Urban Horticulture

Urban Horticulture

Editor: Mary Moreno

R CALLISTO REFERENCE

www.callistoreference.com

Callisto Reference,
118-35 Queens Blvd., Suite 400,
Forest Hills, NY 11375, USA

Visit us on the World Wide Web at:
www.callistoreference.com

ISBN: 978-1-64116-144-2 (Hardback)

Cataloging-in-Publication Data

Urban horticulture / edited by Mary Moreno.
    p. cm.
Includes bibliographical references and index.
ISBN 978-1-64116-144-2
1. Horticulture. 2. Urban gardening. I. Moreno, Mary.
SB453 .U73 2019
635--dc23

Table of Contents

Preface

In my initial years as a student, I used to run to the library at every possible instance to grab a book and learn something new. Books were my primary source of knowledge and I would not have come such a long way without all that I learnt from them. Thus, when I was approached to edit this book; I became understandably nostalgic. It was an absolute honor to be considered worthy of guiding the current generation as well as those to come. I put all my knowledge and hard work into making this book most beneficial for its readers.

Horticulture is the science that is concerned with growing fruits and vegetables. Urban horticulture is a branch of horticulture that studies the relationship between plants and the urban environment. Such studies are applied for the improvement of the surrounding urban area. It is based on the principle that nature has a positive influence on human health, their emotional and psychological wellbeing. In urban horticulture, crops are grown in small gardens and fields, as well as in flowerpots and grow bags. Community gardening is an upcoming practice in this field. It helps in cleaning neighborhoods, restoring nature in industrial areas and creating a relationship between a place and the people. This book is a compilation of chapters that discuss the most vital concepts and emerging trends in the field of urban horticulture. From theories to research to practical applications, case studies related to all contemporary topics of relevance to this field have been included in this book. The readers would gain knowledge that would broaden their perspective about urban horticulture.

I wish to thank my publisher for supporting me at every step. I would also like to thank all the authors who have contributed their researches in this book. I hope this book will be a valuable contribution to the progress of the field.

Editor

SOIL MICROBIAL ACTIVITY IN AN ORGANIC EDIBLE ROSE CROP

Ana Cornelia BUTCARU[1], Florin STĂNICĂ[1], Gabi-Mirela MATEI[2], Sorin MATEI[2]

[1]University of Agronomic Sciences and Veterinary Medicine of Bucharest,
59 Mărăști Blvd., 011464, Bucharest, Romania
[2]National Research-Development Institute for Soil Science, Agrochemistry and Environment - ICPA,
61 Mărăști Blvd., 011464, Bucharest, Romania
Corresponding author email: anabutcaru@gmail.com

Abstract

The paper presents the evolution of the microbial activity analyzed through the evolution of the soil respiration, bacteria and fungi density between March 2015 and November 2016, in an organic edible rose culture under the influence of three ameliorative species and two mulching systems. Beginning with the spring of 2015, with the goal of planting an edible rose culture, in the experimental field of USAMV Bucharest, a special soil preparation was applied using three ameliorative plants, Sinapis alba L., Tagetes patula L. and Phacelia tanacetifolia L. They have a special role to control pathogens in soil and were used in seven different combinations (V1-V7) and a control plot was kept without seeding (V8). After flowering and seed formation, the mature plants were trimmed and incorporated into the soil. They were seeded in the organic roses culture also, same variants between the rose rows in the spring of 2016. In the summer of 2016, two mulching variants were applied for each initial variant (Vn), on the rose's rows: Vn.1. wood chips and Vn.2. wool, while the control Vn.3. was represented by unmulched soil. Microbial activity was stimulated especially in variants with two plant species. The highest potential of soil respiration was characteristic for combinations including Tagetes but also in the variant with Sinapis alone that stimulated the bacterial activity in microbial communities. Generally, the bacteria and fungi density and species number was higher in V1-V7 variants than in the V8 control variant. Microbial species identified included ubiquitous bacteria and fungi with high metabolic capabilities to degrade various substrates such as cellulose from vegetal wastes or keratine from sheep wool added, due to efficient production of cellulase and keratinolytic protease enzymes (bacteria from genera Bacillus, Xanthomonas, Actinomycetes and fungi from genera Trichoderma, Aspergillus, Penicillium, Cladosporium, Paecilomyces, Myrothecium), many of them contributing to biological control of potential plant pathogens and nematodes in rose cultures.

Key words: bacteria, fungi, soil respiration, wool mulch, wood chips mulch.

INTRODUCTION

One of the most important activities in organic agriculture is maintaining and enhancing the soil health respectively the soil organic matter (IFOAM, 2010; Berca 2011; Reeve, 2007).

In the same time, as it is stated by the principle of ecology (IFOAM, 2010), the organic agriculture should be based on living ecological systems and cycles, work with them, emulate them and help sustain them.

An important component for increasing the soil fertility and health can be green manure, cover crops, living mulch (Crossland et al., 2015).

Different kind of organic matter can bring additional positive effects on yield through amelioration of soil life, water retention, humus content (van Opheusden et al., 2012; Butcaru et al., 2016).

Maintaining diversity is another important aspect for perennial cultures in organic agriculture is. Intercropping can be a way of increasing crop diversity (Andersen, 2005; Butcaru et al., 2016).

The present paper presents the results of the microbiological activity in the soil after using an alternative and innovative method for improving the soil activity by using three ameliorative species: *Sinapis alba* L., *Tagetes patula* L. Sparky Mix and *Phacelia tanacetifolia* L., before and after the plantation of an organic edible rose culture.

In addition to the three ameliorative species, from the first year of plantation, two kind of mulch was used: wood chips and wool.

The research analyses the evolution of bacteria, fungi population and respiration coefficient, measured before the establishment of the edible rose culture and after one year and reflect the potential of the ameliorative plants and mulch to develop and maintain the soil activity.

MATERIALS AND METHODS

The research was conducted in the experimental plot at the University of Agronomic Sciences and Veterinary Medicine of Bucharest of a total area of 1,350 m^2 with the purpose of planting three edible rose varieties using an organic technology.

Beginning with spring of 2015 a special soil preparation was applied using three ameliorative plants, *Sinapis alba* L., *Tagetes patula* L. and *Phacelia tanacetifolia* L., with role in soil disinfection (Butcaru et al., 2015; Butcaru et al., 2016).

Crops were sown in late March, by combining the three species in 7 variants: V1 *Sinapis*, V2 *Sinapis + Phacelia*, V3 *Phacelia*, V4 *Sinapis + Tagetes*, V5 *Sinapis + Tagetes + Phacelia*, V6 *Tagetes + Phacelia*, V7 *Tagetes* and a control parcel V8, was kept as black field, without sowing.

The same variants were seeded between the rose rows in the spring of 2016, after the organic roses planting.

After flowering and seed formation, the mature plants were trimmed and incorporated into the soil, all three species in the same time in June 2016.

Beginning with July 2016, the roses, planted on three rows on each variant (V1-V8), were supported by wire trellis and a drip system was installed and operational.

In the summer of 2016, two mulching variants were applied for each initial variant (Vn), on the roses rows: Vn.1. wood chips and Vn.2. wool, while the control Vn.3., was represented by un-mulched soil. Both mulched rows had the same 1 m width with the specific material.

The inter-row was kept grassy through repeated mowing.

In each variant (Vn) was applied the same scheme of treatment, including: fertilizing with manure in autumn 2015 at planting and organic products in 2016; bio stimulatory and caw milk for increasing the immunity system and plant protection with different organic.

For the analysis of the soil microbiological activity, samples were collected before and after planting the organic rose culture (from the total area in March 2015, from each variant Vn in July and October 2015, from each sub-variant Vn.1., Vn.2., Vn.3. in November 2016).

Microbiological analysis studied the number of heterotrophic bacteria determined using dilution plate method - by dispersing soil suspensions on the nutrient agar medium; number of microscopic fungi determined by dispersing soil suspension on PDA medium and soil respiration determined through the substrate induced respiration method according to RS-ISI-14240-1-(2012).

The taxonomical identification were carried out on the basis of the cultural, morphological and / or physiological characteristics in accordance with bacteria Identification Manual (Bergey, 1994) and fungi in agricultural soils (Domsch & Gams, 1972).

It has been used circular chromatograms of soil extracts, with diffusion through absorption on paper Whatmann no. 1, argentic coloring, which generates information on biological quality of the soil due to analytical separation and formation of images whose model of consistency, shape, size, color, texture may indicate the degree of soil health, vitality, fertility, the intensity of biotic activity, soil conditions, the complexity of organic matter and the presence of stable humus.

RESULTS AND DISCUSSIONS

The microbiological analyses proved an increased activity of the soil under the influence of alternative and innovative methods applied.

The bacteria population significantly increased in the March 2015 - Novembre 2016, with a relative stabilisation in the last period.

Soil samples collected in 2016 (phase IV) showed a high density of heterotrophic aerobic bacteria relative to gram of dry soil, which ranged from a minimum of 32 x 10^6 viable cells/g dry soil to V8.2. - Control with wool sub-variant to a maximum of 88 x 10^6 viable cells / g dry soil to the V7.2. - *Tagetes* with wood chips sub-variant (Figure 1).

The application of organic technology by using organic materials (wool, wood chips) for mulching and ameliorative crops alone or combined caused significant increases in the number of heterotrophic aerobic bacteria relative to controls plots in general.

In the fourth phase it is visible the stimulatory effect of *Sinapis* and *Tagetes* on the proliferation of bacteria. The most important values of bacterial density registered under the effect of wool mulch were obtained at variant with *Sinapis*, followed by that of *Tagetes* as ameliorative plants, but generally using wool mulch has led to less numerous bacterial populations than in the rest of variants, especially when were used combinations of two species of ameliorative plants.

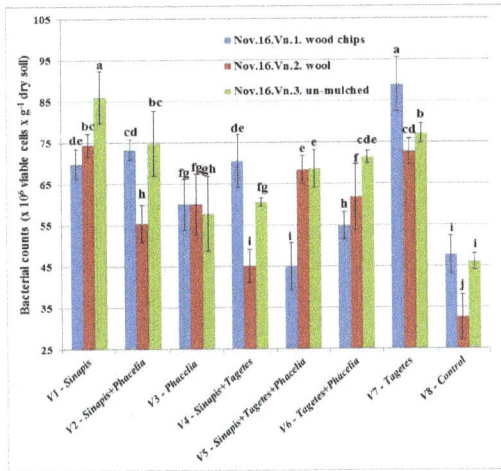

Figure 1. Influence of organic technology on bacterial microflora

Fungal microflora presented moderate values in this stage, below 100×10^3 cfu / g of dry soil at control plots and to all variants with wool mulch and values considered high to a number of variants, from which the sub- variants with wood chips V3.1. *Phacelia*, V7.1. *Tagetes* or V1.1. *Sinapis* and the sub-variants un-mulched V7.3. *Tagetes*, V2.3. *Sinapis + Phacelia* and V6.3. *Tagetes + Phacelia* (Figure 2).

In terms of taxonomy, bacteria and fungi from analyzed soils include ubiquitorius species with high adaptive capacity and species equipped with enzymatic complex equipments, which enable efficient exploitation of a wide variety of substrates with very different origins. There is a considerable number of species capable to degrade and to metabolize organic substrates as wool, wood chips or debris of organic matter due to enzymes such as proteases (keratinases), cellulase and include bacteria belonging to the genera *Bacillus*, *Xanthomonas*, actinomycetes or fungal species from genera *Penicillium*, *Aspergillus*, *Paecilomyces*, *Myrothecium*, *Cladosporium*, *Trichoderma*.

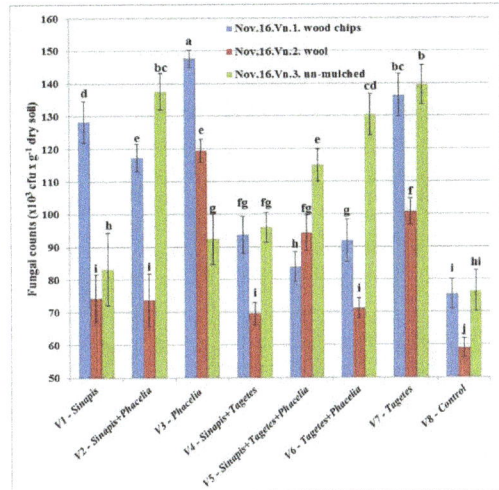

Figure 2. Influence of organic technology on fungal microflora

Many of these microorganisms, such as *Pseudomonas fluorescens*, actinomycetes, *Trichoderma viride*, *Trichoderma hazianum*, *Paecilomyces marquandii* stimulated by the presence of ameliorative plants and organic mulch represented by wool and wood chips act as antagonists against soil borne pathogens of genus *Fusarium*, *Phytophthora* and *Alternaria*, producing a beneficial effect on the health of edible rose culture (Figures 3 and 4).

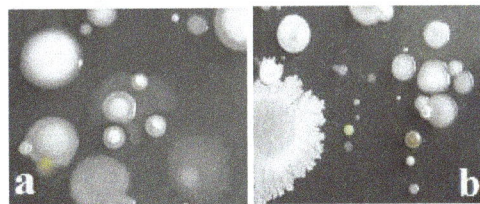

Figure 3. Bacterial microflora to V1.2. *Sinapis* with wool (a) and V7.1. *Tagetes* with wood chips (b)

Figure 4. Fungal microflora to V3.2. *Phacelia* with wool (a) and V7.1. *Tagetes* with wood chips (b)

Using ameliorative plants and different types of mulch in the organic technology for edible roses led to more dynamic global physiological activities of the soil microorganisms as reflected by a significant increase in soil respiration compared to controls, where recorded moderate values of released CO_2/100g dry soil (Figure 5).

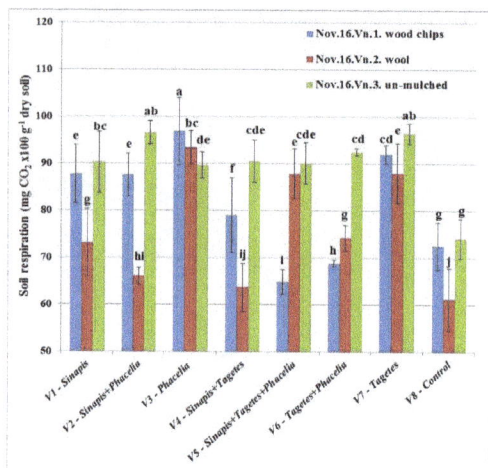

Figure 5. Influence of organic technology on soil respiration

The most intense metabolic activities were recorded at V3.1. *Phacelia* with wood chips sub-variant, mainly due to the activity of cellulolytic fungi, at V7.1. *Tagetes* with wood chips sub-variant mainly on account of bacterial activities (actinomycetes in particular) and cellulolytic fungi, followed by variants with *Tagetes* as ameliorative plant (based on fungal microflora) or combinations of two species of ameliorative plants.

In many of the variants with one or two species of ameliorative plants, soil respiration was more intense when it was used wood chips mulch compared to wool mulch but, in most cases, weaker or similar to the version un-mulched.

Figure 6 presents sectors of Pfeiffer chromate-grams for illustrating changes in soil quality under the influence of ameliorating plants (*Tagetes* and *Phacelia*) and mulch (wool and wood chips) compared to the control.

The analysis of chromatograms reveals an increasing silica organization due to biological activity though embattled shape of the outer edge of the central area in mulched variants, especially those with wood chips.

Clay shows most well-organized at the V3.1. *Phacelia* with wood chips sub-variant, the

remaining variants presenting organization trends in different degrees of evolution, the organization level being correlate, in general, with the high level of chemical complexity.

The content of minerals increased significantly in V7.1. *Tagetes* with wood chips sub-variant, compared with the other experimental variants. Biological activity (bacterial and fungal), mineral diversity and enzyme activity is reflected in the organization of external and middle areas of the chromatograms, organizing corresponding to the increases of protein content, of the nutritional potential, of diversity of sources of carbon, of humic acids formation. Increases integration, connections between particles, due to increased dynamics of some processes compared with the control, but still insufficient to achieve a maximum level in the soil, possibly due to short action time of microbiota on the organic supplements. In the external area appear highlighted varied nutrient sources with an increase degree of stability, colloidal nature in the mulched variants and particularly better highlighted in the *Phacelia* variants. Formation of endings from the external area reflects the favorable evolution of organization of organic matter at variants with ameliorative plants and is more clearly evidenced at the variants mulched with wood chips.

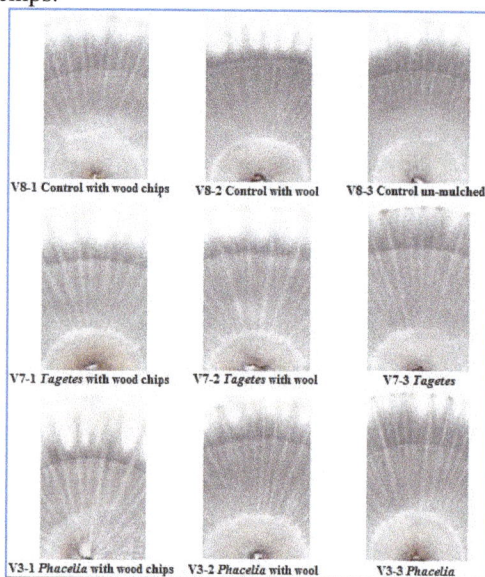

Figure 6. Soils chromatograms at the variants with ameliorative plants and mulch

CONCLUSIONS

Positive results through organic technology applied to edible rose cultures were obtained regarding stimulating the development of fungal and bacterial microflora and increasing the global physiological activities of edaphical microorganisms compared to controls and with the initial phases.

It is remarkable the beneficial effect of *Tagetes* alone or in combination with *Phacelia* on the development of bacterial and fungal microflora. Using wood chips mulch determined a large numbers of bacteria developing on V7. *Tagetes* variant and fungi on V7. *Tagetes*, V3. *Phacelia* variants, were was recorded the most intense soil respiration also.

Using wool as mulch induced a weaker stimulation of soil microbial populations compared with wood chips, the best results being those related to soil respiration increase on V3. *Phacelia* or V7. *Tagetes* variants, or by the stimulation of bacterial increase in V1. *Sinapis* variant.

Analysis of chromatograms revealed favorable effect on soil quality evolution in mulched variants with ameliorative plants of *Tagetes* or *Phacelia*.

REFERENCES

Andersen M. K., 2005. Competition and complementarity in annual intercrops – the role of plant available nutrients. Department of Agricultural Sciences Environment, Resources and Technology, The Royal Veterinary and Agricultural University, Copenhagen, Denmark, 3.

Berca M., 2011. Agrotehnică – Transformarea modernă a agriculturii. Ed. Ceres, București, 173.

Bergey D.H., Holt J.G., 1994. Bergey's manual of determinative bacteriology 9, Wiliams and Wiliams Eds., Baltimore, USA, 787.

Butcaru A.C., Stănică F., Matei G. M., Matei S., 2016. Alternative methods to improve soil activity before planting an organic edible rose crop, Journal of Horticulture, Forestry and Biotechnology, Volume 20(4), 12-17, ISSN: 2066-1797.

Butcaru A.C., Stănică F., Matei G.M., Matei S., 2015. Pregătirea solului în vederea înfiinţării unei culturi de trandafiri de dulceaţă în sistem ecologic, revista Hortus nr.14, 165-168.

Crossland M., Fradgley N., Creissen H., Howlett S., Baresel P., Finckh M. and Girling R, 2015. An online toolbox for cover crops and living mulches, Aspects of Applied Biology - Getting the Most out of Cover Crops, Volume 129, 1.

De Baets S., Poesen J., Meersmans J., Serlet L., 2011. Cover crops and their erosion-reducing effects during concentrated flow erosion. Catena Journal 85, ISSN 0341-8162 DOI: 10.1016/j.catena. 2011.01.009, 237–244.

Dhima K., Vasilakoglou I., Garane V., Ritzoulis C., Vaia Lianopoulou, Eleni Panou-Philotheou, 2010. Competitiveness and Essential Oil Phytotoxicity of Seven Annual Aromatic Plants. Weed Science 58,ISSN 1550-2759 DOI:10.1614/WS-D-10-00031.1, 457–465.

Domsch, K.H., Gams, W., 1970. Fungi in agricultural soils, T&A Constable Ltd., Edinburg, London, 290.

Hooksa C.R.R., Wangb K., Ploegc A., Mcsorleyd R, 2010. Using marigold (*Tagetes* spp.) as a cover crop to protect crops from plant-parasitic nematodes. Applied Soil Ecology 46, ISSN 0929-1393, 307–320.

IFOAM EU GROUP, 2010. Organic food and farming. A system approach to meet the sustainability challenge. www.ifoam-eu.org/workareas/policy/php/CAP.php, 7-8.

Liu J., Khalaf R., Ulén B., Bergkvist G., 2013. Potential phosphorus release from catch crop shoots and roots after freezing-thawing. Plant Soil Journal, electronic ISSN 1573-5036 DOI 10.1007/s11104-013-1716-y.

van Opheusden A.H.M., van der Burgt G.J.H.M., Rietberg P.I., 2012. Decomposition rate of organic fertilizers: effect on yield, nitrogen availability and nitrogen stock in the soil, Louis Bolk Institute, www.louisbolk.org, 33.

Penhallegon R., 2003. Nitrogen-phosphorus-potassium values of organic fertilizers. OSU Extension Service - Lane County Office, Karen Ailor, http://extension.oregonstate.edu/lane/sites/default/files/documents/lc437organicfertilizersvaluesrev.pdf, 4.

Reeve J. R., 2007. Soil quality, microbial community structure, and organic nitrogen uptake in organic and conventional farming systems. Washington State university; Department of Crop and Soil Sciences;http://www.researchgate.net/publication/228542927.

DIFFERENT APPROACHES ON BULBLET FORMATION WITH SCALING IN MADONNA LILY *(LILIUM CANDIDUM)*

Arda AKÇAL[1], Özgür KAHRAMAN[2]

[1]Çanakkale Onsekiz Mart University, Faculty of Agriculture, Terzioglu Campus, 17020, Çanakkale, Turkey
[2]Çanakkale Onsekiz Mart University, Faculty of Architecture and Design, Terzioglu Campus, 17020, Çanakkale, Turkey
Corresponding author email: aakcal@comu.edu.tr

Abstract

The purpose of this study was, to determine the effects of different treatments on bulblet formation with scale propagation in 'Madonna lily' (Lilium candidum). The research was conducted in growth chamber at Çanakkale Onsekiz Mart University, Faculty of Agriculture, Department of Horticulture in 2015-2016. Scales of Lilium candidum bulbs with 22-24 cm circumference, were used as a plant material. Effects of different incubation periods (10,12,14 weeks), incubation temperatures (10-15 ºC, 20 -25ºC), auxin (IBA 100 ppm, IBA 200 ppm) doses and scale positions (outer, middle, inner, center) on bulblet formation were investigated. The experiment was established according to randomised plot design with 3 replications. Some parameters like, bulblet formation ratio, bulblet number per scale, weight, diameter and height of bulblet, scale number, root number and root length of bulblet were also determined. As a conclusion, treatments have not any significant effect on bulblet formation ratio, but there were significant differences between the other parameters for some of treatments. In spite of this, incubation period of 14 weeks gives the highest average value for bulblet number per scale (1.467 piece) and bulblet height (19.105 mm). Also, the highest average value of bulblet weight (0.792 g) and bulblet diameter (13.282 mm) were measured in outer scales. While incubation temperature of 10-15 ºC gives the best result for bulblet scale number (3.511 piece), the highest average value for root number of bulblet (3.900 piece) and root length of bulblet (11.224 cm) were measured in auxin dose of 200 ppm IBA.

***Key words**: Lilium candidum, scale propagation, incubation, bulblet formation, ornamental plants*

INTRODUCTION

Turkey is very rich in terms of a plant diversity, in different three phytogeographic zones in the cross-point of Asia and Europe, along with the climatic change and soil properties. Turkey has about 12.000 taxa species and 3.750 of them are endemic (Avcı, 2005). Endemism ratio is 34.5 % (Uyanık et al., 2013). There are 1056 taxa geophytes and 424 of them are endemic (Özhatay, 2013). Geophytes, are the plants whose above ground parts such as stems, leaves and flowers dries and dies after completing their growth period and which in summer months live on thanks to their under earth storing parts such as bulbs, tubers and rhizomes, are also called natural flower bulbs. These are economically important in the sector of ornamental plants (Aksu et al., 2002; Zencirkıran, 2002). Some of these species have been exported over a hundred years. Bulb exportation of *Lilium candidum* are permited from only bulbs propagated in Turkey

(Anonymous, 2016). *Lilium candidum* is globally known as "Madonna lily" or "white lily". It is a herbaceous, bulbous perennial plant belonging to *Liliaceae* family. It has fibrious roots and the roots are yellowish white colored. Its stem length is between 43-150 cm and has white flowers blooming between the end of May and the end of June, depending on the climate conditions (Özen et al., 2012). *L. candidum* can be propagated from seed and bulblets. However, propagation of seed takes five or more years from seed to develop plant capable of flower production. During the uprooting of the plant which takes such a long time to grow, newly sprouted, not ripened seedlings are also uprooted and therefore the damage increases more and more.

The objective of this study was, to develop some new methods with different approaches for scale propagation in *Lilium candidum* bulbs, which has been uprooted a lot from the nature and exported, also it was aimed on providing bulbs for exporting with these methods.

MATERIALS AND METHODS

This study has been conducted in 2015-2016 period in a growth chamber (Figure 1), where the temperature (^{o}C) and humidty (%) controlled by automatically, in COMU, Faculty of Agriculture, Department of Horticulture.

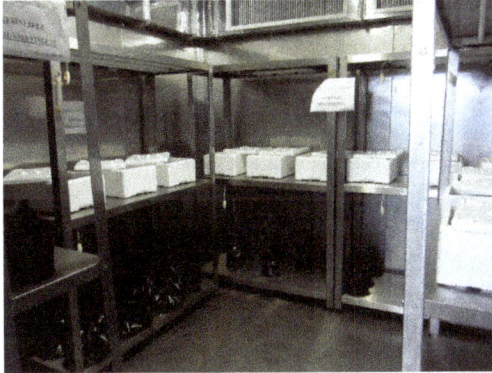

Figure 1. Inside of growth chamber

L. candidum bulbs with 22-24 cm in circumferences and 87.98 g in weight, were used as a plant material (Figure 2).

Figure 2. Bulb of *Lilium candidum*

The bulbs of *L. candidum* (Madonna lily) were provided from a firm exporting flower bulbs. The dry outer scales, any root remains and the bulb tip were all removed prior to scaling (Figure 3).
Scales were separated from basal plate of *L. candidum* bulbs (Figure 4), measured and calibrated for trials (Figure 5). Then washed with distilled water and sterilised with dilute alcohol.

Figure 3. *Lilium candidum* wihtout root

Figure 4. Scales separated from basal plate

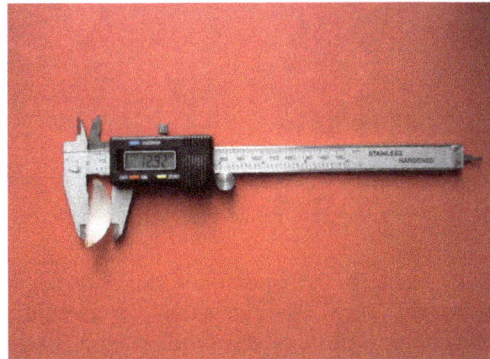

Figure 5. Calibration of lilium scales for trials

The average length of these scales is about 35.60 mm and the average width of the scales is 12.46 mm.
The scales were treated in 1% Captan and 0.5 % Mancozeb for 20 minutes to prevent fungal diseases (Figure 6), left in the shade for 10 minutes to remove the excess water, and kept in a cool place until planting.

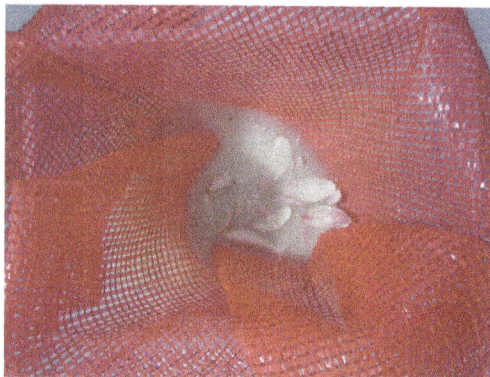

Figure 6. Scales treated with fungicide

Figure 8. Perlit and scale mixture inside of PE bag

15 prepared bulb scales were mixed with 3 liters of damp perlite (Figure 7) and after the mixture was put to black polyethylene bags (5 L) (Figure 7). The bags were fastened tightly leaving some space on top (Figure 8,9), on October, 2nd 2015 (Aksu et al., 2002; Zencirkıran and Mengüç, 2002).

After uprooting bulblets from bags, some parameters like bulblet formation ratio (%), bulblet number per scale (piece), bulblet weight (g), bulblet diameter (mm), bulblet height (mm), bulblet scale number (piece), root number of bulblet (piece) and root lenght of bulblet (cm) were measured.

The data were analyzed statistically by analysis of variance with SPSS 23. Separation of means was by the Duncan's multiple comparison test at $p = 0.05$.

Figure 7. Perlite and scales

Figure 9. Polyethylene bags in growth chamber

This research is consist of four different treatment such as incubation temperature, incubation period, scale position and auxin dose. In incubation temperature trial, the bags were incubated in a growth chamber at 10-15°C and 20-25 °C until January, 14th 2016 to form bulblets. Incubation period trial includes 10 weeks, 12 weeks and 14 weeks period. Scale position trial composed of outer, middle, inner, and center scales of Madonna lily. 100 ppm IBA, 200 IBA and control (without IBA) were used in auxin dose trial.

All experiment were established according to randomised plot design with 3 replications composed of 15 scales each.

RESULTS AND DISCUSSIONS

According to Marinangeli and Curvetto (1997), bulblet formation or number of bulblets per scale were affected by some biotic factors such as cultivar, size, age, physiological status of the bulb, position of the scale in the bulb and by some abiotic factors such as, temperature, humidity, light, physiological and chemical treatments (Matsuo, 1987; Grassotti and Magnani, 1988; Magnani et al., 1988).

In this study it was determined that, incubation periods had a statistically significant effect (p<0.05) on only the bulblet number per scale. Maximum values of bulblet number per scale were obtained from 14 weeks of incubation period (1.467 piece).

On the other hand, there was not any significant difference between the three incubation period (Figure 10) (10,12,14 weeks) for the other bulblet characteristics (Table 1).

Figure 10. Scales and bulblets in incubation periodes

Table 1. The effect of incubation period on bulblet characteristics

Incubation Period	Bulblet Formation Ratio (%)	Bulblet Number per Scale (piece)	Bulblet Weight (g)	Bulblet Diameter (mm)	Bulblet Height (mm)	Bulblet Scale Number (piece)	Root Number of Bulblet (piece)	Root Lenght of Bulblet (cm)
10 Weeks	100.000 a	1.194 b	0.533 a	9.977 a	17.731 a	3.000 a	1.833 a	8.081 a
12 Weeks	96.970 a	1.067 b	0.629 a	10.912 a	18.158 a	2.938 a	1.806 a	9.483 a
14 Weeks	75.000 a	1.467 a	0.702 a	10.659 a	19.105 a	2.705 a	2.091 a	7.727 a

Data having the same letter in a column were not significantly differed by Duncan's multiple comparison test (p< 0.05).

Many researchers stated that bulblet formation was also influenced by various plant growth regulators (Matsuo, 1972; Roh, 1990). In this research, similar results have been observed for auxin doses on some bulblet characteristics.

Auxin doses had a statistically significant effect (p<0.05) on diameter and height of bulblet, bulblet scale number, root number and root length of bulblet (Figure 11).

Figure 11. Bulblets in auxin doses and control

The highest value of bulblet diameter was 6.872 mm with 100 ppm IBA, while the lowest value of diameter was 5.446 mm with 200 ppm IBA (Table 2). Park (1996) reported that, generally the diameter of the bulblets was increased when the scales were treated with the 100 ppm IBA. With a value of 16.127 mm and 15.022 mm, 100 ppm IBA and control bulblets gives the best result for bulblet height. At the same time, the highest value was measured for bulblet scale number, at 100 ppm IBA (3.000 piece) and 200 ppm IBA (2.840 piece), respectively (Table 2).

Table 2. The effect of auxin doses on bulblet characteristics

Auxin Dose	Bulblet Formation Ratio (%)	Bulblet Number per Scale (piece)	Bulblet Weight (g)	Bulblet Diameter (mm)	Bulblet Height (mm)	Bulblet Scale Number (piece)	Root Number of Bulblet (piece)	Root Lenght of Bulblet (cm)
Control	22.932 a	1.083 a	0.268 a	6.398 ab	15.052 a	2.167 b	2.375 b	1.833 b
IBA 100 ppm	66.667 a	1.280 a	0.362 a	6.872 a	16.127 a	3.000 a	2.708 ab	9.152 a
IBA 200 ppm	48.889 a	1.300 a	0.249 a	5.446 b	11.356 b	2.840 a	3.900 a	11.224 a

Data having the same letter in a column were not significantly differed by Duncan's multiple comparison test (p< 0.05)

Also, the highest value for root number of bulblets determined (3.900 piece) in 200 ppm IBA, similarly root length increased with auxin doses, thus, 200 ppm IBA gives the best result with an average value of 11.224 cm (Table 2). Hence, growing structure of scales depends on their positions; it's an important factor for bulblet formation and characteristics. According to some researchers, scales from the outer and middle scales of lily bulb tended to produce more bulblets, which can be correlated with the total carbohydrate content in those scales (Matsuo, 1975; Park, 1996). Except bulblet formation ratio, all parameters affected significantly (p<0.05) by scale positions (Table 3). With 1.167 piece, the highest average value for bulblet number per scale was obtained from outer scales, other positions statistically acted at same level on bulblet number per scale. For weight, diameter, height, scale number and root number of bulblets, similar results observed in scale positions (Figure 12).

Table 3. The effect of scale positions on bulblet characteristics

Scale Position	Bulblet Formation Ratio (%)	Bulblet Number per Scale (piece)	Bulblet Weight (g)	Bulblet Diameter (mm)	Bulblet Height (mm)	Bulblet Scale Number (piece)	Root Number of Bulblet (piece)	Root Lenght of Bulblet (cm)
Outer Scale	83.335 a	1.167 a	0.792 a	13.282 a	18.319 a	3.500 a	2.273 ab	3.955 b
Middle Scale	80.000 a	1.028 b	0.517 b	9.473 b	15.824 b	2.765 b	2.355 a	10.468 a
Inner Scale	90.790 a	1.025 b	0.462 b	8.788 b	14.207 b	2.769 b	1.897 bc	10.000 a
Center Scale	79.840 a	1.000 b	0.236 c	7.107 c	12.036 c	2.000 c	1.500 c	8.813 a

Data having the same letter in a column were not significantly differed by Duncan's multiple comparison test (p< 0.05).

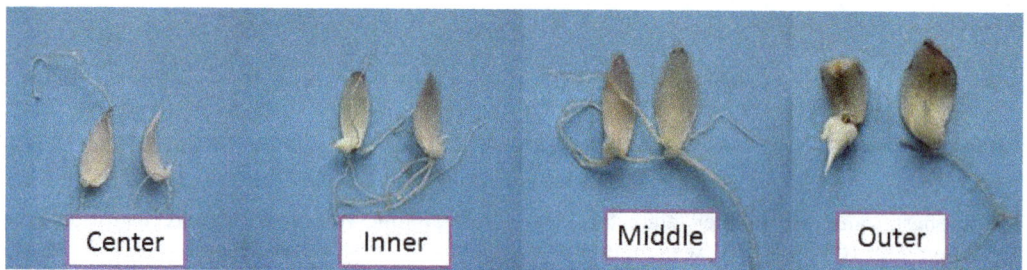

Figure 12. Scales and bulblets in scale positions

When the highest values measured in outer scales, center scales were give the lowest values, generally middle and inner scales statistically acted at the same level. So it was clear that, there was an increase from center to outer scales. On the other hand, it was not investigated for root length of bulblet (Table 3). According to the results of Matsuo et al. (1987), larger scales produce more bulblets in Lily. Hanks (1985) reported that, the number and sizes of differentiated bulblets are influenced by the relative position of starting scales in Narcissus bulbs. Marinangeli et al. (2003) conclude that, middle scales are the best starting materials for experimental uses involving scale propagation and external scales

must be included for production. Also, Padasht et al. (2006) reported that, outer and middle scales at 20 and 25 °C regenerated more bulblets with better properties than inner scales. Many researcher reported that, the growth and development properties of newly formed bulblets depends on the temperature during scaling (Van Tuyl, 1983; Aquettaz et al., 1990). Our results for *L. candidum* indicates that incubation temperatures had a statistically significant effect (p<0.05) on only the diameter and root length of bulblet (Table 4) (Figure 13).

Table 4. The effect of incubation temperature on bulblet characteristics

Incubation Temperature	Bulblet Formation Ratio (%)	Bulblet Number per Scale (piece)	Bulblet Weight (g)	Bulblet Diameter (mm)	Bulblet Height (mm)	Bulblet Scale Number (piece)	Root Number of Bulblet (piece)	Root Lenght of Bulblet (cm)
10-15 °C	95.553 a	1.089 a	0.410 a	7.933 b	17.279 a	3.511 a	2.178 a	5.658 b
20-25 °C	100.000 a	1.133 a	0.427 a	9.056 a	16.378 a	2.889 a	2.133 a	7.351 a
	ns	ns	ns	*	ns	**	ns	*

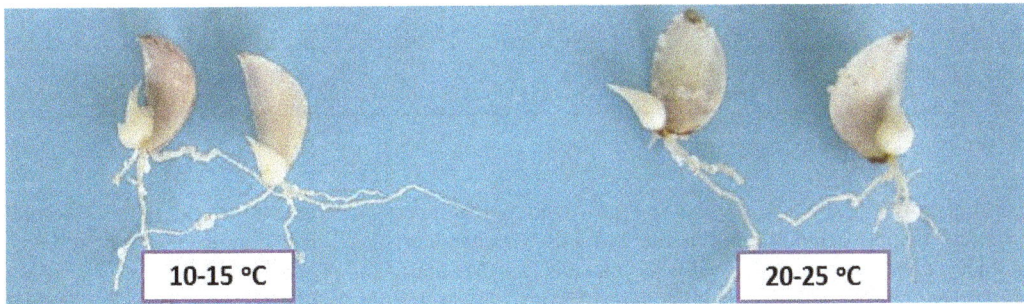

Figure 13. Scales and bulblets in incubation temperatures

The highest value for diameter of bulblet (9.056 mm) and root length of bulblet (7.351 cm) were taken from incubation temperature of 20-25 °C (Table 4). Similar results had been observed by Suh and Lee (2006), the number or bulblets produced per scale in both lilium varieties used in the trial, were not affected by incubation temperatures, however the diameter

of 'Casablanca' bulblets were icreased as temperatures was increased from 20 °C to 30 °C. On the other hand, treatments have not any statistically significant effect on bulblet formation ratio, but there were only some quantitative differences between the treatments (Figure 14).

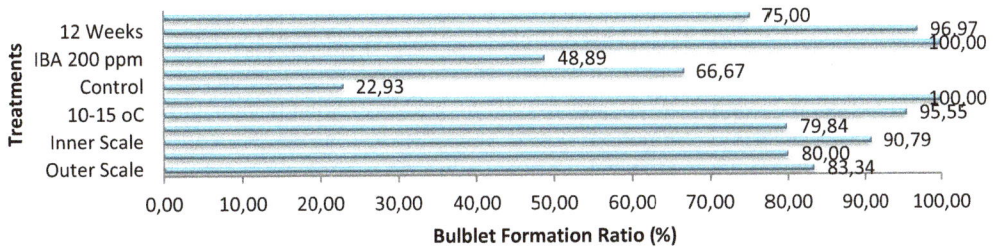

Figure 14. Changes on bulblet formation according to treatments.

The correlation coefficients related in bulblet characteristics are given in Table 5. According to this results, the strongest relationship (r=0.841) in *L. candidum* was found between the bulb diameter and bulb weight. In spite of this, there was a positive correlation between bulblet height (r=0.770) and bulblet weight and also between bulblet height (r=0.639) and bulblet diameter, while there was a negatively weak relationship (r=-0.189) between root length of bulblet and bulblet number per scale (Table 5).

Table 5. Correlation between the bulblet characteristics of *Lilium candidum*

	Bulblet Formation Ratio	Bulblet Number per Scale	Bulblet Weight	Bulblet Diameter	Bulblet Height	Bulblet Scale Number	Root Number of Bulblet	Root Lenght of Bulblet
Bulblet Formation Ratio	1							
Bulblet Number per Scale	-,008	1						
Bulblet Weight	,306	,100	1					
Bulblet Diameter	,247	,138	,841**	1				
Bulblet Height	-,092	,108	,770**	,639**	1			
Bulblet Scale Number	,500	-,028	,631**	,609**	,595**	1		
Root Number of Bulblet	,060	,016	,448**	,436**	,291**	,462**	1	
Root Lenght of Bulblet	,233	-,189*	,180	,093	,040	,162	,364**	1

*. Correlation is significant at the 0.05 level
**. Correlation is significant at the 0.01 level

Consequently, there were some positive relations too between bulblet scale number with bulblet weight (r=0.631), bulblet diameter (r=0.609) and bulblet height (r=0.595) respectively. Similar relations were found between root number of bulblet with bulblet weight (r=0.448), bulblet diameter (r=0.436), bulblet height (r=0.291) and also with bulblet scale number (r=0.462) respectively. It is seen on Table 5 that, there was an another relation observed between root length of bulblet with root number of bulblet (r=0.364).

CONCLUSIONS

This research was focused on the effects of different treatments, such as incubation period, incubation temperature, auxin doses and scale position on bulblet formation and characteristics with scale propagation in 'Madonna Lily' (*Lilium candidum*).
The overall results indicate that, treatments have not any significant effect on bulblet formation ratio, but there were significant differences between the other bulblet characteristics for some of treatments. Especially for the incubation period, 14 weeks gives the highest average value for bulblet number per scale and bulblet height. However, study results shows that 10 week is sufficient for an incubation period. We conclude that, the highest average value of bulblet weight and bulblet diameter were determined in outer scales. So, it was cleared that outer scales are the best propagating material, in addition to, middle scales could be used for propagation. While incubation temperature of 10-15 ^{0}C gives the highest value for bulblet scale number, 20-25 ^{0}C was more effective for bulblet diameter and root length of bulblet. Also, importance of auxins on rooting factors were understood once again with results of the study. As a conclusion it was seen that, different approaches and proper methods were necessary for scaling procedure in *L. candidum*. However, for geophytes there was not any specific information about formation, growth and development of new bulblets.

REFERENCES

Aksu E., Görür G., Çelikel F.G., 2002. Göl soğanı (*Leucojum avestivum*)'nın vegetatif yöntemlerle üretilme olanaklarının araştırılması. II. Ulusal Süs Bitkileri Kongresi. Antalya. s 29-34.

Anonymous, 2016. Doğal Çiçek Soğanlarının Sökümü. Üretimi ve Ticaretine İlişkin Yönetmelik. Resmi Gazete. Sayı: 29556

Avcı M., 2005. Çeşitlilik ve Endemizm Açısında Türkiye'nin Bitki Örtüsü. İstanbul Üniversitesi Fen Edebiyat Fakültesi Coğrafya Dergisi. Sayı:13. İstanbul. s27-55.

Aquettaz P., Paffen A., Delvalle I., Van Der L.P., De Klerk G.J., 1990. The Development of Dormancy in Bulblets of Lilium speciosum Generated In Vitro. I. The Effects of Culture Conditions. Plant Cell Tissue Organ Cult. 22: 167-172.

Grassotti A., Magnani G., 1988. Stato Attuale Eprospettive Della Moltiplicazione In Vivo Del Lilium. Colture Protette 17: 33-42.

Hanks G.R. 1985. Factors Affecting Yields of Advantitious Bulbils During Propagation of Narcissus by Twin Scaling Technique. J. Hort. Sci. 60: 531 – 543.

Marinangeli P.A., Curvetto, N., 1997. Bulb Quality and Trumatic Acid Influence Bulblet Formation from Scaling in Lilium Species and Hybrids. Hortscience 32 (4): 739 -741.

Marinangeli P.A., Hernandez, L.F., Pellegrini, C.P., Curvetto, N.R., 2003. Bulblet Differentiation After Scale Propagation of Lilium longiflorum. J. Amerc. Soc. Hort. Sci. 128 (3): 324 -329.

Matsuo E., 1972. Studies on The Easter Lily (*L. longiflorum*) of Serkaku Retto (Pinnacle Islands) I. Comparative Study on The Growth. J. Jap. Soc. Hort. Sci. 41:383-392.

Matsuo E., 1975. Sudies on The Leaf Development of The Scale Bulblet In The Easter Lily (*L. longiflorum*) IV. Effect of Temperature and Light Conditions on Leaf Emergence of Scale Bulblets. J. Jap. Soc. Hort. Sci. 44:281-285.

Matsuo E., Nonaka, A., Arisumi, K., 1987. Some Factors Influencing The Type of Leaf Development (Plant type) of Scale Bulblets of Easter Lily, *Lilium longiflorum*. Bul. Fac.Agr., Kagoshima University, Japan.

Magnani G., Malorgio, F., Moschini, E., 1988. Influenza Del Livello Termico In Fase De Moltiplicazione Da Scaglie Sulla Produzione Di Bulbetti Di Lilium. Colture Protette 17: 69-74.

Özen F., Temeltaş H., Aksoy Ö., 2012. The Anatomy and Morphology of the Medicinal Plant. Lilium candidum L. (Liliaceae). Distributed in Marmara Region of Turkey. Pakistan Journal of Botany. 44(4): 1185-1192

Özhatay N., 2013. Türkiye'nin Süs Bitkileri Potansiyeli: Doğal Monokotil Geofitler. V. Süs Bitkileri Kongresi. Cilt:1. 06-09 Mayıs. Yalova. s1-12.

Padasht D.M.N., Khalighi A., Naderi R., Mousavi A., 2006. Effects of temperature, Propagation Media and Scale Position on Bulblet Regeneration of Chelcheragh Lily (Lilium ledebouri Boiss.) by Scaling Method. Seed and Plant Improvment Journal(SPII), Karaj, Iran, 22 (3).

Park N.B., 1996. Effects of Temperature, Scale Position and Growth Regulators on The Bulblet Formation and Growth During Scale Propagation of Lilium. Proc. Int. Sym. On Lilium. Acta Hort.414, 257-262.

Roh M.S., 1990. The Effects of Growth Regulators on Bulblet Formation from Easter lily Leaves. Plant Growth Regulator Society of America Quarterly. 18 (3):140-146.

Suh J.K., Lee J.K., 2006. Bulblet Formation and Dormancy Induction as Influenced by Temperature, Growing Media and Light Quality During Scaling Propagation of Lilium Species. Proc. Int. Sym. On Lilium. Acta Hort.414, 251-256.

Uyanık M., Kara Ş.M., Gürbüz B., Özgen Y., 2013. Türkiye'de Bitki Çeşitliliği ve Endemizm. Özet Kitabı. 2-4 Mayıs. Ekoloji Sempozyumu. Tekirdağ. s:197.

Van Tuyl J., 1983. Efffect of Temperature Treatments on The Scale Propagation of Lilium longiflorum 'White Europe' and Lilium x 'Enchantment'. Hortscience 18: 754-756.

Zencirkıran M., Mengüç A., 2002. Parçacık ve ikiz pul yöntemlerinin Galanthus elwesii hook.'de yavru soğan oluşumu üzerine etkileri. II. Ulusal Süs Bitkileri Kongresi. Antalya. s24-28.

3

THE INFLUENCE OF PRECEDING PLANT CULTIVATION ON GROWTH AND PHYSIOLOGY OF AN *OCIMUM BASILICUM* L. CULTIVAR

**Marian BURDUCEA[1], Andrei LOBIUC[1], Naela COSTICĂ[1],
Maria-Magdalena ZAMFIRACHE[1]**

[1]"Alexandru Ioan Cuza" University of Iasi, Carol I Bld., 20 A, 700505, Iasi, Romania

Corresponding author email: alobiuc@yahoo.com

Abstract

The paper aimed to assess some morphometric and physiological parameters of a purple leaved Ocimum basilicum L. cultivar plants grown in substrates in which other plants (a green leaved Ocimum basilicum L. and Armoracia rusticana Gaertn. Mey. & Scherb.) were grown. The number of lateral stems and leaves and the mass of plants were positively influenced by cultivation after the other plants. Photosynthesis and transpiration rates of the same plants decreased, however the chlorophyll fluorescence parameters (Fv/Fm and φPSII) did not reveal a major influence on the photosynthetic apparatus. Chlorophyll and total phenolic contents decreased in plants grown after green basil and increased in plants grown after horseradish. The results show that basil can be grown in the same substrate used by the tested species with positive influences on growth.

Key words: biometry, photosynthesis, chlorophyll fluorescence, substrate, horseradish.

INTRODUCTION

Ocimum basilicum L. known as sweet basil belongs to *Lamiaceae* family, *Ocimum* genus, which includes approximately 60 species. It is a valuable plant species with multiple uses in medicine, cosmetics and gastronomy. Apart from the flavoring properties, the basil has antimicrobial, insecticidal, antioxidant, anti-inflammatory etc. activities (Putievsky and Galambosi, 1999). The properties of this species are mainly owed to the essential oils it synthesizes and to the large infraspecific variability with numerous phenotypes and also chemo-types (Grayer et al., 1996). More than 100 sweet basil cultivars exist that differ in leaf shape, color, height and odor. Originated in subtropical areas, where it can be grown either as an annual or perennial plant, basil is cultivated all over the world as annual plant. Due to its economic importance basil is cultivated on large areas and, thus, the basil crops have to meet the current requirements for agricultural sustainability. Also, as a medicinal and culinary herb, the value of the basil increases when cultivated under organic conditions. Such conditions include crop rotation, intercropping or organic fertilization. The cultivation parameters for basil have been largely investigated, with an emphasis on nitrogen, phosphorus and potassium requirements, time of seeding, distance between plants and rows, soil type, temperature, weeding strategies and pest control (Simon 1996; Meyers, 2003; Arabaci and Bairam, 2004). It is also known that the basil should be rotated once every 4 years and that the most suited preceding crops are cereals and legumes (Pârvu, 2002). However, the influence of the preceding crops on basil physiology and biochemistry has been less studied, rather the species for which basil can be a companion are known, such as tomatoes or marigold (Meyers, 2003; Tringovska et al. 2015).

The effects of preceding crops or intercropping species on other species are mainly due to the substances released in the soil by either leaf litter or roots. Such substances range from high molecular mass compounds such as proteins and polysaccharides, enzymes such as acid phosphatase, nucleases, invertases (Chang and Bandurski, 1964; Chhonkar and Tarafdar, 1981) to low molecular weight phenolic acids (Vaughan et al., 1994). The root exudates, when deposited in soil, a process also known as rhizodeposition, influence the microbiota of the soil and also the growth and development of surrounding plants. The exuded substances may have an effect on proximate plants physiological processes, protein synthesis or enzyme functioning (Bertin et al., 2003; Farooq et al., 2011).

The present paper aimed to test whether basil grown on substrates following other plants is influenced at a physiological and biochemical level in its development. Prior to cultivating a purple leaved basil cultivar, another, green leaved basil cultivar or horseradish (*Armoracia rusticana* Gaertn. Mey. & Scherb.) were grown in the same substrates. *Armoracia rusticana* is a crop specie grown for culinary uses, due to its pungent taste which is given by glucosinolates and their transformation products such as isothiocyanates, but also for medicinal purposes, with proven in vitro anti-inflammatory activity (Yamaguchi, 2012; Marzocco et al., 2015). The specie presents an increasing demand on various markets and is also used in crop rotation for preventing potato and tomato crops pests (Filipović et al., 2015). Horseradish is a specie which is known to exert negative influences on the surrounding plants due to the root exudates it synthesizes (Dias and Moreiro, 1988; Itani et al., 2013). Several parameters were evaluated, to assess the opportunity of cultivating sweet basil in a culture comprising different basil cultivars or other spices which could influence the succeeding crops due to chemicals secreted in soil.

MATERIALS AND METHODS

Plant material
The tested specie was *Ocimum basilicum* L. cv. "Violet de Buzau", which was grown from seeds obtained from the Agricultural Research and Development Station at Buzau, Romania. The species grown prior to this cultivar in the substrate were *Ocimum basilicum* L. cv. "Aromat de Buzau", obtained from the same source and grown from seeds and *Armoracia rusticana*, planted as one year old roots obtained from local suppliers.

Experimental conditions
For the growth of plants, 4 L (15 cm height x 18 cm diameter) plastic pots were used. Pots were filled with 3 L of a mixture composed of commercial soil (60% v/v), peat moss (30% v/v) and perlite (10% v/v). For each variant, 3 pots were used.

Three experimental variants were set up: pots with purple basil seeded in substrate mixture alone (control plants), in substrate mixture in which green basil (4 plants per pot) was previously grown and in substrate mixture in which horseradish (4 plants per pot) was previously grown. Both the green basil cultivar and the horseradish plants were grown for approximately 3 months, period which, in the case of basil, coincided with the fruiting stage. After this period, the green basil and the horseradish plants were removed manually from the pots. Each pot was seeded with 10 purple basil seeds. The plants were thinned at 4 individuals per pot after 1 week.

The pots were kept at constant temperature regimes, between 22° C (night) and 25° C (day). Artificial light was supplied by 4800K fluorescent tubes for 14 h each 24 h. The atmospheric humidity was relatively constant, around 50%. Plants were irrigated twice a week with distilled water. The growth of purple basil lasted for 3 months, until the fruiting stage of plants.

Morphometrical assessments
For each variant, 12 individuals were assessed for stem height, number of leaves, number of lateral stems, number of inflorescences and mass of fresh plants.

Physiological measurements
The photosynthetic activity, transpiration rates and stomatal conductance were measured with an ADC Bioscientific LCi apparatus, on 3 leaves (from the lower, middle and upper regions of the plants) from 3 individuals each per treatment, 5 readings per leaf. Measurements were done during the light regime.

Chlorophyll fluorescence was evaluated by measuring F_0, Fm, Fv, Fv/Fm, Fs, Fm', ϕPSII on 5 leaves per treatment. Fv/Fm was measured following a 20 minutes dark adaptation of leaves using provided clips. ϕPSII was measured at normal light regime.

Biochemical parameters
Assimilatory pigments were extracted from grounded leaves (approx. 0.1 g) using 80% aqueous acetone. The absorbance of extracts were recorded at 470, 646 and 663 nm with a

Shimadzu UV 1240 spectrophotometer in 1 cm light path length glass cuvettes. Pigment contents were calculated using Wellburn (1994) equations.

The total phenolic contents were determined following the method described in Herald et al. (2012). An aliquot of 0.1 ml of 5% (w/v) extracts in 30% ethanol was mixed with Folin reagent, incubated for 5 min, Na_2CO_3 7.5% was added and incubated for further 90 min. Results were calculated using absorbance at 760 nm and expressed as gallic acid equivalents per gram fresh weight.

Total flavonoids were assessed in the same extracts using the $AlCl_3$ method and expressed as quercetin equivalents per gram fresh weight while antioxidant activity of extracts was determined according to the DPPH (2,2-diphenyl-1-picryl-hydrazyl-hydrate) method (Herald et al. 2012).

The statistical analyses conducted were represented by analyses of variance among treatments and the Tukey post hoc test at p<0.05, the results being expressed as means and standard errors.

RESULTS AND DISCUSSIONS

Morphometry
From a morphometric point of view, the analyzed parameters of the purple basil plants recorded certain significant variations. An increase in the number of lateral stems (Fig. 1) and leaves (Table 1) were recorded for plants grown after green basil. For the same plants, a significant decrease in height was recorded, thus the plants presented a more compact habitus.

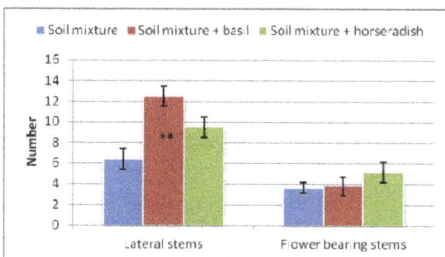

Figure 1. Number of lateral stems and flower bearing stems of purple basil plants
(*-significant differences from control plants at p < 0.05; **-significant differences from control plants at p < 0.01).

Increases in mass and number of leaves, although not significant, were recorded also for plants grown after horseradish. The number of flower bearing stems was not significantly modified in any of variants.

Table 1. Morphometric indices of purple basil plants

Treatments / Parameters	Height (cm)	No. leaves	Mass (g)
Soil mixture	74.5±3.07	119.33±9.36	22.9±1.48
Soil mixture + basil	53.9±2.08**	188±14.11**	27.3±3.61
Soil mixture + horseradish	71.17±4.3**	133.83±9.19**	29.38±3.4

(*-significant differences from control plants at p < 0.05; **-significant differences from control plants at p < 0.01).

Physiological analyses
The photosynthetic rate of purple basil plants significantly decreased when grown after the green basil cultivar and it remained within similar values with control plants when grown after horseradish plants (Table 2). Transpiration rates decreased significantly for both purple basil individuals grown after green basil and for those grown after horseradish plants.

Table 2. Physiological parameters of purple basil plants

Treatment/ Parameter	Photosynthesis rate (μmolls $CO_2/m^2/s$)	Transpiration rate (molls $H_2O/m^2/s$)	Stomatal conductance (molls/m^2/s)
Soil mixture	2.31±0.1	2.9±0.15	0.25±0.02
Soil mixture + basil	1.45±0.14**	1.67±0.15**	0.1±0.02
Soil mixture + horseradish	2.26±0.15	2.41±0.1*	0.35±0.22

(*-significant differences from control plants at p < 0.05; **-significant differences from control plants at p < 0.01).

The stomatal conductance (Table 2), although not statistically significant, was lower in basil plants grown after green basil and higher when grown after horseradish plants compared to control plants.

Chlorophyll fluorescence measured in the light adapted state (ϕPSII) increased in a significant manner in both purple basil plants grown after green basil and also grown after horseradish (Table 3).

The higher values of ϕPSII occurred due to significantly lower values in both Fs and Fm' for the two variants compared to the control.

Table 3. Chlorophyll fluorescence parameters in purple basil plants

Treatments /Parameters	Soil mixture	Soil mixture + basil	Soil mixture + horseradish
F_0	22.4±1.87	23.8±5.11	28±2.35
Fm	246.4±10.92	236.6±27.88	255.8±27.86
Fv	224±9.59	212.8±24.74	227.8±25.89
Fv/Fm	0.91±0.01	0.9±0.02	0.89±0.01
Fs	300.5±17.99	113.5±10.2**	146.67±9.08**
Fm'	1461.08±90.93	623.58±40.27**	790.75±47.75**
PSII	0.79±0	0.82±0.01**	0.81±0*

(*-significant differences from control plants at p < 0.05; **-significant differences from control plants at p < 0.01).

The Fv/Fm parameter, measured under a dark adapted state, recorded lower values, but not significantly, in treatments compared to the controls. These results are the effect of increased values of the F_0 parameter, although not significantly, in both variants compared to control plants.

Biochemical parameters
Among assimilatory pigments, the contents of chlorophyll *a* and carotenoids increased in purple basil grown after horseradish, while in the other cases, the contents remained similar compared with the controls (Fig. 2)

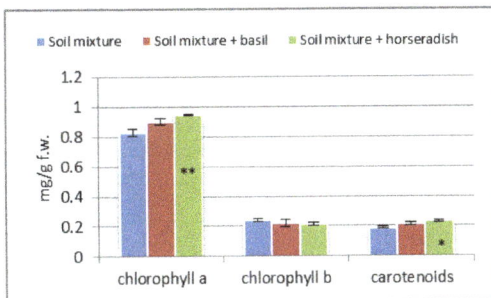

Figure 2. Assimilatory pigments contents in basil plants
(*-significant differences from control plants at p < 0.05; **-significant differences from control plants at p < 0.01).

The total content of phenolic compounds and of flavonoids, determined in ethanolic extracts,

increased in plants grown after horseradish compared to controls, with significance in the case of total phenolics. Purple basil plants grown after green basil registered a decrease of these parameters, significantly so in the case of flavonoids (Fig. 3).

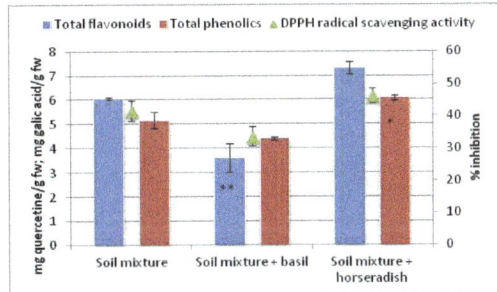

Figure 3. Total phenolics and flavonoid contents and free radical scavenging activity of purple basil plants 2.5% ethanolic extracts
(*-significant differences from control plants at p < 0.05; **-significant differences from control plants at p < 0.01).

These values were reflected also in the scavenging activity of the extracts, which was higher in plants grown after horseradish and lower in plants grown after green basil compared to control plants.

In crops, as in natural populations, the plants can influence each other through various processes. These include direct competition for light, water or nutrients, release of volatile compounds and soil mediated interactions. The soil represents a medium in which numerous substances are released, either by decomposition of plant material or by root exudates. These processes represent the basis for many agricultural practices that are included in organic farming systems, practices such as crop rotation, intercropping, mulching etc. (Farooq et al., 2011).

In the area around the roots, various substances can be found, originating from root exudates. These substances are represented by ions, oxygen, low molecular weight organic compounds (such as arabinose, glucose, oligosaccharides, amino acids such as arginine, asparagine, cysteine, glutamine etc. organic acids such as acetic, ascorbic, benzoic, ferulic, malic and phenolic compounds), as well as higher-molecular-weight compounds such as flavonoids, enzymes, fatty acids, growth regulators, nucleotides, tannins, carbohydrates, steroids, terpenoids, alkaloids, polyacetylenes, and vitamins. Some of these compounds, especially the phenolics, influence the growth and development of surrounding plants and soil microorganisms (Bertin et al., 2003).

The basil roots are known, under certain conditions, to produce rosmarinic acid (Bais et al., 2002). Horseradish roots were proven to secrete, in the substrate, peroxidase (Willey, 2016), probably together with other compounds, as the horseradish root residues are known to be toxic to certain species such as lettuce (*Lactuca sativa*) (Dias and Moreira, 1986; Itani et al., 2013).

In the present study, the fresh mass of basil plants and the number of flowering stems increased (however not significantly), as did number of lateral stems and leaves (significantly) when cultivated after other plants, suggesting a better availability of nutrients in substrates. This type of effect is known to occur as a result of root exudates and biomolecules interacting with compounds present in soil. Root exudates can enhance the solubility of both anions and cations (the latter by the presence of organic acids), can release Al and Fe from compounds, can form nano precipitates and can alter the adsorption of toxic elements and nutrients on substrate particles (Violante and Caporale, 2015). It is considered that certain properties of root exudates can favor agricultural production by increasing nutrient availability (Gianfreda, 2015) and also that proper crop rotation can lead to a 20% increase in yield (Farooq et al., 2011).

A possible stress might have occurred in the case of plants grown after the green basil cultivar, as these plants had reduced height but the highest number of lateral stems and leaves.

Regarding the photosynthetic rates, the decrease in basil plants cultivated after the green basil cultivar may be attributed to the higher number of leaves, but with smaller area per leaf. This kind of effects can occur under the presence of certain organic compounds in substrates such as organic acids which can also determine reduced stomatal opening (Zhou and Yu, 2006), as was found in the present study. The same parameters were not significantly influenced in the case of basil plants grown after horseradish, suggesting that the composition of exudates does not affect the photosynthetic process. The reduced stomatal conductance values of plants grown after green basil may also explain the decrease in transpiration rates (Chapin, 1991). Although apparently the stomatal conductance increased in plants grown after horseradish, the large standard deviations of the mean value may explain the reduction in transpiration rates in the same plants.

Regarding the chlorophyll fluorescence, in our study, the basil plants recorded similar values among treatments for both Fv/Fm and ϕPSII parameters, although the slight increase in the latter was statistically significant. Fluorescence parameters are generally known to decrease under various types of stresses, including the presence in substrates of allelochemicals, as was recorded for *Cucumis sativus* after application of cinamic acid (Zhou and Yu, 2006) or for *Dactylis glomerata*, *Lolium perenne* and *Rumex acetosa* after application of benzoxazolin-2(3H)-one and cinamic acid (Hussain and Reigosa, 2011a) or *Lactuca sativa* when exposed to cinamic acid (Hussain and Reigosa, 2011b). However, other reports have shown that the effect on chlorophyll fluorescence of *Bidens pillosa* and *Lolium perenne* of applied plant leachates was concentration dependent (Rashid et al., 2010). Such concentration dependence is also known as "saw tooth effect" and is explained by the simultaneous influence of chemicals on various physiological processes and the interactions among these (Reigosa et al., 1999). Also, the effects of chemicals on physiological processes depend on the nature of the chemical and on the tested species. For instance, the same compounds (benzoxazolin-2(3H)-one and p-hydroxybenzoic acid) had no effect on the chlorophyll fluorescence of *Polygonum persicaria*, but significantly decrease the values for *Dactylis glomerata*, while ferulic acid had no influence in the case of both species (Reigosa et al., 2001). We could therefore assume that the treatments did not induce a significant stress in tested plants and the concentrations of potentially toxic compounds exuded were low.

Assimilatory pigment content of tested plants increased, significantly for chlorophyll *a* and carotenoids in plants grown after horseradish.

Chlorophyll content is known to increase with the availability of the nutrients as reviewed in Marschner (2011). This is the case for many species, such as wheat (Bojovic and Sojanovic, 2005), melissa (Sharafzadeh et al., 2016) and basil (Politycka and Golcz, 2004), thus further sustaining the idea of improved nutrient availability in the substrates.

The content of phenolic compounds in plants is subjected to biotic and abiotic factors such as light, nutrient, salt stresses or pathogen interaction (Waśkiewicz et al., 2012). Phenolic contents can increase under fertilization of plants, as was observed in the case of basil (Scagel and Lee, 2012). However, a role as growth regulators was suggested for certain phenolics, with positive correlations observed between the amount of phenolics in tissues and plant growth (Kefeli and Kutacek, 1977). Since the basil plants grown after horseradish registered higher amount of phenolics and also increased plant height, while plants grown after green basil registered lower amounts of phenolics and reduced stem height compared to controls, such a hypothesis should be further investigated. Also, the different contents of phenolics in the plants further suggest the different nature of chemicals present in the substrates and their differential effect on basil plants. As a consequence of the amount of phenolics, the antioxidant activity of basil extracts varied accordingly, with higher scavenging ability for plants grown after horseradish, a correlation between these two parameters being proved for many species, including for basil (Juliani and Simon, 2002).

Overall, the data obtained for the growth of purple basil in the present study, indicate that basil can be successfully grown both after other basil cultivars as well as after horseradish. The positive effects on the growth of basil can be attributed to the presence in the substrate of exuded organic compounds such as organic acids and enzymes. Organic acids can improve nutrient availability (Violante and Caporale, 2015) while enzymes such as peroxidase can use organic toxic compounds as a substrate, degrading them and thus decreasing toxicity in the substrate (Vaughan et al., 1994). Although some exuded substances may influence physiological processes (photosynthesis, transpiration, stomatal conductance), cell division and elongation, membrane fluidity, protein synthesis, enzyme regulation (Farooq et al., 2011), the cultivation of several species simultaneously or in succession can positively contribute to the content of organic matter and nitrogen in the soil, improve water and nutrient availability and suppress weeds (Tringovska et al., 2015). *Armoracia rusticana* is a specie already used in crop rotation systems, which protects other crops from pests (Filipović et al., 2015) and it may exert other types of beneficial effects.

CONCLUSIONS

Ocimum basilicum L. plants can be grown on substrates where other basil or horseradish plants were previously grown with some beneficial effects, especially in the case of the latter. Positive effects are represented by increased morphological parameters values and bioactive compounds content. Further analyses are required to determine the type and the concentration of compounds excreted in substrate by roots and a more complete assessment of agronomic, physiological and biochemical parameters of plants grown in succession.

ACKNOWLEDGMENTS

Some of the analyses in this paper were performed using infrastructure provided by CERNESIM project, grant number 257/28.09.2010, SMIS/CNMR code 13984/901.
We thank PhD Floarea Burnichi and PhD Costel Vînătoru from Agricultural Research and Development Station at Buzau, Romania, for kindly providing the *Ocimum basilicum* seeds.

REFERENCES

Arabaci O., Bayram E., 2004. The Effect of nitrogen fertilization and different plant densities on some agronomic and technologic characteristic of *Ocimum basilicum* L. (basil). Journal of Agronomy, 3(4), 255-262.

Bais H.P., Walker T.S., Schweizer H.P., Vivanco J.M., 2002. Root specific elicitation and antimicrobial activity of rosmarinic acid in hairy root cultures of *Ocimum basilicum*. Plant Physiology and Biochemistry, 40(11), 983-995.

Bertin C., Yang X., Weston L.A., 2003. The role of root exudates and allelochemicals in the rhizosphere. Plant and Soil, 256, 67–83.

Bojović B., Stojanović J., 2005. Chlorophyll and carotenoid content in wheat cultivars as a function of mineral nutrition. Archives of Biological Sciences, 57 (4), 283-290.

Chang, C.W., Bandurski, R.S., 1964. Exocellular enzymes of com roots. Plant Physiology, 39, 60.

Chapin F.S., 199, Integrated Responses of plants to stress source. In: Willey N., Environmental Plant Physiology, BioScience, New York: Garland Science, 41(1), 29-36.

Chhonkar P.K., Tarafdar J.C., 1981. Characteristics and location at phosphatases in soil-plant system. Indian Society of Soil Sciences, 29, 215-219.

Dias L.S. Moreira I., 1988. Allelopathic interactions between vegetable crops and weed. In: Cavalloro R., Titi A. El (Ed.) Weed Control in Vegetable Production, CRC Press, 197-211.

Farooq M., Jabran K., Cheema Z.A., Wahid A., Siddique K.H., 2011. The role of allelopathy in agricultural pest management. Pest Management Science, 67 (5), 493–506.

Filipović V., Popović V., Milica Aćimović M., 2015. Organic production of horseradish (Armoracia rusticana Gaertn. Mey. and Scherb.) in Serbian Metropolitan Regions. Procedia Economics and Finance 22, 105 – 113.

Gianfreda L., 2015. Enzymes of importance to rhizosphere processes. Journal of Soil Science and Plant Nutrition, 15(2), 283-306.

Grayer R.J., Kite G.C., Goldstone F.J, Bryan S.E., Paton A., Putievsky E., 1996. Infraspecific taxonomy and essential oil chemotypes in sweet basil, Ocimum basilicum. Phytochemistry, 43(5), 1033-1039.

Herald T.J., Gadgil P., Tilley M., 2012. High-throughput micro plate assays for screening flavonoid content and DPPH-scavenging activity in sorghum bran and flour. Journal of the Science of Food and Agriculture, 92(11), 2326-31.

Hussain M.I., Reigosa M.J., 2011a, Allelochemical stress inhibits growth, leaf water relations, PSII photochemistry, non-photochemical fluorescence quenching, and heat energy dissipation in three C3 perennial species. Journal of Experimental Botany, 1-13.

Hussain M.I., Reigosa M.J., 2011b, A chlorophyll fluorescence analysis of photosynthetic efficiency, quantum yield and photon energy dissipation in PSII antennae of Lactuca sativa L. leaves exposed to cinnamic acid. Plant Physiology and Biochemistry, 49, 1290-1298.

Itani, T., Nakahata Y., Kato-Noguchi H., 2013. Allelopathic activity of some herb plant species. International Journal of Agriculture and Biology, 15, 1359-1362.

Juliani, H.R., J.E. Simon. 2002. Antioxidant activity of basil, In: Janick J., Whipkey A. (Eds.), Trends in new crops and new uses. ASHS Press, Alexandria, VA. 575–579.

Kefeli V.I., Kutacek M., 1977. Phenolic substances and their possible role in plant growth regulation, In: Pilet

P.-E. (Eds.) Plant Growth Regulation, Part of the series Proceedings in Life Sciences. Springer-Verlag Berlin Heidelberg, 181-188.

Marschner H., Mineral Nutrition of Higher Plants. Second Edition, Academic Press, 299-312.

Marzocco S., Calabrone L., Adesso S., Larocca M., Franceschelli S., Autore G., Martelli G., Rossano R., 2015. Anti-inflammatory activity of horseradish (Armoracia rusticana) root extracts in LPS-stimulated macrophages. Food and function, 12(6), 3778-3788.

Meyers M., 2003. Basil a Herb of America Society Guide, The Herb of America Society. Ohio, 10-12.

Pârvu C., 2002. Enciclopedia plantelor. Vol. I, Ceres, București, 289-290.

Politycka B., Golcz A., 2004. Content of chloroplast pigments and anthocyanins in the leaves of Ocimum basilicum L. depending on nitrogen doses. Folia Horticulturae, 16(1), 23-29.

Putievsky E., Galambosi B., 1999. Production systems of sweet basil, In Hiltunen R., Holm Y., (Eds.) Basil: The genus Ocimum. Harwood academic publisher, 39-66.

Rashid, M.D.H., Asaeda T., Uddin M.D.N., 2010. Litter-mediated allelopathic effects of kudzu (Pueraria montana) on Bidens pilosa and Lolium perenne and its persistence in soil. Weed Biology and Management, 10(1), 48–56.

Reigosa M.J., Sánchez-Moreiras A., González L., 1999. Ecophysiological approach in allelopathy. Critical Reviews in Plant Sciences, 18(5), 577-608.

Reigosa M.J., González L., Sánchez-Moreiras A.M., Durán B., Puime O., Fernández D.A., Bolaño J.C., 2001. Comparison of physiological effects of allelochemicals and commercial herbicides. Allelopathy Journal, 8, 211-220.

Scagel C.F., Lee J., 2012. Phenolic composition of basil plants is differentially altered by plant nutrient status and inoculation with mycorrhizal fungi. Hortscience 47(5), 660–671.

Sharafzadeh S., Khosh-Khui M., Javidnia K., 2011. Effect of nutrients on growth and active substances of lemon balm (Melissa Officinalis L.). Acta Hort, 925, 229-232.

Simon, J.E., 1996. Basil. New crop factsheet. Purdue Unviersity, West Lafayette, Indiana.

Tringovska I., Yankova V., Markova D., Mihov M., 2015. Effect of companion plants on tomato greenhouse production. Scientia Horticulturae, 186, 31–37.

Vaughan D., Cheshire M.V., Ord B.G., 1994. Exudation of peroxidase from roots of Festuca rubra and its effects on exuded phenolic acids. Plant and Soil, 160, 153-155.

Violante A., Caporale A.G., 2015. Biogeochemical processes at soil-root interface. Journal of Soil Science and Plant Nutrition, 15(2), 422-448.

Waśkiewicz A., Muzolf-Panek M., Goliński P., 2013. Phenolic content changes in plants under salt stress, In: Parvaiz A., Azooz M.M., Prasad M.N.V., (Eds.) Ecophysiology and Responses of Plants under Salt Stress. Springer, 283-314.

Wellburn A.R., 1994. The spectral determination of chlorophyll *a* and *b*, as well as total carotenoids, using various solvents with spectrophotometers of different resolution. Journal of Plant Physiology, 144, 307-313.

Willey N., 2016. Environmental plant physiology. Garland Science, New York, 315.

Yamaguchi M., 2012. World Vegetables: principles, production and nutritive values. Springer Science and Business Media, 236.

Zhou Y.H., Yu J.Q., 2006. Allelochemicals and photosynthesis, In: Reigosa M.J., Pedrol N., González L. (Eds.), Allelopathy - A Physiological Process with Ecological Implications, Springer, 127-140.

4

COMPARATIVE STUDY ON GROWTH AND DEVELOPMENT IN *TAGETES* GENUS

Ana Maria BĂDULESCU[1], Florina ULEANU[1]

[1]University of Pitesti, Department of Science Romania, 1 Târgu din Vale str., 110040, Pitesti, Romania

Corresponding author email: uleanuflorina@yahoo.ro

Abstract

Tagetes species are well known and grown both in cities and in villages. They are appreciated for their long flowering from early summer to late autumn. Improvement of technological links to the species Tagetes in solar is one of the main concerns of flowers growers that yields high quality products and low production costs. As a result of widening the range of Tagetes varieties is constantly necessary of their study in comparative cultures to see how expressing the genetic heritage in areas where are grown.

Key words: Tagetes, technological links, solarium.

INTRODUCTION

About the total area planted with flowers in Romania today it is hard to say. In 1987 there were about 135 ha planted with flowers, greenhouses belonging to the state sector. Other 1,500 ha were recorded with flower field crops and nurseries. Areas planted with flowers, both in protected crops and out, were much bigger if we consider the private sector that work in this area of activity, but these did not recorded in official statistics (Şelaru, 2008).

Currently, the trade balance is negative (import is increasing and exports almost nonexistent). The main causes are the production decline due to outdated technologies, overcome assortment and, above all, a favoured commercial policy.

For Romania, the main problem is to align external quality standards, recover old markets and break into new ones through diversification of the assortment, upgrading, using seedlings. Floriculture future depends on how are provided: quality, quantity and production continuity.

Tagetes genus includes about 30 species of which only three annual species shows more interest in ornamental plants culture (Drăghia and Chelariu, 2011). They are appreciated for their long flowering period from early summer to late autumn, but also to control and combat *Pratylencus* nematodes in vegetable gardens in terms of environmental technologies without the use of insecticides. The results show the possibility to fight *Pratylenchus* nematodes with 40-70% depending on the used species (Prohab and Borcean, 2009).

Improvement of technological links to the species Tagetes in solar is one of the main concerns of flowers growers that yields high quality products and low production costs. As a result of widening the range of *Tagetes* varieties is constantly necessary of their study in comparative cultures to see how expressing the genetic heritage in areas where are grown.

MATERIALS AND METHODS

The experience was conducted in solarium conditions, in 2015, in Pitesti area, Argeş county. As biological material were used seven *Tagetes* hybrids, that represented the experimental variants as follows:

V_1-*Tagetes patula nana „Aton Spry"* (Figure 1)

V_2-*Tagetes patula nana „Aton Flamed"* (Figure 2)

V_3-*Tagetes patula nana „Aton Bee"* (Figure 3)

V_4-*Tagetes patula nana „Aton Orange"* (Figure 4)

V_5-*Tagetes patula nana „Durango Bolero"* (Figure 5)

V_6-*Tagetes erecta „Antigua Yellow"* (Figure 6)

V_7-*Tagetes erecta „Antigua Orange"*(Figure 7)

No studies have been made on these hybrids of *Tagetes*. During the growing season, 16.04-04.06.2015, there were applied proper care works. Biometric measurements have been made weekly from planting to sale: the height of plants, number of leaves, leaf length and width, also the number of flowers.

Figure 1. V₁— *Tagetes patula nana ,,Aton Spry"*

Figure 2. V₂— *Tagetes patula nana ,,Aton Flamed"*

Figure 3. V₃— *Tagetes patula nana ,,Aton Bee"*

Figure 4. V₄— *Tagetes patula nana ,,Aton Orange"*

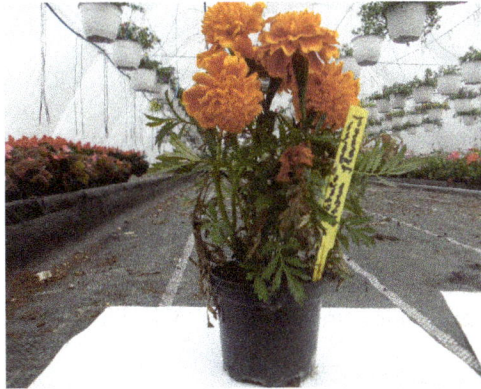

Figure 5. V₅— *Tagetes patula nana ,,Durango Bolero"*

Figure 6. V₆— *Tagetes erecta ,,Antigua Yellow"*

Figure 7. V₇— *Tagetes erecta „Antigua Orange"*

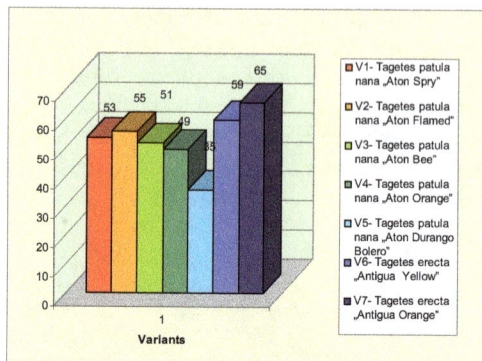

Figure 9. The total number of leaves per plant

RESULTS AND DISCUSSIONS

It was determined the rate of growth in height of plants by taking measurements weekly from planting to sale (Figure 8). Growth in *Tagetes* varieties was different from one variety to another. The best was developed as expected a representative of the *Tagetes erecta* species namely *Tagetes erecta* "Antigua Orange" -V7 with 26.3 cm and the weakest *Tagetes patula nana* "Aton Durango Bolero"-V5 with 20.7 cm.

The total number of leaves per plant was different from one variety to another and ranged between 35 leaves at V5- *Tagetes patula nana* — "Durango Bolero" and 65 leaves at V7- *Tagetes erecta* "Antigua Orange".

The number of flowers per plant varied from one variety to another (Figure 10). On 16.04.2015 flourished varieties of *Tagetes patula nana* and on 23.04.2015 flourished *Tagetes erecta* varieties, with flowers obviously higher than those of the species mentioned above.

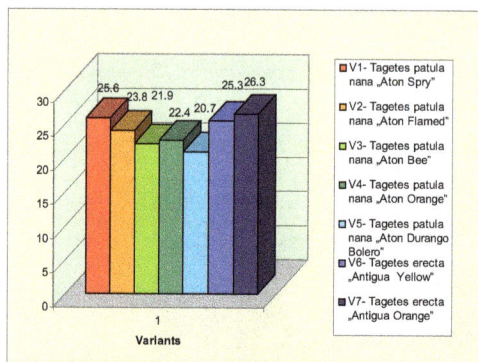

Figure 8. Growth in height of plants

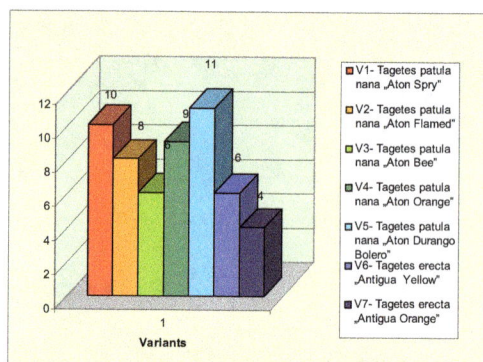

Figure 10. The number of flower per plants

The number of leaves per plant is shown in Figure 9.

At transplanting date (14.03.2015) all plants had the same number of leaves (2), but at the end of the measurements (4.06.2015) as we can see the number of leaves per plant varied depending on the variety.

The best results were registered for the variety *Tagetes patula nana* "Durango Bolero" by 11 flowers per plant, and the worst results were recorded in *Tagetes erecta* "Antigua Orange" with a maximum of 4 flowers per plant.

It is noted that representatives of the *Tagetes patula nana* had a larger number of flowers per plant compared with representatives of *Tagetes erecta*.

CONCLUSIONS

Comparative study of new varieties of *Tagetes* is required to recommend the most suitable varieties to growers according to their destination.

Rhythm of growth in plant height proves again that the species *Tagetes erecta* show higher growth compared to *Tagetes patula*.

It appears that at transplanting moment all plants have the same number of leaves, but at the end this character differs depending on the variety.

Antigua orange variety shows the best results in terms of this character, while the lowest values recorded in Durango Bolero variety.

The best results in case of flower number were registered for the variety *Tagetes patula nana* "Durango Bolero" by 11 flowers per plant, and the worst results were recorded in *Tagetes erecta* "Antigua Orange" with a maximum of 4 flowers per plant.

If you are looking for the highest plant with larger flowers we recommend the cultivation of the species *Tagetes erecta*, and if desired cultivation of varieties shorter but more abundant flowering we recommend the choice of the species *Tagetes patula nana*.

All varieties studied were characterized by good growth rate, high percentage of flowering and looked very attractive.

The study undertaken can confidently recommend cultivating of new varieties because they have high productivity, particularly commercial aspect and resistance to handling and transport.

REFERENCES

Drăghia L., Chelariu E. L., 2011. Floricultură, Ed. „Ion Ionescu de la Brad", Iași

Prohab I, Borcean I., 2009. Tagetes: nematode unfriendly medicinal,ornamental plant tagetes Research Journal of Agricultural Science DADR Timișoara, USAMVB Timișoara, 41.

Șelaru E., 2008. Cultura florilor de grădină. Ed. Ceres, București.

5

THE AESTHETIC CRITERION – COMPONENT FOR DESIGNING URBAN PUBLIC GREEN SPACE. STUDY BASED ON TWO ALTERNATIVES FOR IZVOR PARK, BUCHAREST

Anca STĂNESCU[1]

[1]University of Agronomic Sciences and Veterinary Medicine of Bucharest,
59 Marasti Blvd, District 1, Bucharest, Romania
Corresponding author email: anca_stanescu2001@yahoo.com

Abstract

The design of urban green space is based on a series of principles, criteria and determinant factors for the quality of the urban public space. These criterion and principles are defined by multiple aspects: physical, functional, ambient, aesthetic and ecological, each of which are important in the configuration of the landscape arrangement project. The current work addresses the problem of the aesthetic criterion in landscape design by studying comparatively two project alternatives, each having a distinct compositional style and each representing a certain type of aesthetic vision for the organization of the physical space. These alternatives belong to a mixt composition style specific for the urban landscape design of the 20th century.

Key words: urban green space, landscape design, aesthetic criterion

INTRODUCTION

Izvor Park covers a surface of approximately 17 hectares and is located in the central area of Bucharest close to the People's Palace, while being framed by the roads: Splaiul Independenței, Izvor Coșbuc Street, B. P. Hașdeu Street and Mihai Vodă Street.

The park was partially developed between 1987 and 1988 as the first version of land vegetal "furnishing", followed by the selection of one of the 22 designing alternatives created for the park in the same period of 1988.

The 1989 revolution has brought to a stand the final design development and construction of the park. For the completion of the study alternatives, both the landscape function and the general compositional shape of space organization were considered, while the aesthetic criterion represented an essential component in designing this urban green entity. In landscape design the aesthetic criterion has direct final correspondence in the visual and ambient quality of the physical space. Therefore the importance of approaching the aesthetic criterion when developing a design is of maximum interest for the landscape architects.

MATERIALS AND METHODS

The research method used within the present work is that of comparative analysis of the two alternatives V_1 and V_2 in order to highlight the main aesthetic characteristics of both projects. These alternatives represent the actual study material of the work. It pursues the analysis of aesthetic composition elements which match two different aesthetic trends. Each of the two alternatives is the result of practical compliance within the project towards the landscape functionality attributed to this space.

The prevalent landscape function is that of promenade-rest, alongside the function of pedestrian transit and pedestrian connection between the adjacent roads of the site. The composition elements are alleys, water and vegetation, the latter being represented by ensemble arboreal vegetation, lawns, arboreal alignments and floral decorations (in alternative V_1). The park entrances and promenade alleys build the base compositional structure of the projects, which is further sustained by the other elements – water and vegetation – in the final aesthetic configuration of each alternative. In alternative V_1 water has a considerable presence, the surface of the water mirror holding approximately 40000 m^2 while representing an important visual element of the designed space. In alternative V_2 water is present on much smaller surfaces in two circular basins located at alley intersections (small circuses); within this alternative water holds a surface of 630 m^2.

RESULTS AND DISCUSSIONS

Stylistically, the two study alternatives each belong to a distinct trend: alternative V_1 (Fig. 1) is composed of a geometrical style (Stănescu, 2011). All the elements – alleys, water, vegetation – take regular geometrical shapes, resulted from the association of straight lines as they highlight the main characteristics of this particular style (Mostaedi, 2004): tracks of straight alleys, alley intersections under 90° angles market with two decorative water basins which take circular shape. The surface of the water mirror is the dominant element of the composition and it also has rectangular shape with slightly smoothed edges. Vegetation is arranged wither in simple or double alignments along the alleys or in groups and bulks towards the outer areas of the park. Vegetation also includes floral decorations with geometrical shapes located alongside the main alleys of the park, which they enrich aesthetically and visually.

Alternative V_2 (Fig. 2) is composed of a free landscape style characterized mainly by network of alleys and the layout method for the vegetation (Stănescu, 2011). The alleys comprise of free shapes which are sinuous, comfortable, with a pleasant aspect; the two circuses representing the alley intersections are marked by two decorative water basins, circularly shaped, each having the surface of 315 m^2. The presence of water is much more discreet that in the other alternative, leaving more space for vegetation, which is the dominant characteristic of this design alternative. The central area of the park is reserved for wide spaces covered in lawn, while the outskirts are covered in groups and bulks of broadleaves and coniferouses (Kluckert, 2005). There are also present alignments in multiple sequences in order to create a real protection screen for the park against the rest of the urban area.

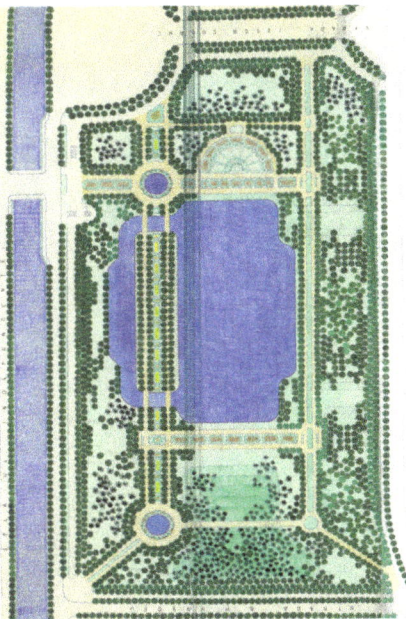

Figure 1. Alternative V_1

Figure 2. Alternative V_2

CONCLUSIONS

The current study was based on the analysis of two landscape design alternatives – V_1 and V_2. Alternative V_1 was considering the geometric architectural style specific for the 80' in this urban area of Bucharest, which is located near the People's Palace and is considered to be a representative area of the city. Alternative V_2 is characterized by a free landscape style which is much less drastic and strict.

It can be concluded that the compositional aesthetic principles applied in these projects prevail over the functional elements, but do not cancel them, but on the contrary sustain them.

The alternatives studied fall under two distinct style trends which are essential components of the aesthetic criterion in landscape design.

ACKNOWLEDGEMENTS

I would like to express my sincere gratitude to the landscape architect Valentin Donose, the author of the two project alternatives, and to the Design Institute "Proiect Bucureşti" in which he carried out his professional activity.

REFERENCES

Kluckert E., 2005. European Garden Design. hf ULLMANN - Tandem Verlag GmbH, Königswinter, 484.

Mostaedi A., 2004. Landscape Design Today. Carles Broto & Josep Maria Minguet, Barcelona, 158-164.

Stănescu A., 2011. Peisagisică urbană vol. II. Editura Printech, Bucharest, 40-47.

6

EARLY MODERN GARDEN DESIGN ILLUSIONS AND DECEPTIONS. TWO DIFFERENT QUESTS FOR PARADISE - VILLA LANTE AT BAGNAIA AND VILLA ORSINI AT BOMARZO

Alexandru MEXI[1]

[1]"Vasile Goldiş" Western University of Arad, 94-96 Revoluţiei Avenue, Arad, Romania
Corresponding author email: alx.mexi@gmail.com

Abstract

Considered some of the most fascinating examples of Mannerist gardens in Italy, Villa Lante at Bagnaia and Villa Orsini at Bomarzo, were conceived in the late 16th century as a kind of landscape and architectural expressions of the Man's quest for the Lost Paradise. Although they were designed roughly in the same period of time, they differ completely in almost every sence, but, in the same time, they complete each other in many ways.
The following paper aims to compare the architectural layouts, planting designs and allegorical programmes of both gardens as to emphasize the way Paradise was seen in the late 16th century in Western Europe and to show two different landscape design mechanisms that try to mimic the quest for the (re)discovery of Eden.

Key words: Paradise, garden philosophy, Mannerist, 16th century, Villa Lante, Bomarzo

INTRODUCTION

If we can consider the garden to be an image – a figure, then, walking through the garden, through its paths may often be a real process of allegories - a process of figuration and most of all, especially in the Late Reneissance – a quest for finding God. Perhaps the most representative examples of this figurative process of some of the ideas the garden is trying to convey are the famous Italian Mannerist gardens of the sixteenth century; and out of these, two stand out - they both complement but also cancel each other; they intertwine, but are fundamentally different, they revolve around almost identical ideas and symbols, but convey different messages (apud. Lazzaro, 1980). The two gardens, Villa Lante from Bagnaia and Villa Orsini or Sacro Bosco from Bomarzo, build both on their own and together a universe translated into a microcosm of symbols, alchemical allegories and meta-linguistic messages – they tackle the theme of Man's return in Paradise but offer distinct solutions and variants:
"The Renaissance Universe is hierarchical, with God at the summit, human beeings in the center, nature below, and each part related to the other. The natural world was perceived in terms of its usefulness for human needs: plants and animals provide food and medicine. They also reflect human traits, virtues and beliefs and therefore serve as symbols – heraldic, moral, philosophical and religious. At the same time, the visible world correspondes with the divinely cosmos – the microcosmo reflects the macrocosm. To know this world is therefore to know God. " (Lazzaro, 1990)

MATERIALS AND METHODS

In order to understand the process of figuration and the quest to find the Lost Paradise that the two gardens promise, one must first closely observe all of the components of Villa Lante and of Villa Orsini. To this end, the following chapter aims to focus on reorganizing the gardens' elements in order to recompose the allegorical messages they intended to express when they were built.

Villa Lante and the Retrieval of Paradise
Villa Lante is probably the most interesting landscaping complex of the sixteenth century in Italy; it amazes through the architectural structure adopted in the garden, the use of slopes, water, by joining a *Hortus Conclusus* with a *bosco* etc. But the most spectacular part of this villa is precisely the philosophical story

it is trying to share through the arts, architecture and natural elements.

The landscaping complex brings together two complementary compositions – the mathematical garden or the tamed nature and the *bosco* or the wild nature – *"Nature, the very embodiment of ultimate divinity, was terrifying not only in its absolute power but also in its lack of content. Only mankind, the political animal, intended something"* (Scully, 1991). If the landscaping mathematics suggests divine reason and talks about a Heaven created by a mathematical God (*Deus geometer*) then the *bosco* or the forest represents the original chaos, the Fall and also the antithesis of Paradise.

The garden uses artistic metaphors and takes the visitors on an allegorical route that includes the main historical stages or key moments of the human life on its continuous path to absolute reason, to finding Paradise again – to illumination. This figural process starts from the chaos of the forest and ends in "geometry and perspective", but at the same time, it goes through several "checkpoints" that define the human condition from its appearance on earth until it reaches the supreme goal – reaching illumination or finding the lost Eden.

To be more exact, the route begins through a secondary gate and not at the primary (as one might think) that brings visitors face to face with the fountain of the mythical Pegasus. The image, through the legend that revolves around this sacred animal (Chevalier and Gheerbrant, 1995), suggests the fact that this place is the origin of all things and of the adventure that aims to obtain the ultimate divine illumination. The place where the winged horse's hoof hits the ground becomes the starting and the inspiration point that will help man find the realm of reason – the mathematical Heaven! (Don, 2008).

Located at the entrance of the complex, the fountain of Pegasus can be found right next to the very landscaping expression of Heaven – the mathematic garden; but the visitor is not allowed to get directly to it and neither does he know this garden exist. This garden takes the form of a *Hortus Conclusus* and a *Garden of paradise* at the same time – *"Originally, fences were a practical land symbolical necessity for enclosing sacred places, and indeed the very concept of the enclosure had been deeply rooted in the vocabulary of the garden ever since the medieval term Hortus Conclusus identified its special character. Enclosing even a small bed with a fence or even just a small border was a means of making it sacred, so that its symbolic value took priority over any practical purpose"* (Vercelloni and Vercelloni, 2009).

The fountain has therefore a double role - invites the visitor to enter the garden, but also represents a hyphen which both separates and links the wild garden with the rational one.

Leaving the fountain of Pegasus behind, in front of the visitor's eyes appears the forest (symbol of the unconscious and of anti-reason, chaos) from which point he can choose many paths, each sending him to a new adventure and a new discovery. Among the most significant such points of the bosco one might remember a few elements such as: Il Barco, the maze, the beaver's fountain, the mascarons' fountain, the lions' fountain, the fountain of Bacchus etc.

Il Barco. Il Barco is actually a hunting annex built by Cardinal Ottaviano Riario Visconti, grandson of the original owner of the complex, the Cardinal Raffaele Sansoni Riario (Ehrenfried, 2007). The latter was the one who in 1498 closed the entire perimeter of Villa Lante with walls, making it a hunting park. The building, which still bears the symbols of Cardinal Ottaviano Riario Visconti (a snake coiled around a flower), was used as a hunting pavilion and later was incorporated into the bosco's structure (https://villalante.wordpress.com/il-barco/).

The lions' fountain. The Lions' fountain represents a statuary set, hidden deep in the woods, and it is made up of a circular dip in stone with 4 symmetrically displayed lions with water flowing out of their mouths. There is no known accurate information on the origin and/or its significance, but this element is present even in the sixteenth century famous plan of Villa Lante. (https://villalante.wordpress.com/il-barco/).

The mascarons' fountain. The mascarons' fountain is also hidden deep in the woods, and, just as in the Lions' fountain, there is not much information on its origin or meaning.

Moreover, in the centre of the rectangular basin that surrounds it, one can find an almost identical fountain with that on the right of the Diluvial Fountain on the upper terrace of the geometric garden (https://villalante.wordpress.com/il-barco/). These parallels between the two distinct microcosms (the forest harden and the mathematic garden) seem to realize connections between them, joining the two separate parts of the villa both physically and symbolically, making them indissoluble.

Beaver's Fountain. This fountain also follows the same pattern, as it is not yet described in any of the researched documents. Perhaps it also aims to highlight the natural wilderness and it contributes to shaping that image and atmosphere of chaos.

The Labyrinth. Unlike the other elements mentioned above, this one no longer exists today. Given the fact that the symbolic significance of the maze is made up of a search process, it is perfectly plausible for this part of the garden to be located in the *bosco* just to emphasize the chaos and incessant search of the individual for a higher world. Perhaps its purpose was to entice visitors, making them believe that going through the maze would get them out of this forest with sacred powers, while it did nothing else but to bring man even more to the Fall, deceiving and tiring him. Moreover, the metaphor of the labyrinth was doubled by the metaphor of the forest (in itself a maze), the route becoming increasingly more difficult and dangerous through the bosco, the exit out of this place getting a much brighter aura.

Having escaped from the forest adventure, visitors can step into perhaps the most spectacular garden of the Italian Mannerism. This new landscaping construction suggests liberation from the terror of the unconscious and takes its visitors on a straight path that runs together with the tumultuous journey of a watercourse. So the water becomes the guide and leads man towards his ultimate goal.

"The Early Reneissance garden was primarily static and could be viewed in its entirelly from a fixed point of view. To borrow from literary terminology, it had a unity of space and time. The gardens after the 1520s consisted of a series of successive spaces, isolated from each other physically and visually. They could only be experienced through movement, and the relationship between spectator and garden became active rather than passive. " (Graafland, 2003 quoting MacDougall, ***)

The water, which must be followed closely, stems from the so-called "Fountain of the Diluvial" (Iliescu, 2014). It symbolizes the biblical flood that cleansed the world of "impurities" and that provides through purification the chance of a new order of life. This allegorical fountain separates therefore the garden of knowledge from the infamous *bosco*. It is composed of three major elements, each creating a new image and outlining a new symbol that completes the allegories of this ensemble. The fountain leads the way to thinking and reason, representing in the philosophical semantics of the garden, the "archaic harmony between man and nature" (http://www.lazio.dk/villa_lante_di_bagnaia.htm) – the relationship of interdependence and mutual respect between a rational being and his life environment.

So the main source of water comes from a cave, a symbol of the maternal womb and uterus, highlighting both the purifying role of the water and the symbolic role of the cave. This cave is flanked by two loggias that Cardinal Gambara adorned with paintings of the muses from the Greco-Roman mythology – thereby suggesting the divine role of the arts in the man's process of search and finding of the Paradise. From the roofs of these loggias spring up, in the spirit of Late Renaissance or Mannerism, a series of fine water jets designed to capture and sprinkle visitors – a typical farce of the sixteenth century.

The last piece of this fountain is composed of a container made of carved stone with four symmetrically arranged human figures on the outside. Perhaps, just like its counterpart in the *bosco*, this too forms a link between the human being and the route that he must follow in life and the entire vocabulary of symbols associated to number four (4 seasons, 4 cardinal points, 4 human states, 4 dominant winds, 4 sacred rivers of Paradise, 4 phases of the moon, 4 elements etc). Moreover, this element makes, through repetition, a new link between the forest and the geometrical garden, pointing out the connections, the rivalry, but also the interdependence between the two seemingly antithetical components of the landscaping complex from Bagnaia – *"a garden manifests the rivalry between man and nature, not the victory of one over the other"* (apud. Lazzaro, 1990).

The allegorical route continues along a strong symmetry axis subordinated to a straight route of a water course. Following its course, water takes the form of a hexagonal fountain (the Dolphin Fountain- which incorporates the heraldic symbols of Cardinal Gambara) where, the path narrows and opens in a depth perspective, along a water staircase which opens and ends with one architectural item representing Cardinal Gambara's emblem – the crayfish. From here, the water flows further to the Giants' waterfall, an architectural waterfall flanked by the statues of the Tiber and Arno rivers. This, with its two statues arranged symmetrically as to the way water falls, symbolizes the friendship between the Papacy in Rome (Tiber) and the Medici family in Florence (Arno). From the allegorical fountain of the two rivers, the water keeps flowing through the Cardinals' table (object made out of stone that served as entertainment and dining space) (apud. https://villalante.wordpress.com/mensa-del-cardinale/).

According to Kluckert Ehnrenfried, the Giants'fountain and its layout on the central axis in the middle of the geometrical garden can be interpreted as a metaphor not only for the friendship between the Medici and Vatican, but also as a metaphor for the road from Bagnaia to Bomarzo (Ehrenfried, 2007), or from Searching for Heaven to Avoiding Eden.

From the Cardinals' Table, the water leads to an "Enlightenment" which is more and more visible. The course takes the form of a new waterfall – Waterfall of Lights, and this one, unlike the other decorative pieces, does not include in its design statues of human or animals figures, but terraces with pots of flowing water. However, although it doesn't incorporate human figures, statues of Greek and Roman deities, Venus and Neptune, can be found in caves arranged symmetrically to

the central axis (Iliescu, 2014). The Waterfall of Lights can be seen as representing boiling water or the purifying force of water and fire combined (https://villalante.wordpress.com/fontana-dei-lumicini/).

Leaving behind this fine work of art, the view widens and opens up to the last two parts of the garden. However, before concluding the allegorical route, the visitor can observe the appearance of two identical buildings located on one side and the other of the watercourse: Pallazina Gambara and Pallazina Montalto.

Pallazina Gambara was the first building erected in the "Garden of knowledge/of reason". Built between 1568-1578 after a square plan, it incorporated the Gambara family symbol, the crayfish (symbol of the road to Salvation, symbol of loyalty), and a torch (symbol of martyr fire) and an inscription (SOL ALIIS) that can be interpreted as saying eighter "Only for others" - meaning that the project was conceived as a spiritual message addressed by the Cardinal to the visitors or "Light for others" - meaning that the sun, symbol often used by heretics and pagans, is for all and not just for the Church of Rome (https://villalante.wordpress.com/palazzina-gambara/).

Moreover, it seems that these symbols are related one way or another with a comet which coincided with the time the pavilion was constructed (https://villalante.wordpress.com/palazzina-gambara/); but so far, this information could not be verified. Inside, the building has painted scenes which almost always show ideas linked to Salvation and the (re)discovery of Paradise.

Palazzina Montalto, although it appeared in the original design of Cardinal Gambara, it was built between 1590 and 1612 by Alessandro Peretti Montalto Damascena, nephew of Pope Sixtus V.

The only difference between the two pavilions consists in symbols that decorate the exterior of the two buildings. Thus, in Pallazina Montalto's case, the building's frieze is adorned with the following symbols: mountains representing the Montalto dynasty, a branch with pears, referring to the Peretti dynasty, the star with eight corners - Christian symbol of the eternal salvation. On the other hand, the pavilion's interior was decorated in the same iconographic spirit as Pallazina Gambara (https://villalante.wordpress.com/palazzina-montalto/).

Passing through the two pavilions that served as the owners and their servants' house in the past, the visitor ends the allegorical route of taming and of the absolute control of nature. This mastery of nature emerges as a geometric floor in which even water stops being wild, and it becomes subject to the human will. So, it is no longer churning and it no longer falls, but takes the form of a still water and stands in a geometric pool – the Fountain of the Moors.

This fountain is a reinterpretation of the Persian model of Heaven and is often perceived as embodying a calm sea that can be crossed in all directions. At its centre is not a piece of water or a pot with plants, as it happens in the Persian design and later in the Arabic and even Christian gardens, but we discover the statues of four young people (called Moors because of the dark color of the volcanic rocks they are made of) who support the heraldic symbols of the Montalto family (pears, mountains and the star with eight corners).

It seems that this reinterpretation of the Garden of Paradise is actually much more complex than it may seem. It appears that the perfect geometric shape of the garden was inspired by the legend of the martyrdom of St. Lorenzo. He was burned on a grill shaped like a grid because he refused to provide money from the Christian community to the prefect of Rome.

Moreover, Cardinal Gambara was apparently very attached to this saint because one might even see that, in his pavilion, there is a fresco depicting the martyrdom of St. Stephen and Lorenzo. It is considered by some authors that this legend became the main theme of the garden of reason - "*the grid is, in other words, the sacred symbol representing Saint Lorenzo, one of the most celebrated holy Catholic men, who devoted his entire life to Jesus Christ and the Cross.*" (https://villalante.wordpress.com/graticola-di-san-lorenzo/).

Regarding the circular shape of the basin, it seems to have been inspired by the architecture of the church of Santo Stefano Rontondo in Rome and together with the floors with rich embroideries of buxus represented, in Gambara's view, symbols of absolute faith. Even the water gushing from the well was used as a metaphor of the road, purification and salvation that the visitor will receive when he will reach the heavenly Jerusalem (https://villalante.wordpress.com/fontana-dei-mori/).

In the four rectangular pools arranged symmetrically around the central statue, there are four stone boats, each containing a soldier carved in stone. These were intended to represent four soldiers throwing water at the heraldic symbols, suggesting both the Protestant attacks of the Latin Church and especially the Turks' attacks stopped by the fleet led by some members of the Gambara family at the Battle of Lepanto (https://villalante.wordpress.com/fontana-dei-mori/). At present, however, for unknown reasons, the water jet produced by the statues of the four soldiers goes the opposite direction - toward the ground floors with the box embroideries.

Beyond the geometric ground there is a narrow perspective that connects the central axis of the garden and the main street of Bagnaia. But to reach the city, the visitor must pass through a monumental gate disposed axially to the garden. This gate is present even in the sixteenth century plans, but its current configuration is most certainly due to the Lante della Rovere family (last owners of the Villa, XVII century) who decided to adorne this facade with their personal emblem. (https://villalante.wordpress.com/passeggiando-nella-villa/ingresso-foto/)

The allegorical route of this garden ends therefore, next to the central gate (Ingresso). It ends a process of figuration taking the visitor through a world that takes secular, mythological and biblical meanings, and that can talk to and relate both the divine purpose of man (retrieving Paradise) and his own passage through life. On the other hand, the same route can be travelled in the opposite direction, suggesting a path of retreat and meditation:

"We start in the town at the foot of the mountain: the good, proud, sharp-edged,

manmade town. From there, we enter a gate in a mythical wall. What can lie behind it? In fact, it opens upon a broad, geometric parterre with a deep basin of water lifted in its center, above which the stemma of the Montalto is flaunted in the air. Beyond the parterre, the villa is divided into two pavilions, one on the left, one o the right, eith the ramps of Praeneste and Tivoli mounting between them toward a dark, heavily wooded garden. This climbs the mountainside toward secret pools and springs an dis continued by a wooded park that climbs farther up the slopes – who knows, perhaps into the unknown heart of the mountain itself.

What a sinister courtesy the villa opens up to invite us into the wild. How dark it i sunder the trees. As we press onward, the shaggy shapes of the forest begins an driver gods, water-worn, emerge, half human, half animal, covered with moss. Human shapes are merging back into the nature, perhaps beyond the animal to the vegetable world. [...] until suddenly the villa opens into its two parts once again, this time, miraculously, to let us aut, to let us see the civilised parterre, but more than that, far more than that, to show us the city out there in the light beyond the forest, the work of man, our refuge and our only hope. " (Scully, 1991)

Regardless of the route one takes, the landscaping from Bagnaia emphasizes the sanctity and divine meaning of the garden and creates through it, a microcosm which actually represents an Eden at a human scale, made posible through architectural and horticultural processes that mimic the human reasoning processes.

The garden being covered, and the message revealed, there is only one question left concerning the projection of the Villa Lante.

It is known that the Renaissance brought a series of physical and mathematical discoveries, including perhaps the most famous one, namely the linear perspective. This geometric construction will govern art, architecture, and gardens of the Renaissance, Mannerism and Baroque, becoming one of the main features of landscape design done in this period of time.

"The harmony of the universe was ehoed in the harmony created through huma nart in buildings, pictures, households, governments and gardens. [...] The order in the garden, achieved through similar means (mathematics), likewise reflects and reproduces a cosmic order. [...] Because the ordered microcosm reflects the macrocosm, the garden was the ideal vehicle to aquire knowledge of the divine order, a step by step process all things in the visible world were understood as links in a chain leading to the divine." (Lazzaro, 1990)

But if art, architecture and the gardens are subordinate to the perspective, how is it possible that the plan of the Villa Lante is drawn partly in a rising perspective and partly in elevation? The answer can be given by the general shape of the plan. Thus, by further abstracting its image, one may notice a certain similarity between the general outline of the plan and the shape of a human skull.

If we consider that the fundamental idea of Villa Lante is the retrieval of Paradise through reason, then, in terms of specificity and particularities of studies; especially those concerning the allocation of imagination, reason and memory in the human brain (apud. Clarke and Dewhurst, 1996) made in the Renaissance; through alchemy and through a complex hermeneutical approach, it can be deduced that the carefully drawn plan of the garden itself suggests the message that it wants to convey through the proposed routes – Retrieval/ rediscovery of lost Eden by mathematical reason. Moreover, if we compare the plan of the gardens from Bagnaia with Robert Fludd's drawings from *Utriusque cosmi maioris scilicet et minoris metaphysica, physica atqve technica historia*, we can notice that there are many correlations between design, theme and the garden.

Although it is probably not the best example, the correlation between the orriginal plan of Villa Lante and Robert Fludd's drawing is used as to emphasize the fact that the shape of the garden was itself a means of expressing the fact that this particular garden was designed as to represent the true path that leads to Paradise.

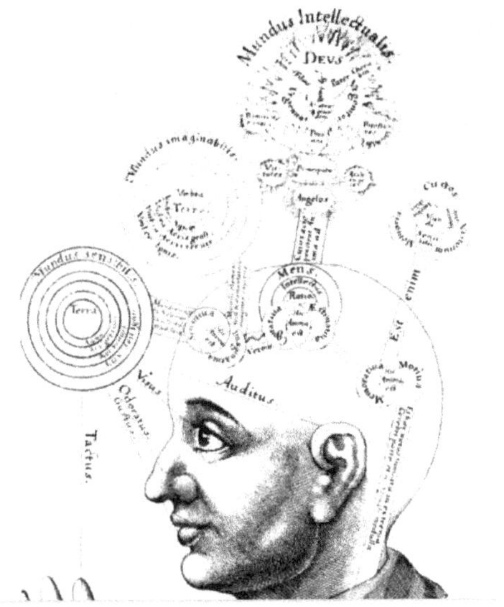

Figure 1. Utriusque cosmi maioris scilicet
et minoris metaphysica, physica atqve
technica historia
(Clarke E., Dewhurst K., 1996, p. 42)

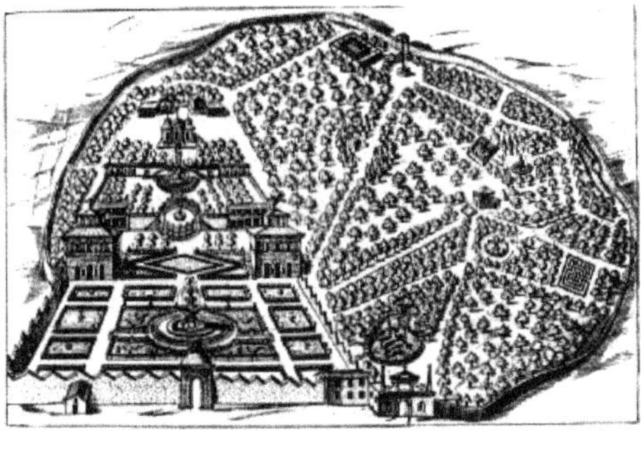

Figure 2. Villa Lante, 16th cenury plan
(www.gardenvisit.com)

In conclusion, Villa Lante can be considered an architectural landscaping microcosm which uses all the artistic means specific to the Italian Mannerism to convey a complex message that hides in fact an allegorical story of man searching for the lost Paradise. Unlike the Villa Lante, Villa Orsini from Bomarzo tries to trick visitors, guiding them through a sacred grove but which, unlike the *bosco* site at Bagnaia, leads the visitors in circles and offer them a single image - the image of the Fall. Villa Orsini does not offer a way out of the sacred forest and no route through which one can access the Lost Paradise, even if it gives the impression that it will reveal itself to man and that it will be decisive and strong to overcome all the wonders of the Tartarus found on the Villa's route.

" [...] an attempt to banish the melancholy by subjecting onself to cheerful stimuli (the case of Villa Lante). The other is to do the opposite, namely to surround oneself with sad and gloomy things, thus giving oneself a kind of homeopatic dose of melancholy to stimulate a counter-reaction (Villa Orsini)." (McIntosh, 2005)

Sacro Bosco or Paradise Lost
"Tu ch'entri qui con mente Parte a parte Et dimmi poi se tante Maraviglie Sien fatte per ingano O pur per arte – You who enter this place, observe it piece by piece and tell me afterwards whether so many marvels were created for deception or purely for art." (McIntosh, 2005)
Just like the message posted at the entrance to the garden says, Villa Orsini will introduce visitors to an unfamiliar world, in an abstract but real, kinetic and aggressive universe.
Unlike Villa Lante, although the Sacro Bosco from Bomarzo has the same subject - namely the route to the retrieval of Paradise, it offers a new vision of this initiation process. Most likely, the route was inspired by the visionary's own life experience, the garden being designed as a monument of commemoration of the tragedies he has surpassed.
Unlike Villa Lante, Villa Orsini was designed entirely as a complex maze (Ehrenfried, 2007). This structure enabled the visionary to use different registers of shapes, sizes and symbols to recreate a false initiation road where nothing that exists seem real, and nothing that is promised or expected will ever exist. In this garden, even nature is no longer subject to its own rules, it is strongly deformed and decomposed – *"In Bomarzo, the rules of the world are no longer valid, it is as if this garden wished to escape from the laws of nature."* (Ehrenfried, 2007)
As with Villa Lante, the Bomarzo garden offers visitors several possible routes through which they can discover both the universe

transformed through landscaping art and the promised Heaven. Irrespective but the chosen path, going through the labyrinth will never lead to the expected outcome.

"It is a sacred wood, full of disquieting monuments, some seeming to grow aut of the rock itself. The original entrance and axes of mouvement along its slopes can only be conjectured. The intended sequence of experiences is therefore not clear, perhaps was never meant to be so. The path through the modern entrance is perhaps as good as any other. It leads downward toward the trees, crosing a little watercourse leading farther into the depths of the forest. Directly ahead, standing in the light across a gentle ope field, a good, rationally abstract chapel can be seen, columned and domed, but the path does not lead toward it. Instead, it turne away from it down the darkening slope. Soon, hewn aut of the natural rock, Hercules rises before us, tearing the giant Cacus to pieces in his hands. He is guarding the garden for us but is markedly alarming, nonetheless. Finally, we come to the deepest part of the forest, and the darkest. There the stream runs into a cleft in the earth and disappears with a gurgling throat. The horrible gargle is surely enhanced by the whale's mouth, all teeth and gullet, into which the living rock at the mouth of the crevice has been carved.

Right there, looming over the cavern, an enormous round-eyed tortoise has been carved aut of the rock. On its back, the figure of a woman, apparently sounding a trumpett, is placed. We know from the emblem book of teh Cavaliere Ripa that she is an image of Fame – Fame lost down here in the depths of the wood, sounding her trumpet in the wild, while high above her the Orsini Castle can be seen shining in the sun. There is no connection between the two, no apparent route from his place tot hat. The effect is again bestial – the woman is, after all, right aut of the Apocalyptic Vision of St. John, mounted as if in ecstasy upon a beast. [...] just beyond Fame, in the darkness of the forest, a bright light gleams. It is Pegasus, the winged horse, symbol of hope, touched by a ray of sun, rearing up in the darkness. He show sus the only route to follow aut of here. It is a sinister-enough path through the wood: Nel mezzo del

cammin di nostra vita/ mi ritrovai per una selva oscura.

As we follow it upward, we are led further into dream. Goddeses of earth recline like Etruscan matrons in the rock, heavily and somnolent, bearing urns upon their heads. A house appears before us in a sunlit glade. It is leaning steeply into the hill. We mount ever higher beyond it. A war elephant looms up: a castle crowns its back, and a mahout sits upon its head. It is lifting the broken body of a soldier in its trunk. An enormous lizard flares beside it. We are climbing aut of the depths, but the images around us are becoming more alarming all the while. At last our dream, the guardianof our sleep, is broken by a figure o true nightmare: a colossal screaming face, as big as a house, demolishing the censor, awaking us as if to our own scream. And then, awake, we are aut, standing in the sunshine beside the mercefully abstract chapel we saw before and looking beyond it toward the palace, bathed now in clear white light, but remembering still the woman on the beast deep in the wood of our unaccountable yearning of fame. " (Scully, 1991)

From the description given by Vincent Scully, one can observe the mystical character of the garden and how it is translated into reality as a complex maze.

Although it belongs to the same historical period as Villa Lante, Villa Orsini at Bomarzo offers a new model for the use of the principles of construction of Mannerist gardens and although it approaches the same theme of rediscovering Paradise Lost, it uses many metaphors hidden in the carved stone, conveying a message which is different from the one in Bagnaia and opposite atmosphere to the balanced landscape of Bagnaia.

Seen by Christopher McIntosh as a complex landscape which combines several different themes: "part of it a memorial to Vicino's deceased wife, part therapy for melancholy, part autobiography in stone, part collection of alchemical symbols, part mannerist experiment" (McIntosh, 2005) Villa Orsini from Bomarzo must be seen as a whole, as a sum of elements that make up one story and one life experience - Orsini's tragedies.

RESULTS AND DISCUSSIONS

As already shown, both Villa Lante and Villa Orsini were designed as to represent the quest for Paradise. The final result of the allegorical processes that make up the two villas offer though two different views over what the road to Paradise and what Eden should look like.

Although very different in this aspect, they tend to be closely related to one another because they tend to use similar images and visual expressions as to emphasize the lanscape design philosophy.

To this end, both gardens use similar natural, mythological and/or philosophical themes and elements but interpret their images completely different:

Pegasus. While at Villa Lante, the mythological winged horse is inviting the visitor to walk the paths to Paradise and offers him a glimpse of the promised Eden, at Bomarzo, the same animal is placed probably only to annoy the visitor, offering him a false hope. Pegasus is placed in the *bosco* in both gardens, but his presence does not deliver similar messages.

The watercourse. Both gardens invite their visitors to walk their paths following a watercourse. At Villa Lante, the water is a phisical substitute for the human reason, its "taming" reflects the man's own road to illumination by following the path of reason. This path, and the water that makes it blends itself with a handfull of mythological, profane and religious elements, emphasizing the fact that achieving divine reason is a very complex and delicated process that covers all areas of sciences and humanism. On the other hand, the forever untamed watercourses at Bomarzo drive the visitor further and further away from what he hopes he will achive by following them.

The *Bosco*. While both gardens use woods as mazes and symbols of the chaos and the unknown, the Bagnaia *bosco* is part of the allegorical path from chaos to reason, thus being just part of a visitor's experience. At Bomarzo on the other hand, the forest represents the action itself. Here, the *bosco* is the perfect foreground for the entire spectacle of chaos and deceit.

The Paradise. The final expected outcome of the two villas is the (re)descovery of Eden. To this end, Villa Lante offers a more or less straight path that leads to the Paradise, while Villa Orsini only promises it and deceit its visitors by showing them a false hope under the image of the Pegasus.

CONCLUSIONS

The figuration processes described in the two examples discussed above, provides a comprehensive view of the role of the gardens in the sixteenth century, on the relationship between man and the divine in the same period of time, on the relationship between nature, art and science, etc.

Both Villa Lante and Villa Orsini represent two architectural and horticultural experiments belonging to the Mannerist era in Italy; extremely complex and delicate topics related to the history of art, the art of gardens. They address common themes and even common elements (the sacred forest, Pegasus' statue, etc.) but contain different actions and outcomes, regardless of how they are covered. The two gardens mimic the way to Paradise, but, depending on the chosen route, they can delight, disappoint or madden.

ACKNOWLEDGEMENTS

Research for this paper has been supported by UEFISCDI, PN-II-RU-TE-2014-4-0694, *Collaborative research, technological advancement and experimental philosophy in the seventeenth century: The Hartlib Circle and the rise of „the new science".*

REFERENCES

McIntosh C., 2005. Gardens of the Gods – myth, magic and meaning, I.B. Tauris, London and New York

Clarke E., Dewhurst K., 1996. An illustrated history of brain function – imaging the brain from Antiquity to the present, Norman Publishing, San Francisco

Conan M., 1997. Dictionnaire historique de l'art des jardins, Fernand Hazan Editions, ***

Graafland Arie, 2003. Versailles and the Mechanics of Power – the subjugation of Circe, An essay, OIO Publishers, Rotterdam

Iliescu A. F., 2014, Istoria artei grădinilor, Ceres, Bucharest.

Ehrenfried K., 2007. European Garden Design from Classical Antiquity to the present day, Tandem Verlag GmBH, Oxford şi Abignon

Lazzaro C., 1990. The Italian Reneissance garden – from the Conventions of Planting, Design and Ornament to the Grand Gardens of the Sixteenth-Century Central Italy, Yale Universitz Press, New Haven and London

Vincent S., 1991. Architecture – The natural and the manmade, St. Martin's Press, New York

Turner T., 2005. Garden History – Philosophy and Design 2000BC – 2000AD, Spoon Press, London and New York

Vercelloni M., Vercelloni V., 2009. The invention of the Western Garden – the history of an idea, Jaca Book S.p.A., Milano

Monty Don, *Around the World in 80 Gardens,* episodul 7, documentar BBC, 2008

www.xroads.virginia.edu/~ma99/hall/Dumbartonoaks/garden_dum.html

www.gardenvisit.com

www.villalante.wordpress.com

www.bildgeist.com

7

THE CONSERVATION PROCESS
ADAPTING THEORY TO A NEW CONTEXT

Smaranda COMĂNESCU[1], Violeta RĂDUCAN[2]

[1]Landscape architect Freelance
[2]University of Agronomic Sciences and Veterinary Medicine of Bucharest,
59 Marasti Blvd, District 1, Bucharest, Romania
Corresponding author email: violetaraducan@gmail.com

Abstract

In Romania historical garden conservation is a new discipline. Often the professionals involved in this process have to rely on their own intuition or on case studies of projects undertaken in countries with a richer tradition in this field. It is obvious that a unitary and professional approach is needed. We propose in the present study a possible methodology for approaching the conservation process, based mainly on the rich experience of the English National Trust.
In the first part, this study will attempt to draw from this accumulated knowledge a set of principles, which is by no means exhaustive, and does not represent a guarantee for successful conservation. Rather, this is a list of procedures which have become widely established in England. They have been verified by experience, and can be adapted to a new context. The approach to conservation can and does vary, depending upon the subject of study, and its context as the practice of the National Trust proves. The second part of the study deals with the way these principles have been adopted, adapted and applied on a school project during the Historic Garden Restoration classes at the USAMV Landscape Architecture department. The methodology of working with the students on a conservation project involving the regeneration of the Floreşti Estate focused on those procedures which would help the students develop the basic skills needed when dealing with a heritage asset.

Key words: heritage, historical gardens, garden conservation, restoration, regeneration.

INTRODUCTION

When heritage is mentioned, most people would probably think about buildings and monuments, art collections, maybe literature and music. However, the value and importance of historical gardens[1] as part of the common heritage is increasingly being recognised, hence a new and unprecedented interest in garden protection, conservation and regeneration has emerged. In Romania **garden conservation** is a very new discipline, and the professionals pioneering this path often have to rely on their own intuition and common sense when dealing with it. Publications on this subject are few, the legislation is lacking, and there is little unity of approach. In this study 'garden conservation' refers to a much more complex process than the

term 'conservation' seems to suggest. The Burra Charter defines conservation thus: 'Conservation means all the processes of looking after a place so as to retain its cultural significance' (Burra Charter, 1999, Article 1.4).[2] Simply put, cultural significance means 'aesthetic, historic, scientific, social or spiritual value for past, present or future generations' (Ibid, Article 1.2.). Thus, conservation has come to mean, especially for professionals, retaining the meaning and importance of a place, and not only preserving its physical matter, or fabric. Conservation can encompass a wide variety of interventions, ranging from maintenance to repair, restoration, reconstruction,[3] or more complex processes of regeneration.

---

[1] In this article a **historical garden** will be defined according to the Florence Charter: 'A historic garden is an architectural and horticultural composition of interest to the public from the historical or artistic point of view.' (Florence Charter, 1982, Article 1). It includes notions like private and public garden or park, country estate, etc.

[2] The definitions proposed by the charter have become established, at least in England, having been adopted by Historical England and other institutions dedicated to heritage protection; they also have the advantage of being simple and short. (Watkins and Wright, 2007);

[3] '**Maintenance** means the continuous protective care of the fabric and setting of a place, and is to be distinguished from repair. **Repair** involves restoration or reconstruction. **Preservation** means maintaining the

In countries like England, garden conservation has become a well-established practice, indeed, a tradition.

The National Trust, founded by some of John Ruskin's disciples at the end of the nineteenth century (Waterson, 1995), and at present the greatest owner of historical gardens in Europe (cf. National Trust website), has been a major pioneer of garden conservation, and its experience has helped define today's standards of good practice within this field. As the history of the trust illustrates, approaches to garden conservation have undergone many changes, from restorations 'in spirit,' or just creating 'appropriate' gardens for historical buildings, restoring to 'the last significant phase' according to thorough research, or the approach of 'conserve as found,' to ample regeneration projects (Cook, A., 2004).[4] Although approaches to garden conservation are likely to continue changing, this on-going process has led to the accumulation of a valuable mass of knowledge and practical experience, to the establishment of standards of good practice, and to the formation of a dedicated vocabulary.

In **the first part**, this study will attempt to draw from this accumulated knowledge a set of principles, which is by no means exhaustive, and does not represent a guarantee for successful conservation (understood in the wider sense stated above). Success depends on other factors as well, not least on the competence of all the people involved, from specialists to workmen, and their dedication

---

fabric of a place in its existing state and retarding deterioration. **Restoration** means returning the existing fabric of a place to a known earlier state by removing accretions or by reassembling existing components without the introduction of new material. **Reconstruction** means returning a place to a known earlier state and is distinguished from restoration by the introduction of new material into the fabric. **Adaptation** means modifying a place to suit the existing use or a proposed use. [...] **Setting** means the area around a place, which may include the visual catchment. [...] **Meanings** denote what a place signifies, indicates, evokes or expresses. Meanings generally relate to intangible aspects such as symbolic qualities and memories. **Interpretation** means all the ways of presenting the cultural significance of a place.' (Ibid, Article 1.4.-1.17.)

[4] For a brief account of the history of the changing approaches to garden conservation see also 'Hartwell House and Apafi Manor: Conservation through conversion' (Comanescu, 2013).

and commitment. Rather, this is a list of procedures which have become widely established in England.[5] They have been verified by experience, and can be adapted to a new context. The approach to conservation can and does vary, depending upon the subject of study and its context, as the practice of the National Trust proves.

The second part of the study deals with the way these principles have been adopted, adapted and applied during the *Historic Garden Restoration* classes at the USAMV Landscape Architecture department.

The methodology[6] of working with the students on a conservation project involving the regeneration of the Floreşti Estate was based on the stages listed below, but focusing on those procedures which would help the students develop the basic skills needed when dealing with a heritage asset: site survey, documentary research, analysis and reconstruction of the site's design and history, assessing the present condition, and developing a project based on the results of their research.

Part 1. The conservation process

The aim of a conservation project is to retain the 'cultural significance' of the heritage asset, in this case the historical garden. This means understanding what is important and valuable about it, and deciding what to do in order to preserve it. After assessing the significance of the place, its current condition and the issues involving it, one should decide the level of intervention needed in order to preserve this significance.

Some well-preserved places might require only maintenance, others might be threatened by loss of significance due to decay, and might require works of repair and restoration. In other cases, revealing and highlighting the significance of the place might require reconstruction.

Often, historical gardens need to undergo a process of revitalisation and regeneration in order to be integrated into the contemporary context. This might mean being assigned new viable functions and uses, or allowing new development within the protected areas, which, whilst sensitive to preserving the character and

---

[5] See below footnote 7 and the accompanying text.
[6] See Materials and methods.

significance of the garden, will help bring it to life in the new environment. Most often, a number of kinds of intervention will be applied on the same site.

Thus, the stages of the conservation process could be surmised as follows[7]:

- **Understanding the site**: its complete history, what it is today, and its current condition.
- Assessing its **significance**: why is it important and for who?
- **Risks and opportunities**:
- Identifying issues and vulnerabilities: this should result from the above two stages. Of particular importance are the factors that may endanger the significance of the place.
- Defining a vision: aims and policies. Explaining what should be done; this section may include recommendations for procedures like maintenance, restoration or reconstruction, as well as setting out directions for more complex processes like regeneration or revitalisation.
- Developing a **project** and an action plan: this section details the proposed interventions, and sets out the stages in which the proposed work should be undertaken. It may include a master plan, a management programme, a maintenance checklist, etc.
- The **implementation stage**.

A. Understanding the site. Survey and research.

The first step when dealing with a heritage asset is understanding what it is. This means knowing as much as possible about its history, from the earliest times to the present date, about the people who contributed to its creation

and subsequent evolution, about the ideas it might embody, about what it is today and the problems and issues which might threaten it. The first and absolutely necessary steps toward understanding the heritage asset are survey and research.

It is important to bear in mind that in the case of historical gardens surveying techniques will be a little different than for buildings. They include specific procedures like vegetation surveys, ecological assessments, hydrological and geological surveys, garden archaeology, as well as identifying each hard feature of the garden (paths, bridges, garden buildings, water features, etc.), mapping them and assessing their condition. The type and number of surveys undertaken will depend upon the site, its importance, complexity and state of preservation.

A. 1. Site visits

Site visits. When starting the survey and research stage, the first step is visiting the site, in order to form initial impressions, and to get a 'feel' of the place. Subsequent site visits will be needed for detailed surveys, and later on for confrontations with the results of documentary research. Important points to be kept in mind on site visits include: the coherence and unity of the place, or the lack thereof, the condition of the garden, the relationship between the house or other buildings and the garden, identifying significant features and their condition, views, blocked views, things that have a negative impact, planting, the condition of the trees, how the place is used and by whom, accesses, etc.

A. 2. Documentary research

The next important step is **documentary research**, which, combined with site survey, should result into a **history of the place**, a chronological, complete scheme of the site's development from the earliest times to the present day. It will also set the garden into a **wider context**, answering questions like: are there similar gardens? What are the other works of the garden's author(s)? Is it a rare or early example of a garden of this type? Documents to be consulted include: maps, design proposals, pictures, photos, aerial photos, drawings, descriptions, journals of the owners, chronicles, lists of materials and plants to be bought for the garden, and also already published studies

---

[7] This list is largely based on and adapted after recommendations by Historical England, the Heritage Lottery Fund and the National Trust on how to develop a Conservation Management Plan (CMP). The CMP is a widely used document, an instrument which in essence describes what a heritage asset is, its significance, its current condition, issues and vulnerabilities, and sets out long term management policies, as well as short term (3-5 years) prescriptions, including maintenance and restoration project works. The CMP is required for funding and development applications, and is extensively used by the National Trust in order to provide continuity of management for their properties. The CMP usually represents the bases for a project. See: National Planning Policy Framework, Heritage Lottery Fund, (Watkins and Wright, 2007, pp. 25-39);

about the place, articles, and other records. These are only some of the documents to be gathered and examined. There is desktop research, at the local and district town hall, at the records office, at libraries, archives, museums, private collections. Finding documents requires time and skill, and sometimes travelling. All these documents should be organised into an accessible data base, which should then be permanently updated with results of new research or records of new work. From this research, a history of the place will be built. It will help identify the main phases of the site's development, and divide the site in areas with a specific character. One of the most recommended procedures at this stage is map overlay and comparison. Documentary research will always be confronted with site surveys.

A. 3. Site surveys

Site surveys include identifying, assessing and mapping all the elements on site: buildings, garden buildings, water features, earthworks, terraces, paths, landmarks, walls, fences, vegetation, hydrological and geological surveys, tree surveys (besides mapping the existing trees, drawing up files for each outstanding specimen), ecological and wildlife surveys, where necessary archaeology and garden archaeology. Surviving and lost historical views and borrowed landscapes will be identified in order to explain the **local context**.

Surviving features assessments, as well as tree surveys specifying the species, age, condition, aesthetical value, and importance, are of particular importance. Site survey and documentary research are the basis for reconstructing the significant phases of the site's design, and for later works.

B. Assessing significance

It is essential to specify why the heritage asset is significant, for whom, and how this significance is linked to the actual fabric of the place. In some cases assessing the significance of a place can be pretty straightforward and simple. However, in more complex cases there can be many layers of significance, and all of them should be considered when proposing a project which might have impact upon them. Significance can refer to historical, evidential importance, artistic qualities, spiritual associations, importance due to association with an outstanding person or event, rarity, age, condition, superiority to objects of a similar kind (an outstanding example of a garden of a particular style, an impressive collection of rare trees), wildlife and ecology, archaeology, etc. The importance to the local community or other groups of people such as enthusiasts should not be forgotten.

C. Risks and opportunities

Once the significance of the place has been established, real or potential threats to it can be identified in view of the research previously done. Thus, **issues and vulnerabilities** concerning the site will be assessed. They might be related to decay, danger of loss of fabric, fragmentation, loss of character and meaning, danger from development, lack of finances to maintain the place, lack of visitors, or conflicts between different types of heritage, but also loss of authenticity, lack of sustainability, over-commercialization.

As the risks concerning the site are analysed, and solutions are sought, a certain **vision** will emerge. Thus certain **general aims and policies** will be established: the kind of interventions that are necessary in order for the significance of the place to be preserved, and, if possible enhanced. These interventions may range from works of maintenance and repair, to restoration, reconstruction, and the integration of new features such as cafes, souvenir shops, cultural centres, or others.

The general attitude towards change when dealing with heritage assets should be reserved; however in some cases the regeneration of a place requires a creative, but sensitive and respectful approach and it is always necessary to make the place functional, responding to contemporary needs.

D. The project

The project will be based on the above research and conclusions, and can include a **master plan**, an **action plan** with specifications regarding the **stages of the project**, how it should be implemented, which procedures have priority. The proposal should take into consideration things like how the project will be financed, what qualifications are required of the staff, once the main stages of the project are completed, how will the property be maintained and financed in the future.

Like the previous stages, the project will most likely be the result of collaboration between experts. At this point it is very important to make sure that all the participants have a clear understanding of the aims of the project viewed as a whole. The coordinator of the team, in particular, should integrate the input of other specialists into a coherent scheme, making sure that the resulting garden is a harmonious whole.

E. The implementation stage

As part of the conservation process, the implementation stage is of crucial importance for the success of the conservation project, and should be addressed, especially in Romania, where the staff employed for on-site works is usually not trained in work on heritage sites. The manner in which the proposed interventions and works are executed is of great importance, and if inappropriately done, can ruin not only the project, but the historical garden itself. This is why it is recommended that the execution should be supervised by the person who was in charge of the project. Likewise, the staff and other professionals should be familiar with the aims of the project, the significance and character of the garden, with the specific terminology employed in garden conservation, and should have the skills and competences for this type of work. The same should be true of the people who will be in charge of the future maintenance and management of the garden.

Part 2. The conservation project

MATERIALS AND METHODS

For a thorough understanding of the methodology of historical garden conservation (in the broad sense specified above), we propose applying it to a specific, complex, and for many reasons significant case study: the Cantacuzino Estate in Floreşti, Prahova. The subject of the regeneration of this site was addressed in a school-project during the 2014-2015 *Historic Garden Restoration* classes at the Landscape Architecture Department at the USAMV, Bucharest. The *Historic Garden Restoration* classes take place during the first semester of the 3rd year of study (14 weeks) and are usually organised in 2 taught course hours and 2 hours of practical activities per week. For the Floreşti case study, the students had 9 weeks for research and 5 weeks for project work.

The abovementioned methodology was adapted to the school-project, some of the points being necessarily omitted, being outside the sphere of tasks that the landscape architecture students could accomplish. The activity of the students was organised in two stages: research and project work. For the research stage, due to the multiple and diverse research directions which had to be covered, the students were organised into groups of two to five. For optimal involvement, they were given the opportunity to approach the directions of research of their choice, according to their own preferences. At the end of the first stage, an indispensable data base was created, comprised of the results of the research work. For the second stage, the students were organised into larger groups of eight to eleven members. Although the number of students in a group was determined by the professors, the members were not. We opted for this approach to favour good communication in each group. The groups were encouraged to develop different solutions for their projects. These would encompass various types of interventions, including: preservation, repair and restoration, reconstruction and, on a broader scale, regeneration. The projects were meant to organise the proposed interventions into stages, which would allow the concomitant use of the site for cultural, sportive or other activities, which in turn, would financially support the future works. To avoid mistakes, these stages of the school-project were closely guided by supervisors competent in the field of historical garden protection, conservation and restoration.

RESULTS AND DISCUSSIONS

The research stage was preceded by a presentation of the already known information about the Cantacuzino Estate in Floreşti, Prahova: a topographical survey, the surveys of the 'Little Trianon' palace, photographs from various historical periods, data on the original owner, about the architect of the palace and about the supposed designers of the garden, and a historical study of the palace.

In order to become directly acquainted with the object of study, the students' first activity was a site visit. The students were organised in groups after that visit. They were assigned tasks according to their own preferences regarding research directions. The students undertook research at the National Archives, at the Academy Library, at the History Museum of the Ploieşti Municipality, at the Floreşti Village Hall. This endeavour was really successful. A whole archive of documents reflects daily life on the Floreşti estate, although these documents are apparently dry and uninteresting. Historical plans dating from before the construction of the present palace have been found, identifying the main areas of the estate: the pleasure grounds and the hunting park. The 1905 plan already shows a clear division into specific areas: the pleasure grounds, the hunting park with the mills' pond and the river meadow, as well as the Cap Roşu Park at the northern end of the estate (Figure.1).

Figure 1. Floreşti Estate plan, 1905, Detail
Source: Arhivele Naţionale, Planuri, Judeţe lit. O-V, Inventar 2343, Cota 248.

At the National Archives, a 1906 'Boundray Book for the Floreşti Estate' has been found.[8] It encompasses a complete inventory of the

[8] Arhivele Nationale_Hotărnicii_inv.2473_Jud. PH_Cota 53 - "Cartea de Hotărnicie pentru moşia Floreşti" din 1906, publicata in 1908.

estate, and shows a clear division into specific areas: the pleasure grounds, the hunting park with the mills' pond and the river meadow, as well as the Cap Roşu Park at the northern end of the estate.

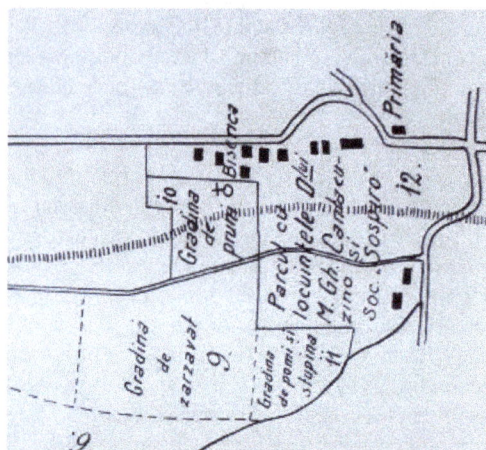

Figure 2. Floreşti Estate plan, 1924, Detail
Source: Arhivele Naţionale, G.Gr.Cantacuzino, Inventar 1829, Cota 608.

A 1924 plan shows a plum tree orchard and a vegetable garden, bee hives in an orchard, a wilderness, hayfields, and poplar and alder woodland (Figure 2.). Other documents mention: buildings in the 'garden in the Park', two glasshouses, beehives and fruit bearing trees, a mill, a cattle farm and 'an orderly dairy.' The most important plan was found at the Central Archives of the Bucharest Municipality, and it represents a restoration proposal for the park, signed 'Pinard' and dated July 1912. Worthy of mention are the important views marked on this plan (Figure 3.).

Other students have elaborated a site survey recording all the trees and the built elements, like buildings, walls, bridges, ponds and other water features. Comparative studies regarding the wider context of the 'Petit Trianon' as archetype were also undertaken, by analysing places that are also named after and likened to the French original.

The studies showed that most of these places were situated in urban areas, with evident consequences upon the dimensions of the gardens. A comparative study on the Cotroceni ensemble highlighted a series of similarities concerning the decorative features, such as a rectangular pond and the balustrades from the

Palace garden, which were erected at the beginning of the twentieth century, like the ones at Floreşti estate.

The studies revealed the fact that the ensemble at Floreşti is a late example of a nineteenth century garden, with a geometrical area around the main building, transitioning into a landscaped pleasure garden, and then into the wider parkland. It should be underlined that, except for the 'Petit Trianon' itself, no other Versailles feature was used as model for the Floreşti estate. The eighteenth century gardens of the Petit Trianon present an idyllic view on village and pastoral life. They have no connection with the public parks of the nineteenth century, like Buttes Chaumont, Monceau or Montsouris, which, on the other hand, have a great number of elements in common with the pleasure grounds at Floreşti.

Figure 3. Plan of the park of the Floreşti Palace, belonging to I. G. Cantacuzino, no. 90, 1912
Source : Arhivele Centrale ale Municipiului Bucureşti, Inventar 2343 vol. II, Planuri O-V, Judeţul Prahova.

The most important and useful information about the site was found in two articles published in contemporary periodicals: 'A day at Floresti,' published in *România Ilustrată* magazine, (Antemireanu, 1905) and 'Disposition en terrasse. Aménagement d'un Jardin régulier, d'une large facture, à flanc de coteau, devant une demeure de style Trianon (Domaine de Floresti, au Prince Cantacuzène, Roumanie)', published in *La vie à la Campagne*, (Maumené, 1914). The first article describes the estate in detail, mentioning specific areas, and providing photographs. It also mentions the author of the first landscaped layout. 'The Pleasure grounds at the artfully crafted Floreşti estate date from around 1830. They were laid out, in their present form, by

Meyer, the famous gardener who was also commissioned by General Kiselef to realise the eponymous, and most admirable boulevard in Bucharest, the most exquisite adornment of the Capital.' (Antemireanu, 1905). The plan published in 'La vie à la Campagne' shows the superior terrace, the geometrical garden around the palace, and also the link to the 'pleasure grounds,' to the edge of the lake (Maumené, 1914).

The current heritage legislation has also been studied, as well as the List of Historical Monuments, according to which the estate is a category A listed heritage asset, that is, of national importance. The Cantacuzino Estate (PH-II-a-A-16490) is an ensemble of national importance, which lies parallel to the Prahova River, from North to South, on a distance of 3 km. The main elements of the ensemble are: the buildings, the most valuable of which is the Palace called the 'Little Trianon' (PH-II-m-A-16490.01), built between 1910 and 1916, designed by the architect Ion D. Berindey, the water tower (PH-II-m-A-16490.02), built between 1910 and 1916, the enclosure wall (PH-II-m-A-16490.04), and the 'Holy Trinity' and 'Nativity' Church, with the Governor Grigore Cantacuzino's family crypt (PH-II-m-A-16491), 1887. The park (PH-II-m-A-16490.03) has naturally been the main object of our study.

The students surveyed the existing vegetation (Figure 4), and analysed the important views for the general composition and for emphasizing both the palace and the grounds.

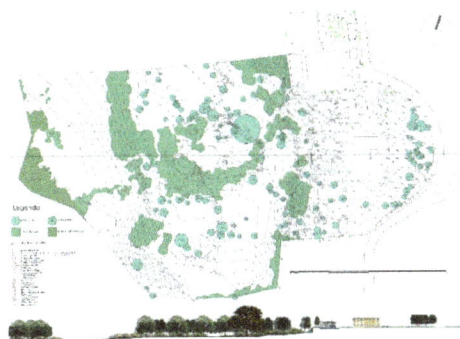

Figure 4. Vegetation Survey of the formal gardens and the pleasure grounds.
Authors: the students from the third year of study

All this documentary research was corroborated with all the other information provided by plans, other documents, and most importantly, the survey of the site. Other surveys such as excavations or other archaeological works, and ground investigation, which are in principle recommended, were not undertaken, this being a school-project. At present, the condition of the buildings in the ensemble is poor, the palace being in an advanced stage of degradation; it is a ruin in fact. The enclosing wall has also collapsed in various places. Some of the gates have disappeared, while the water tower needs to be consolidated and restored.

Assessing Significance

According to the Historical Monuments List (2010 and updated in 2015), the whole ensemble at Floreşti, as well as its main features are of national importance. Even some of the unlisted features, such as the buildings of the present sanatorium are important due to their association with the Governor Grigore Cantacuzino (1800-1849) and his wife Luxita Kretzulescu (Figure 5). He is also the founder of the Floreşti church (1826-1830), which was later rebuilt by his wife. It is said that within the present tuberculosis asylum buildings, previously the villas of Gh. Gr. Cantacuzino's children, there are murals by Gh. M. Tattarescu, who also painted the church built by Luxita Kretzulescu – Cantacuzino in 1887.

Figure 5. The Little Trianon mirrored in the lake. Inset: Vornicul Grigore Cantacuzino and Luxita Kretzulescu, the parents of Gh. Gr. Cantacuzino, called 'the Nabab' (Ion, 2010)

The estate has belonged to one of the most prominent and interesting figures of the beginning of the twentieth century, Gh. Gr. Cantacuzino (Figure 6), called 'the Nabab', due to his enormous fortune. He was one of the most appreciated political figures, Member of Parliament and Prime Minister. 'The Nabab' was renowned for his authentic patriotism, which is remarked upon in the article 'A day at Floresti,' (Antemireanu, 1905).

Figure 6. Mr. and Mrs. Gh. Gr. Cantacuzino in the park at Floreşti (Antemireanu, 1905, p. 259)

Even today, there are many legends about this charismatic man, with a strong but warm personality. The Floreşti estate has been a favourite place for many personalities, including King Mihai (Fabra Bratianu, 2012) (Figure 7).

Figure 7. King Mihai I, Ileana Brătianu and two cousins on the deck of the lake in the pleasure grounds. Source: (Fabra Bratianu, 2012, p. 103).

Apart from its association with the Cantacuzino family, and especially with Gh. Gr. Cantacuzino, the Floreşti estate is important due to the exceptional quality of the palace architecture by I. D. Berindey (Figure 8), and to its relationship with the designed landscape, which has survived to a great degree.

The site is associated both with W. F. C. Meyer, and with E. Pinard, two of the most prominent garden designers in Romania. Further site surveys are needed to determine more precisely what input each of them had and how much of their designs survives.

Figure 8. Details of the 'Little Trianon' Palace
Photo: Mihaela Radu

Most of the parts of the ensemble have survived, including: the pleasure grounds, the hunting park, the villas, the utility areas, and various important features: earthworks, water features, the general planting scheme, and some of the main views. All these elements are still in place, and although deteriorated, they are identifiable and can be restored. Thus, the ensemble is valuable as an example of late nineteenth century and early twentieth century country estate.

Although this type of estate is fairly common in Europe, in Romania they have become rare, which adds to the site's importance on a national level. The refinement and luxury of Pinard's design for the formal terraces are underlined by Maumené, in his presentation of the gardens.[9]

Moreover, the reinforced concrete features testify to Emile Pinard's intervention, whose project for the terraces (Figure 10) was praised at the end of the article 'La vie à la Campagne'.[10] Pinard was familiar with the appreciation of the contemporaneous French landscape architects. The project of the Bibescu (today Romanescu) park in Craiova was awarded the Golden Medal at the 1900 Paris International Exhibition and its authors were Edouard Redont, Jules Redont and his brother, and Emile Pinard.

Although the pleasure grounds at Floreşti are not a veritable arboretum, they do accommodate a collection of rare trees, and a plane tree, remarkable for its age, dimensions and aesthetic value (Figure 9).

Figure 9. The Pleasure Grounds, October 2014
Photo: Andreea Soare

Risks and opportunities

The most noteworthy feature of the ensemble is the palace called the 'Little Trianon,' which is at present in a ruinous state, and in danger of collapsing. Urgent consolidation works are imperiously needed so that this most important element of the park should not be lost. The whole composition revolves around this central element, and depends upon its presence. Works undertaken in the immediate proximity of the palace can induce vibrations which may affect and further deteriorate the monument. As it is, major and irreversible deterioration of the palace's fabric has already taken place. The retaining wall, the staircases, the inferior pond are also in a poor state, while only dispersed fragments of the balustrade have survived.

The lack of funds for a complete restoration has led to the need to find alternative solutions: for the park maintenance works a contract between USAMV Bucharest and the Cantacuzino Floreşti Foundation was signed.

---

[9] « On descend sur la deuxième terrasse par des escaliers latéraux accompagnes, comme l'est le mur de soutènement, de balustrades qui ont été prolongées latéralement. Un nouveau bassin est dispos contre le mur de soutènement et sur toute la largeur de la partie saillante. Il est alimenté par l'eau du bassin supérieur. Le mur de soutènement de cette seconde terrasse, qui se retourne en pan coupé, sera maintenu bas avec des caisses à fleurs posées sur les pilastres, cela pour éviter la répétition de la balustrade supérieure ; dans les pans coupes s'encastreront des bancs de pierre, abrite chacun par un portique recouvert de plantes grimpantes. » (Maumené, 1914, p. 188).

[10] « Par la dominante de ses grandes lignes, sa facture sobre et élégante, son encadrement libre de massifs et de grands arbres, cet ensemble doit parfaitement

s'harmoniser avec le Parc paysager dans lequels il s'encastre. Il est digne en tout point des créations de l'école française des Jardins contemporains, dont, en Roumanie M. Pinard est l'excellent représentant. » (Maumené, 1914, p. 188).

For the consolidation and restoration of the palace, an idea competition was organised, which will be followed by developing a project and applying for EU funding.

The Floreşti estate is full of life even in its present state. The international horsemanship competition, Karpatia Horse Trials is annually organised here and enjoys great popularity. Although it is a great opportunity to bring people on the site, it has some drawbacks too: a few huge trees from the hunting park have been cut, new land works were undertaken in order to build water obstacles for the horse races, and, not least, new works involving reinforced concrete were undertaken on the geometrical pond in front of the palace.

The grounds can be visited anytime. One of the main problems is that at present no effective security can be provided for the site. This leads to further deterioration of the built edifices, as well as of the poplar woodland, through uncontrolled cuts. This situation can lead to loss of authenticity.

Another threat is uncontrolled young tree growth. Thus, clearing works are needed, as well as maintenance works for old, rare and spectacular specimens. Likewise, the hard elements of the pleasure grounds should be restored: ponds, staircases, bridges, and water features. The research undertaken by the students revealed the fact that the area around the palace is situated on the crest of the *Floreşti Anticline*, on a salt massive, which can provoke landslides. This is important to know, because it will influence the types of future work which will be undertaken on the superior terrace, where the palace is situated.

For the conservation of an ensemble as complex and valuable as the Floreşti estate, a vast variety of interventions are required. Apart from the classical maintenance, restoration and revitalization works, a creative and sensitive approach will be needed in order to make sure that the ensemble will be functional in the future. This type of approach, called regeneration, allows for new functions to be introduced, and for new features such as: new accesses, parking lots, cabins for security staff, restrooms, resting places, belvederes, and event dedicated areas. These features should be integrated so as to affect neither the substance nor the spirit of the place.

The project

By studying the materials accumulated during the research stage, both documentary, and site surveys, we concluded that we have the possibility to elaborate a simplified classical conservation project. In the future these materials will be completed with archaeological surveys, which are needed for uncovering lost artefacts, as well as for finding the fragments of features that have been destroyed in time.

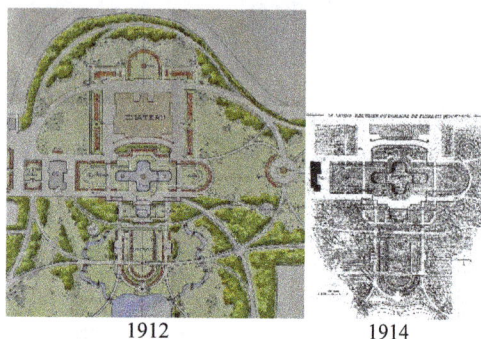

1912   1914
Figure 10. Floreşti Estate, *the geometric gardens*
Plans by Emile Pinard

The owners have expressed a few requirements concerning the conservation project: they would like the restoration of the pleasure grounds to be as exact as possible, but with the addition of a parking area; the project should be sustainable and easy to implement; they are looking for proposals of activities which should take place both on the superior terrace, and on the pleasure grounds. These activities should bring in revenues which would then be used for further restoration of the park and palace. These sensible suggestions transform the project in something more than just revitalization. It will become a regeneration project, which involves not only a resuscitation of the place, but its rebirth. This is why the students have been organised in large groups of eleven, eight, and respectively ten members. As in the case of establishing the teams for the research stage, the preferences of the students were taken into consideration, keeping in mind a certain vision of the project. The supervisors adopted this attitude with the aim of obtaining the best possible results and of inducing the students the pleasure of working in this field. The students were encouraged to elaborate diverse projects, starting from the same data. Each group has had full access to the materials resulting from research, as well as to the requirements of the owners. Each group has drawn their own set of conclusions, which led them to diverse solutions. The plans of each stage of the site's development were juxtaposed, in order to be examined and analysed, and on this basis a strategy of approach to the project was decided. Each group was encouraged to elaborate stages of the implementation of the project, so that the park would function continuously, bringing in revenue and attracting visitors.

The aim of this school-project was helping the students develop the basic skills needed when dealing with a historical garden as a heritage asset. This includes: becoming familiarised with undertaking research at libraries, archives, etc. in view of understanding the asset and developing a project, organising a database with all the accumulated information, understanding the importance of a sensitive and sensible approach, adopting an 'in spirit' intervention, but avoiding pastiche, dealing both with teamwork and individual work, inducing a positive, empathic attitude toward the condition of heritage assets in general, and also the actual involvement in salvaging endangered assets.

CONCLUSIONS

The resulting three projects have many points in common, but they also present substantial differences. The spectacular trees will be retained and highlighted, while the valuable surviving features such as bridges and water features will be restored. The differences between the projects revolved mainly around the way the area around the palace was resolved, the connection between the palace and the gardens, and he connection between the terrace and the lake (Figure 11). Only the second team proposed a formal access from the east. This proposal was unfortunately not sustained by convincing arguments. The project work of the students has not been sustained financially either by the Cantacuzino Foundation or by the owners of the estate, but it has been facilitated by the convention between the University and the Foundation, which allowed all the students, from every year to conduct their practical activities on the site.

The initiative of approaching this subject during the Historic Garden Restoration classes belonged to the professor of this subject.

Group 1

Group 2

Group 3
Figure 11. The projects of the students

The research stage has been difficult due to the distance of approx. 85 km to the site and also due to the unfavourable weather (October, November, and December). Another difficulty was linked to the students' timetable and the programme of the archives, libraries, museums where the research was undertaken. The students were given the opportunity to have intercourse with the owner of the estate and to participate in the 'Karpatia Horse Trials' event. During the project work the students have become affectively involved in their work, which has greatly contributed to the outcome of the projects. We strongly believe that a scholastic approach to conservation is less efficient, and cannot benefit from the same level of involvement, without which exceptional results are impossible.

ACKNOWLEDGEMENTS

This research work was carried out with the cooperation of the representative of the owners of the Floreşti estate, the students and their supervisors. We thank the students for the passion with which they got involved in the project and undertook the research work, which is like a veritable and fascinating detective work.

REFERENCES

Antemireanu Al., 1905. O zi la Floreşti, in *România Ilustrată*, Year III, Nr. 9, Sep. 1905.

Arhivele Centrale ale Municipiului Bucureşti (Central Archives of the Bucharest Municipality), INV. 2343 vol. II, Planuri O-V, Judeţul Prahova, Planul parcului castelului de la Floreşti, proprietar I. G. Cantacuzino, nr. 90, 1912.

Arhivele Naţionale (National Archives), Gh. Gr. Cantacuzino, Inventar 1829, Cota 608.

Arhivele Naţionale (National Archives), Hotărnicii, Inventar 2473, Jud. PH,_Cota 53.

Arhivele Naţionale (National Archives), Planuri, Judete lit. O-V, Inventar 2343, Cota 248.

Athens Charter - for the Restoration of Historic Monuments, 1931.

Batey M., 2000. Ruskin Morris and the early campaigns, in *Indignation*, Kit-Cat Books, London.

Burra Charter - the Australia ICOMOS charter for the conservation of places of cultural significance, 1979, rev. 1981, 1988, 1999.

Cantacuzino-Enescu M., 2005. Lights and shadows, the memories of a Moldavian princess, Aristarc.

Comanescu S., 2013. Hartwell House and Apafi Manor: Conservation through conversion, A Dissertation for the Diploma in Conservation of Historic Gardens and Cultural Landscapes available from https://gardenwalks.wordpress.com/.

Cook A., 2004. Changing Approaches to Historic Gardens with a case study of Chiswick House & Grounds. A Dissertation for the Diploma in Conservation (Landscapes & Gardens), unpublished.

Department for Communities and Local Government, 2012, 'National Planning Policy,' London, available from www.communities.gov.uk

Elliot B., 2010. Changing fashions in the conservation and restoration of gardens in Great Britain *Bulletin du Centre de recherche du château de Versailles*.

English Heritage, 2008. Enabling Development and the Conservation of Significant Places, English Heritage, London.

Fabra Bratianu Marie-Hélène, 2012. Memory of Dead Leaves, Humanitas, Memorii/Jurnale collection, Translation: Emanoil Marcu.

Florence Charter - for the Preservation of Historic Gardens, 1982.

Heritage Lottery Fund, 2008. Conservation Management Planning, available from www.hlf.org.uk

Historical England website https://www.historicengland.org.uk/

Hoinărescu C., 1985. Ctitoriile cantacuzine din Prahova – premisă fundamentală a arhitecturii brâncoveneşti, in *Revista Muzeelor şi Monumentelor. Monumente Istorice şi de Artă*, year XVI, no. 1, 1985, p. 69. http://crcv.revues.org/10764, accessed 2015. http://www.cultura.ro/page/17 http://www.monumentul.ro/pdfs/Narcis%20Dorin%20Ion%2009.pdf. https://www.nationaltrust.org.uk/

Ion N. D., 2009. The destiny of some aristocratic residences during the first decade of the communist regime (1945-1955), available from

Ion N. D., 2010. Residences of Romanian aristocratic families, Editura Institutului Cultural Român.

Lista Monumentelor Istorice - 2010.

Maumené A., 1914. Disposition en terrasse. Aménagement d'un Jardin régulier, d'une large facture, à flanc de coteau, devant une demeure de style Trianon (Domaine de Floresti, au Prince Cantacuzène, Roumanie), in *La vie à la Campagne*, Hachette et Cie, Vol XV, no.180, 15 March 1914.

Quest-Ritson C., 2003. The English garden: a social history, Penguin, London.

Răducan V., 2005. The Court of Gheorghe Grigore Cantacuzino in Floreşti - Prahova, în *Scientifical Papers*, Seria B, *Horticulture*, vol. XLVIII, chapter *Ornamental Plant & Landscape Architecture*, ISBN 973-7753-18-6, Bucureşti, p. 187-193.

The National Trust, 2001. Rooted in History Studies in Garden Conservation, The National Trust, London.

Venice Charter - for the Conservation and Restoration of Monuments and Sites, 1964.

Waterson M., 1995. The National Trust. The first Hundred Years, BBC Books, London.

Watkins, J. and Wright, T., 2007. The Management and Maintenance of Historic Parks, Gardens and Landscapes. The English Heritage Handbook, Frances Lincoln, London.

8

TEMPERATURE AND PH INFLUENCE ON ANTAGONISTIC POTENTIAL OF *TRICHODERMA* SP. STRAINS AGAINST *RHIZOCTONIA SOLANI*

Cristina PETRIȘOR[1], **Alexandru PAICA**[1], **Florica CONSTANTINESCU**[1]

[1]Research and Development Institute for Plant Protection, Ion Ionescu de la Brad Blvd., No.8, District 1, Bucharest, Romania

Corresponding author email: crisstop@yahoo.com

Abstract

Species of the genus Trichoderma sp. are considered as potential biocontrol agents (BCA) for many plant diseases. The effectiveness of biocontrol agents depends on several parameters therefore their application showed consider climatic factor that could affect their biocontrol capacity.The present study examined the antagonistic potential of two Trichoderma sp. strains (Td85 and Td50) against Rhizoctonia solani depending on pH (4.5;5.5) and temperature (25°C, 30°C). Both strains of Trichoderma sp. studied inhibit stronger growth of Rhizoctonia solani at 30°C compared to 25°C. Also our results revealed that both strains of Trichoderma have maximum antagonistic ability to Rhizoctonia solani strain at pH=4.5.

Key words: climatic factors, biocontrol, inhibition percent

INTRODUCTION

Rhizoctonia solani is a soil pathogen that causes diseases in a wide range of hosts of agricultural, horticultural and ornamentals crops (Cundom et al., 2003). It can cause severe damage specially during the seedlings pre-emergence and post-emergence stages and is a limiting factor in the production of crops. Diseases caused by *Rhizoctonia solani* are difficult because this pathogen survives for many years as sclerotia or as mycelium in soil or organic matter under various conditions and has an extremely wide host range. (Osman et al., 2011).

Many important agricultural and horticultural crops worldwide are mostly affected by *R.solani* including tomato, bean, potato, strawberry, soybean, tobacco, tulip (Rahman et al., 2014; Seema and Devaki, 2012; Osman et al., 2011; Lahlali and Hijri, 2010, Grosch, 2007; Elad et al., 2006; Singh and Chand, 2006; Schneider et al., 1997; Tu et al., 1996).

Species of the genus *Trichoderma* sp. are considered as potential biocontrol agents (BCA) for many plant diseases (Galarza et al. 2015; Singh et al. 2013; Kohl and Schlosser 1989).

The effectiveness of biocontrol agents depends on several parameters, that include specific pathogen, soil texture, water content, pH, temperature and crop history (Berg et al., 2005; Kredics et al., 2003), therefore their application should consider the environmental stress that could affect their ability to maintain their biocontrol capacity.

Although there are several publication on the antagonistic activities of *Trichoderma* sp., there is little information on the effects of pH and temperature on its antagonistic proprieties against *Rhizoctonia solani* (Daryaei et al., 2016; Lahlali and Hijri, 2010; Montealegre et al., 2009; Santamarina and Rosello, 2006).

The goal of this work was to evaluate the *in vitro* effects of different environmental conditions (temperature and pH) on the antagonism of *Trichoderma* sp. towards *Rhizoctonia solani*.

MATERIALS AND METHODS

One *Rhizoctonia solani* strain and two *Trichoderma* sp. strains (Td85 and Td 50) obtained from RDIPP culture collection were used in this experiment. These fungal strains were maintained at 4°C on Potato Dextrose Agar (PDA) with periodical subculturing on the same medium at 25°C.

Trichoderma sp. strains used in this experiment were identified at species level as *Trichoderma*

asperellum according molecular analysis(Paica et al., 2015).

Fungal isolates of *Trichoderma* sp. were *in vitro* screened for their ability to suppress the mycelial growth of *R. solani* in dual culture assays (Morton and Stroube, 1955).

Mycelial blocks (5mm) were cutted from the periphery of 5 days old culture of both *Trichoderma* sp. and *R.solani*. Two mycelial blocks one from *Trichoderma* and other from *R.solani* were placed in a same time on PDA (Potato Dextrose Agar) plate in opposite directions and incubated at two temperatures (25°C, 30°C) for 4 days. The radial growth of each colony was measured at 48.72 and 96 hours interval.

PDA with different pH levels (4.5 and 5.5) were poured into Petri dishes and a 5 mm plug from the margin of actively growing colony of *Trichoderma* sp. strains and *R.solani* were placed in opposite direction and were incubated at 28°C for 4 days.

Controls were also set up with the pathogen alone so that a growth without interactions could be precisely measured. Three replicates for each antagonist-pathogen combination and for the controls were considered. Percent inhibition of mycelial growth of targeted fungal pathogen over control was calculated by following equation:

I% = C - T/C where:

I%-percent inhibition in mycelial growth

C-colony diameter of pathogen in control plates

T -colony diameter of pathogen in dual culture plates

RESULTS AND DISCUSSIONS

A broad range of temperature tolerance for growth and sporulation of *Trichoderma* sp. is a very interesting feature for suitability of the antagonism.

A clear zone of interaction between antagonist and pathogen was observed after 48 h of incubation. The mycelium of both *Trichoderma* sp. strains grew abundantly on *R. solani* after 4 days of incubation.

Both *Trichoderma* sp. strains were able to significantly decrease the radial growth of *R.solani* mycelium within 4 days at both

temperature conditions. *Trichoderma* Td85 was more active against *R.solani* strain at 30°C compared to 25°C. Td85 limited gowth of *R.solani* mycelium more than 57% at 30°C (fig.1). Also, Td85 strain was more effective at 25°C with a inhibition percentage of 54.48% compared to Td50 with a percentage of 53.33% (fig.1 and 2). Our results are in agreement with Grosch et al, (2007) who reported that most strains of *Trichoderma* sp. studied by them showed better antagonistic activity against *R.solani* at higher temperature (30°C).

Also, our results are in conformity with those of Montealegre et al., (2009) who suggested that low and high temperatures (between 5°C and 22°C) do not changes the biocontrol capacity of different *Trichoderma* sp. strains on *R.solani*.

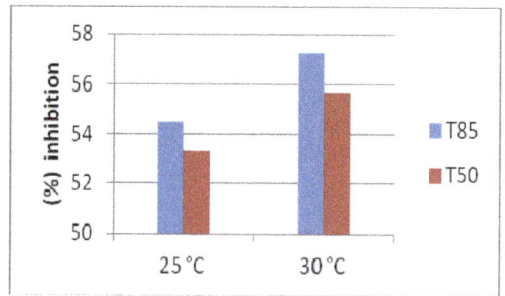

Figure 1. Effect of the temperature on the inhibition of the mycelial growth of *R.solani* in dual culture assay with *Trichoderma* sp. after 4 days of incubation

Figure 2 Antagonistic effect between *R.solani* and *Trichoderma* sp. at 30°C (left) and 25°C (right) after 4 days of incubation

Our results showed that both *Trichoderma* sp. strains were antagonistic to *R.solani* at both pH values although differences were found among strains. The data presented in fig.3 indicate that both *Trichoderma* sp. strains studied were effective in suppressing *R.solani* at pH=4.5 compared to pH=5.5.

At pH=4.5, no significant differences between the two antagonistic strains was observed regarding the percentages of inhibition (56%) of the mycelial growth (fig.3, fig.4). However at pH=5.5, Td85 strain had a slight increase of inhibition procentage (53.72%) compared to Td50 (52.15%)

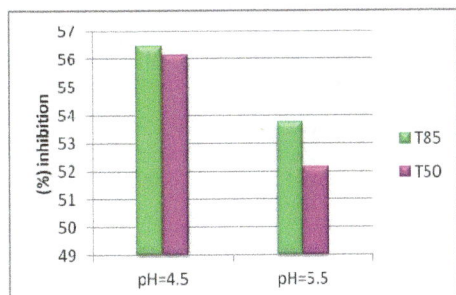

Fig. 3 Effect of pH on the inhibition of radial growth of *R. solani* in dual culture with *Trichoderma* sp. strains after 4 days of incubation

Fig.4 Antagonistic effect between *R.solani* and *Trichoderma* sp. at pH=4.5(left) and pH=5.5 (right) after 4 days of incubation

Results of Daryaei et al., 2016 suggest that at different pH values *Trichoderma atroviride* gave significantly various amount of inhibition and overgrowth activity against R.solani in dual culture assays with the strongest inhibition(76%) at pH=7.5. However Bagwan, 2010 reported that most favourable pH for maximum antagonistic potential of *Trichoderma viride* against *S. rolfsii* and *R.solani* ranged between 5.5 to 6.5.

CONCLUSIONS

The interaction between *Trichoderma* sp. and *R.solani* was dependent on temperature and pH. *Trichoderma* Td85 strain proved to be most effective with the highest percentage of inhibition at 30°C whereas Td50 strain showed lower inhibition at this temperature.
Our results supported that the most appropriate pH for maximum antagonistic potential of tested strains was 4.5.

REFERENCES

Bagwan N.B., 2010. Influence of temperature and pH on antagonistic potential of *Trichoderma viride* in vitro.*International Journal of Plant Protection* 3 (2): 165-169

Cundom M.A., Mazza S.M., Gutierrez S.A., 2003. Selection of *Trichoderma* spp. isolates against *R.solani* .Spanish Journal of Agricultural Research1(4):79-82

Daryaei A., Jones E.E., Glare T.R., Falloon R.E., 2016. pH and water activity in culture media affect biological control activity of *Trichoderma atroviride* against *R.solani*. Biological Control 92:24-30

Elad Y., Chet I, Henis Y., 2006. Biological control of *R. solani* in strawberry fields by *Trichoderma harzianum*. Plant Soil 60:245-254

Galarza L., Akagi Y., Tkao K.., Kim C.S., Maekawa N., Itai A., Peralta E., Santos E., Kodam M., 2015. Characterization of *Trichoderma* species isolated in Ecuador and their antagonistic activities against phytopathogenic fungi from Ecuador and Japan. J Gen Plant Pathol 81:201–210

Grosch R., Lottmann J., Rehn V.N.C., Rehnk G., Mendoca-Hagler L., Smalla K., Berg G., 2007 Analysis of antagonistic interactions between *Trichoderma* isolates from Brazilian weeds and the soil –borne pathogen *Rhizoctonia solani* .J of Plant Diseases and Protection 114(4):167-175

Köhl J., Schlösser E., 1989, Decay of sclerotia of *Botrytis cinerea* by *Trichoderma* spp. at low temperatures. Journal of Phytopathology, 125(4): 320-326.

Kredics L., Antal Z., Manczinger L., Szekeres A., Kevei F., Nagy E., 2003. Influence of environmental parameters on *Trichoderma* strain with biocontrol potential. Food Technol. Biotechnol. 41(1):37-42

Lahlali R., Hijri M., 2010. Screening, identification and evaluation of potential biocontrol fungal endophytes against *R.solani* AG3 on potato plants. FEMS Microbiol Lett 311:152-159

Montealegre J., Valderama L., Herrera R., Besoain X., Perez L.M., 2009. Biocontrol capacity of wild and mutant *Trichoderma harzianum* (Rifai) strains on *R.solani* 618: effect of temperature and soil type during storage. Electronic J. of Biotechnology 12(4):1-11

Morton D.T., Stroube N.H., 1955, Antagonistic and stimulatory effect of microorganism upon *Sclerotium rolfsii*. Phytopathology 45:419-420

Osman M.E.H., El-Sheekh M.M., Metwally M.A. Ismail A.E.A., Ismail M.M., 2011. Antagonistic activity of some fungi and cyanobacteria species against *R.solani*. Int. J of Plant Pathology 2(3):101-114

Paica A., Petrisor C., Constantinescu F. 2015. Influence of abiotic factors on biological control ability of different *Trichoderma* spp. strains. Annals of the University of Craiova. XX(LVI):543-550

Rahman M., Ali M.A., Dey T.K., Islam M.M., Naher L., Ismail A., 2014. Evolution of disease and potential biocontrol activity of *Trichoderma* sp against *R. solani* on potato. Biosci J. Uberlandia 30(4):1108-1117

Santamarina M.P., Rosello J., 2006. Influence of temperature and water activity on the antagonism of the *Trichoderma harzianum* to *Verticilium* and *Rhizoctonia* . Crop protection 25:1130-1134

Schneider J.H.M., Schilder M.T., Dust G., 1997. Characterization of *R.solani* AG-2. isolates causing barepatch in field grown tulips in the Netherlands. Eur.J.Plant Pathol.103:265-279

Seema M., Devaki N.S., 2012. In vitro evaluation of biological control agents against *Rhizoctonia solani*. Jof Agric. Techn. 8(1):233-240

Singh A., Mohd S., Srivastava M., Biswas S.K., 2013. Molecular and antagonistic activity of *Trichoderma atroviride* against legume crop pathogens in Uttar Pradesh, India. Int. Journal of Bioresource and Stress Management.4(4):582-587

Singh S., Chand H., 2006. Screening of bioagents against root rot of mung bean caused by *R.solani*. Mol. Plant Microbe Interact.18:710-721

Tu C.C., Hsieh T.F., Chang Y.C., 1996. Vegetable diseases incited by *Rhizoctonia* spp in Sneh B., Jabaji-Hare S., Neate S., Dijst G.(eds):*Rhizoctonia* species: taxonomy, molecular biology, ecology,pathology and disease control, pp 369-377, Kluwer Academic Publishers, Dordrecht.

9

EFFECTS OF CHRONIC TOXICITY INDUCED BY CADMIUM ON THE GAMETOPHYTE OF TWO FERN SPECIES

Oana-Alexandra DRĂGHICEANU[1], Liliana Cristina SOARE[1]

[1]University of Piteşti, Târgu din Vale Street, No 1, 110040, Piteşti, Argeş County, Romania

Corresponding author email: o_draghiceanu@yahoo.com

Abstract

The aim of the present study was to determine the effect of chronic cadmium (Cd) action on the germination of spores and gametophyte differentiation in species Athyrium filix-femina (L.) Roth and Dryopteris filix-mas (L.) Schott, on different culture media (Knop solution, soil) for a period of 3 months. Cadmium was used in the following concentrations: $C=0$ mg $Cd \cdot L^{-1}$ Knop solution/kg^{-1} soil, $V_1=25$ mg $Cd \cdot L^{-1}$ Knop solution/kg^{-1} soil, $V_2=50$ mg $Cd \cdot L^{-1}$ Knop solution/kg^{-1} soil, $V_3=100$ mg $Cd \cdot L^{-1}$ Knop solution/kg^{-1} soil, $V_4=150$ mg $Cd \cdot L^{-1}$ Knop solution/kg^{-1} soil. The percentage of germinated spores was found to decrease with the increasing Cd concentration in the environment, while germination is delayed in time. Unlike the Knop solution variants, in the soil variants gametophyte development was not significantly affected; for V1-2 soil concentrations the sporophyte appears in Athyrium filix-femina, a stage that was also noted for the Knop solution control, for the same species. In the case of the variants grown on Knop solution, although the spores did germinate and the gametophyte began to differentiate, Cd-induced chronic stress cannot be compensated by the gametophyte, so that the cells lose their membrane integrity, and their survival is compromised.

Key words: heavy metals, Athyrium filix-femina, Dryopteris filix-mas.

INTRODUCTION

Cadmium is one of the most important heavy metals, and it is usually encountered: on the International Agency for Research on Cancer. list due to its carcinogenic properties, on toxic substance list of Agency for Toxic Substances and Disease Registry (CAS ID #:7440-43-9), among the top 126 priority pollutants, according to the United States Environmental Protection Agency (Flora, 2014).

It is a transition metal (block d), which presents chemical similarities to zinc (Zn) – in fact, they both belong to the same group (12). These similarities can cause the toxicity of Cd: replacement of Zn, a trace element, by Cd affects metabolic processes (Wuana and Okieimen, 2011).

According United States Geological Survey, to estimate Cd reserves, Zn reserves are checked, while taking into account this aspect: Cd is approximately 0.003% of Zn ores.

Cadmium pollution is due to natural sources, the contribution of which varies between 10-50% of total emissions, and also to anthropogenic sources. For example, the mining of zinc is estimated to release approximately 6 million tones, as a byproduct

of Cd (Raza et al., 2015). In order to present the main anthropogenic sources International Cadmium Association proposes a classification that takes into account the presence of Cd as impurity – non-Cd products: iron and steel, fossil fuels, cement, phosphate-based fertilizers, and, as a necessity: NiCd batteries, pigments, Cd alloys, electronic Cd compounds, etc.

Determining Cd toxicity on living beings is performed using acute and chronic toxicity tests. Acute toxicity refers to short-period exposure of an organism to the action of one or more toxic agents. Within this framework, lethal concentration (LC_{50}) is determined, i.e. the concentration that leads to the death of 50% of the test organisms. In nature, most pollutants manifests their action after a long time, and pollution is usually in non-lethal concentrations. Chronic toxicity is the "capacity of a substance or a solution to induce adverse effects for a long time, after repeated or continuous exposure, sometimes over the whole lifetime of an organism" (United States Environmental Protection Agency).

The best-known and most severe form of chronic exposure to the action of Cd is that occurring in Japan: consumption of rice contaminated with Cd leads to the disease called "Itai-Itai", which is characterized by kidney damage and disorders of the bone system (osteomalacia and osteoporosis) (Nordberg et al., 2015).

According to Pavlik (1997) 90% of the Cd taken up by plants comes from the ground, and only 10% from the atmosphere, as the main paths of penetration are the roots and leaves.

Catalá et al. (2011) recommended using ferns in toxicity tests, both acute and chronic, because the results can be extrapolated to wild plants or cultivated plants, they are found in different habitats (ecological or organic relevance), and growing spores and development gametophyte can be made on different media (solution, soil, etc.).

In order to know pteridospore sensitivity in a chronic toxicity testing of different substances and environmental samples should be used (Catalá and Rodriguez-Gil, 2011).

The aim of this paper was to determine the chronic effect of the action of Cd on the germination of spores and gametophyte differentiation in species *Athyrium filix-femina* (L.) Roth (*Aff*) and *Dryopteris filix-mas* (L.) Schott (*Dfm*) on different culture media.

MATERIALS AND METHODS

In order to obtain the spores of the two species, the author took several study trips along the Vâlsan Valley over the period August 2015. Mature leaves were collected from several individuals in different sites in order to ensure genetic diversity. After releasing the spores in the sporangia, there followed their collecting and preserving in a refrigerator at 4°C.

Testing media:

Two test media were used: Knop solution [$Ca(NO_3)_2$:1.00 $g \cdot L^{-1}$; $MgSO_4$: 0.25 $g \cdot L^{-1}$; KH_2PO_4: 0.25$g \cdot L^{-1}$; KNO_3: 0.25$g \cdot L^{-1}$] and flower earth Florisol obtained by processing from the deposit in Dersca-Dorohoi, with a pH between 6.5-7, humidity 60-70% N: 410 ppm, P: 192 ppm, K: 1350 ppm; organic substance min 70% dry product. The soil was sterilized at 60°C.

Tested substance: The substance tested was Cd acetate in various concentrations; reporting was done per L for the samples in Knop solution, and per kg for the soil variants: Control (C)=0 mg $Cd \cdot L^{-1}$ kg^{-1}, V_1=25 mg $Cd \cdot L^{-1}$ kg^{-1}, V_2=50

mg $Cd \cdot L^{-1}$ kg^{-1}, V_3=100 mg Cd L^{-1} kg^{-1}, V_4=150 mg $Cd \cdot L^{-1}$ kg^{-1}.

To ensure optimal conditions for development, the culture vessels were kept in growth chamber at 25°C in the daytime, and 15°C at night, with constant humidity and illumination (photoperiod: 16 hours of light, and 8 hours of dark). The soil variants were placed in Petri dishes and periodically watered with distilled water.

The experiment had 3 repeats. For the Knop solution variants quantitative determinations were made: the percentage of germinated spores was determined, and to do the statistical interpretation the SPSS program, version 16 was used, with which the average and the standard deviation were calculated. Comparisons were made using Duncan's test. To monitor the differentiation of the gametophyte in all variants, observations were made at regular intervals, and photomicrographs were made under an OPTIKA B275 microscope with an A630 Canon Power Shoot camera and under a OPTIKA stereo-microscop.

RESULTS AND DISCUSSIONS

Germination of spores is influenced by a number of factors such as light, phytohormones, ions of metal, temperature (Suo et al., 2015).

As far as the cultures of spores are concerned, which used the Knop solution, Cd significantly affected germination, primarily by reducing the percentage of germinated spores. All experimental variants were affected, except V_2Cd *Dfm*, where there were 7 percent more spores than in the controls, and in *Aff* – between V_3Cd and the control there were no significant differences (see Table 1). Also, spore germination was delayed in time in the V_4Cd variant, in both species: germination was reported after a month compared to the control. Also, in this variant the lowest percentage of spores germinated was obtained: 4 for *Dfm* and 15 for *Aff*.

Time delay and a lower percentage of spores germinated due to the presence of various concentrations of Cd were also reported by Gupta and Devi (1992), and Biswas et al. (2015) in several species of ferns.

Gupta et al. (1992) found that, in concentrations of 2.5 and 5 mg Cd L^{-1} spore germination was inhibited, and the development of the gametophyte was discontinued at the stage of prothallium blade in the species Ceratopteris thalictroides.

Table.1. Influence of heavy metals on the germination of spores

Species	C	V$_1$Cd	V$_2$Cd	V$_3$Cd	V$_4$Cd
	Percentage of germinated spores (mean ± standard deviation)				
Athyrium filix-femina	90.3±2[a]	26.6±5[c]	45±1[b]	92.3±1[a]	15±1[e]
Dryopteris filix-mas	81.6±2[b]	75.3±4[c]	89.3±5[a]	54±3[d]	4±1[e]

Legend: The values are the means of 3 replicates ± standard deviation; a, b, c, d, e – the results obtained from the Duncan test: the comparisons were made between control and V$_{1-4}$ for each metal.

In Table 2 and Table 3 the gametophyte differentiation after one month is shown, both on the soil and on the Knop solution, in both species.

Table 2. Gametophyte differentiation of *A. filix-femina* (one month after experiment initiation)

Variants	Knop solution	Soil
Control	blades differentiation, antheridia	young chordate prothallia, antheridia
V$_1$Cd	filaments differentiation, germinated spores	prothallium blade, antheridia
V$_2$Cd	prothallium blade, three-dimensional cell masses, antheridia	chordate prothallia
V$_3$Cd	prothallium filament, three-dimensional cell masses	chordate prothallia, rare prothallium blade, antheridia
V$_4$Cd	filaments differentiation	prothallium blade, antheridia

Table 3. Gametophyte differentiation of *D. filix-mas* (one month after experiment initiation)

Variants	Knop solution	Soil
Control	prothallium blade	young chordate prothallia
V$_1$Cd	damaged filaments and blades differentiation, short rhizoid, three-dimensional cell masses	chordate prothallia
V$_2$Cd	blades differentiation, three-dimensional cell masses	chordate prothallia
V$_3$Cd	filaments, blades differentiation	prothallium blade and filament
V$_4$Cd	germinated spores	prothallium blade and filament

It was found that gametophyte development was much faster in the soil-grown variants, where the following stages were noted:

chordate prothallia in the controls of both species (Figure 14, 17) V$_2$Cd (Figure 23) and V$_3$Cd in *Athyrium*, V$_1$Cd and V$_2$Cd in *Dryopteris*, blade in V$_1$Cd (Figure 20) and V$_4$Cd in *A. filix-femina*, and for the second species, filaments in V$_{3-4}$Cd. In *Athyrium*, antheridia with viable antherosoids were observed, in all cases, except V$_2$Cd. In the Knop solution variants the most advanced stage of development was the prothallian blade one, which occurred in the controls of the two species (Figure 1, 4, 5) and in V$_2$Cd *Athyrium* (Figure 7). The filament stage usually occurred at high concentrations V$_{3-4}$ (Figure 10).

Due to the influence of Cd, the gametophyte development was affected: the filaments (V$_1$Cd in both species) and blades were partially damaged (V$_2$Cd in *Dryopteris* - Figure 12) and three-dimensional cell masses were formed (Figure 8). The abnormal growth of the prothallium blade was reported by Gupta and Devi (1994), as well, in the species *Pteris vittata*, where the gametophyte is much more sensitive to Cd action than the sporophyte.

Table 4. Gametophyte differentiation of *A. filix-femina* (three months after experiment initiation)

Variants	Knop solution	Soil
Control	chordate prothallia with sporophyte	chordate prothallia with archegonia
V$_1$Cd	damaged filaments, germinated spores	chordate prothallia and sporophyte with embryonic leaf
V$_2$Cd	damaged prothallia	chordate prothallia and sporophyte
V$_3$Cd	damaged filaments	mature chordate prothallia with archegonia, young chordate prothallia with antheridia, fecundation
V$_4$Cd	damaged filaments, germinated spores	prothallia, antheridia

Table 5. Gametophyte differentiation of *D. filix-mas* (three months after experiment initiation)

Variants	Knop solution	Soil
Control	young elongated prothallia	chordate prothallia, antheridia
V$_1$Cd	damaged filaments	chordate prothallia,
V$_2$Cd	damaged filaments and blades	chordate prothallia, archegonia
V$_3$Cd	germinated spores, damaged filament	young chordate prothallia
V$_4$Cd	few germinated spores	young chordate prothallia (small)

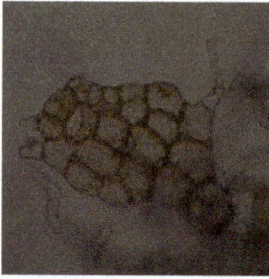

Figure 1. *Dfm,* C, one month (x100).

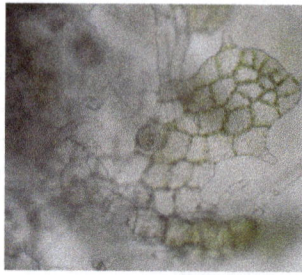

Figure 2. *Dfm,* M, 3 months (x100).

Figure 3. *Dfm,* C, 3 months (x400).

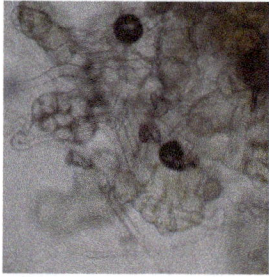

Figure 4. *Aff* C, one month (x100).

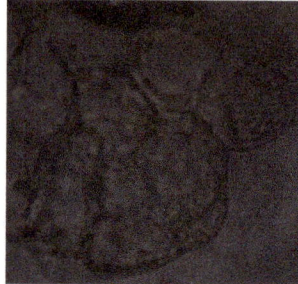

Figure 5. *Aff* C, one month (x400).

Figure 6. *Aff* C, 3 months (x10).

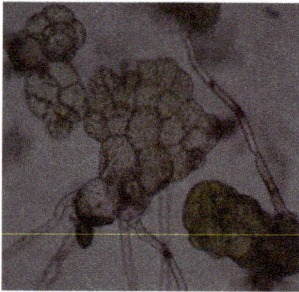

Figure 7. *Aff* V$_2$, one month (x100).

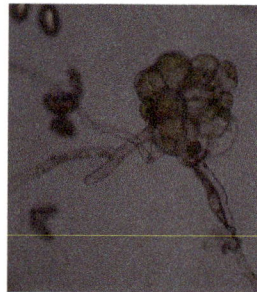

Figure 8. *Aff* V$_2$, one month (x100).

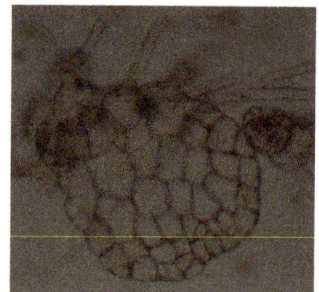

Figure 9. *Aff* V$_2$, 3 months (x100).

Figure10. *Dfm* V$_3$, one month (x100).

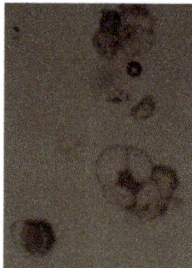

Figure11. *Dfm* V$_3$, 3 months (x100).

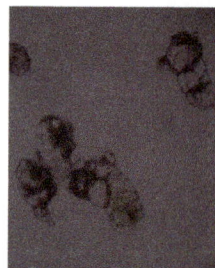

Figure12. *Dfm* V$_2$, one month (x100).

Figure13. *Dfm* V$_2$, 3 months (x100).

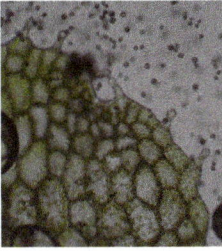

Figure14. *Dfm* C, soil, one month (x100).

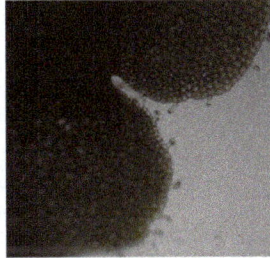

Figure15. *Dfm* C, soil, 3 months (x40).

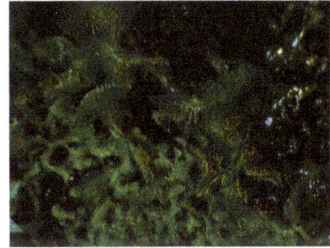

Figure16. *Dfm* C, soil, 3 months (x10).

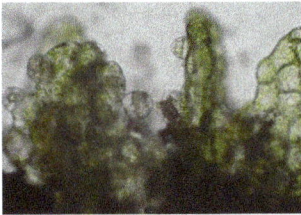

Figure17. *Aff* C, soil, one month 10x10 (x100).

Figure18. *Aff* C, soil, 3 months (x40)

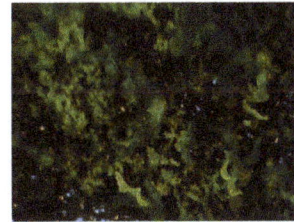

Figure19. *Aff* C, soil, 3 months (x10).

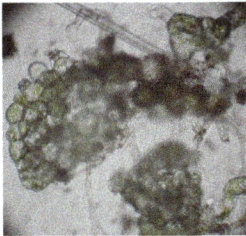

Figure 20. *Aff* V$_1$, soil, one month (x100).

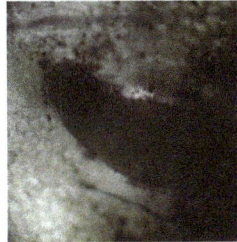

Figure 21. *Aff* V$_1$, soil, 3 months (x100).

Figure 22. *Aff* V$_1$, soil, 3 months (x10).

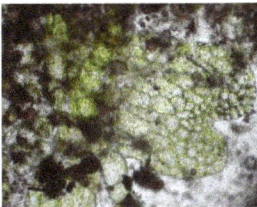

Figure 23. *Aff* V$_2$, soil, one month (x100). Figure 24. *Aff* V$_2$, soil, 3 month (x10).   Figure 25. *Dfm* V2, 3 months (x40).

After 3 months from the initiation of the experiment, in *A. filix-femina* as control, the Knop solution variant, the sporophyte was formed (Figure 6), while *D. filix-mas* was still in the gametophyte stage: prothallium blade (Figure 2) with antheridia (Figure 3). According to Table 4 and Table 5, regardless of species and experimental variant, after 3 months the gametophyte development was stopped in the stage of chordate prothallium (Figure 9), with damaged blades and filaments (Figure 11, 13). For low concentrations (V$_{1-2}$), in the soil-grown variants, Cd stimulated the development so that within 3 months the sporophyte was formed on the gametophyte (Figure 22, 24), in which the juvenile leaves can be noted (Figure 21). In the remaining Cd variants, and also in the control, the gametophyte was still in the stage of heart-shaped prothallium of different sizes (Figure 15, 16, 19) with/without antheridia and/or archegonia (Figure 18, 25).

CONCLUSIONS

Cd affected the percentage of spores germinated and gametophyte development in the experimental variants grown on Knop solution.
The influence of Cd on gametophyte development was far more significant in the solution variants (damaged blades and chordate prothallia) as compared to those grown on soil, and between species, in *Dryopteris filix-mas* as compared to *Athyrium filix-femina*.
In V_{1-2} concentrations for soil-grown *Athyrium* the occurrence of the sporophyte is noted. In the case of the variants grown on Knop solution, although the spores did germinate and the gametophyte began to differentiate, Cd-induced chronic stress cannot be compensated by the gametophyte, so that the cells lose their membrane integrity, and their survival is compromised.

REFERENCES

Biswas M., Khare P., Kumari N., 2015. Effect of heavy metals like mercury, cadmium and lead on the spore germination of *Pteris vittata* L. the common road side fern. International Journal of Science, Technology & Management, 4 (1):194-198.

Catalá M., Esteban M., Quintanilla L.G., 2011. Mitochondrial Activity of Fern Spores for the Evaluation of Acute Toxicity in Higher Plant Development. In: Fernández H., Kumar A., Revilla M.A. (Eds.), Working with Ferns, Issues and Applications, Springer, New York, Dordrecht, Heidelberg, London, 237-247.

Catalá M., Rodriguez-Gil J.L., 2011. Chronic Phytotoxicity in Gametophytes: DNA as Biomarker of Growth and Chlorophyll Autofluorescence as Biomarker of Cell Function. In: Fernández H., Kumar A., Revilla M.A. (Eds.), Working with Ferns, Issues and Applications, Springer, New York, Dordrecht, Heidelberg, London, 249-260.

Flora S.J.S., 2014. Metals. In: Gupta R.C. (Ed.), Biomarkers in Toxicology, Agents Toxicity Biomarkers, Academic Press., San Diego, 485-519.

Gupta M., Devi S., Singh J., 1992. Effects of long-term low-dose exposure to cadmium during the entire life cycle of *Ceratopteris thalictroides*, a water fern. Archives of Environmental Contamination and Toxicology, 23(2):184-189.

Gupta M., Devi S., 1992. Effect of Cadmium on spore germination and gametophyte development in some ferns. Bull Environ Contam Toxicol, 48:337-343.

Gupta M., Devi S., 1994. Chronic toxicity of cadmium in *Pteris vittata*, a roadside fern. Ecotoxicology, 3(4): 235-247.

Nordberg G.F., Nogawa K.., Nordberg M., 2015. Cadmium. In: Nordberg G.F., Fowler B.A., Nordberg M. (Eds.), Handbook on the Toxicology of Metals (Fourth Edition), Academic Press, Amsterdam, Boston, Heidelberg, London, New York, Oxford, Paris, San Diego, San Francisco, Singapore, Sydney, Tokyo, 667-716.

Pavlik Y., 1997. Ecological Problems in Industrially Exposed East Slovakian Regions in Relation to Agriculture (In Slovak). VtjZV Nitr.

Raza S.H., Shafiq F., Rashid U., Ibrahim M., Adrees M., 2015. Remediation of Cd-Contaminated Soils: Perspectives and Advancements. In: Hakeem K.R., Sabir M., Ozturk M., Mermut A.R. (Eds.), Soil Remediation and Plants: Prospects and Challenges. Academic Press, Amsterdam, Boston, Heidelberg, London, New York, Oxford, Paris, San Diego, San Francisco, Singapore, Sydney, Tokyo, 571-597.

Suo J., Chen S., Zhao Q., Shi L., Dai S., 2015. Fern spore germination in response to environmental factors. Front. Biol., 10(4):358-376.

Wuana R.A., Okieimen F.E., 2011. Heavy metals in contaminated soils: A review of sources, chemistry, risks and best available strategies for remediation. ISRN Ecology. Vol 2011: 20 pages.

International Agency for Research on Cancer, IARC Monographs, Classifications, List of Classifications, volumes 1-113, 2016, http://monographs.iarc.fr/ENG/Classification/latest_classif.php.

Agency for toxic Substances and Disease Registry, Most viewed toxic substances, Cadmium, Toxic Substances Portal, 2011 http://www.atsdr.cdc.gov/substances/toxsubstance.asp?toxid=15.

International Cadmium Association, Cadmium Applications, http://www.cadmium.org/cadmium-applications.

U.S. Environmental Protection Agency 1997. Terms of Environment: Glossary, Abbreviations and Acronyms (EPA Publication No.175-B-97-001). U.S. Environmental Protection Agency, Washington D.C.

U.S. Geological Survey, USGS Minerals Information: Cadmium, Mineral Commodity Summmaries, 2015 http://minerals.usgs.gov/minerals/pubs/commodity/cadmium/mcs-2015-cadmi.pdf

10

LOPHANTHUS ANISATUS, A MULTI – PURPOSE PLANT, ACCLIMATIZED AND IMPROVED AT VRDS BUZAU

Costel VÎNĂTORU[1], Bianca ZAMFIR[1], Camelia BRATU[1], Adrian PETICILA[2]

[1]Vegetable Research and Development Station Buzău, No. 23, Mesteacănului Street,
zip code 120024, Buzău, Romania
Email: costel_vinatoru@yahoo.com; zamfir_b@yahoo.com; botea_camelia2007@yahoo.com
[2]University of Agronomic Sciences and Veterinary Medicine of Bucharest,
59 Mărăşti Blvd, District 1, 011464, Bucharest, Romania
Email: apeticila@gmail.com
Corresponding author email: costel_vinatoru@yahoo.com

Abstract

VRDS (Vegetable Research and Development Station) Buzau has tradition for acclimatization of new vegetable species, there being obtained new varieties of Momordica charantia, Cucumis metuliferus, Cichorium crispum, Cichorium latifolium, Momordica conchinchinensis, etc. Biodiversity conservation and crop extention for the new species through acclimatization and breeding have become a major necessity nowadays. The aim of this study was to give special attention to the acclimatization of a new species, e.g. Lophanthus anisatus. It is a native to Asia, is spread in almost all world crops, known by other names (Agastache foeniculum, Lophantus agastache) and its food, medicinal and melliferous properties are widely recognized by scientists but, however, until now there has not been cultivated in Romania. The research started in 2010 with the documentation and purchase of the basic genetic material (seeds, seedlings). After completing these steps, in 2012 were cultivated the first purchased genotypes. After the first year of study in the crop was observed that there were no major phenotypic differences between cultivars. Genotype L3, from Bulgaria, demonstrated higher uniformity in terms of the main characters expressiveness and high adaptability to environmental conditions. In descending, 2013-2014 was cultivated and studied only this genotype, to avoid contamination by pollination with other cultivars. The results are positive, the species has adapted very well, can be grown successfully in Romania. Research will continue for genetic stabilization and marked characteristics for distinctibility in order to approval and registration at SIVTR (State Institute for Variety Testing and Registration).

Key words: adaptability, biodiversity, genotype, melliferous, medicinal.

INTRODUCTION

Lophanthus anisatus study was taken from VRDS Buzau in 2010. In the world is known under several names (*Agastache Foeniculum*, *Lophantus Agastache*) or lofant popular.
"*Agastache* is a small genus of *Lamiaceae*, comprising 22 species of perennial aromatic medicinal herbs. In this article, we review recent advances in phytochemical, pharmacological, biotechnological and molecular research on *Agastache*." (Zielinska, 2014) This is a multi-purpose plant, in the world it is known as a medicinal plant, aromatic, spicy, and even ornamental and melliferous.
"This species is a candidate for large scale, domestic cultivation as a source of nectar for honey bees and as aromatic plant with wide variation in the composition and content of its essential oils." (Fuentes Granados, 1995)
Because of its genetic capacity to adapt to environmental conditions, is cultivated and known worldwide.
"The genus has gained importance in America, Asia, and Europe as a component of tea mixtures and as a flavouring in confections." (Fuentes Granados,1997)
The main objectives of the Laboratory of Genetics Breeding and Biodiversity Conservation from VRDS Buzau are getting new biological creations, competitive, as required by growers and consumers; rehabilitation of neglected plants in culture; acclimatization of new species and promote their culture.
The research undertaken in this species is within the target three priority of the research laboratory.

MATERIALS AND METHODS

The research for this species has been carried out according to an established plan, covering four main stages. The first phase focused on documentation and studies and to obtain basic biological material or seeds from reliable sources and distant geographical areas. The seeds used were from America, Asia and Europe.

In the second stage we worked to acclimate the species in the climatic conditions of our country. In this stage were detained genotypes that have shown adaptability and genetic uniformity descent and those who have demonstrated higher sensitivity and variability of the characters have been removed from the breeding program. Genotypes have passed acclimatization entered the third stage of work aimed at improving the species in order to obtain varieties of genetically stabilized according to international norms.

In the fourth stage has developed specific technology for acclimatization and improved genotypes.

RESULTS AND DISCUSSIONS

Following evaluation, the process of acclimatization acquired genetic material found to L3 coming from Bulgaria best adapted to the climatic conditions of our country and also showed uniformity and genetics. The other origins were removed from the acclimatization process, because this species is entomofila very much preferred by insects, especially bees, contamination risk and prolongation of acclimatization and improvement.

After this genotype has successfully passed acclimatization phase, has undergone extensive improvement works in order to obtain a new variety.

In the process of improvement to follow the main character restriction variability and meeting international standards distinctibility, uniformity and stability (DUS).

Genotype L3 obtained the best results and improvement works ended with getting a new variety and the main features of genotype L3 (table1)

Table 1. The main features of L3 genotype

Studied feature		Variability limit	Average value
Plant height		87-105	90 cm
Shrub diameter		47-55	51 cm
Main shoot no.		10-16	13 buc
Secondary shoot no.		29-38	34 buc
large inflorescences no.		17-24	22 buc
Small inflorescence no.		22-30	26 buc
Inflorescence lenght	large	15-21	18 cm
	medium	9-15	12 cm
	small	4-6	5 cm
Leaf stalk lenght		1.5-2.5	2 cm
Sesil leaf lenght		5-7	6 cm
Leaf diameter		3-4	3.5 cm
Large leaves no/main shoot		11-17	14 buc
Small leaves no./main shoot		35-41	38 buc
Stem lenght		3-5	4 cm
Stem diameter at the basis		13-17	15 mm
Shoots diameter		2.5-3	2.8 mm
Inflorescence diameter		2-2.4	2.2 cm
Floral floors no.	large	14-18	16 buc
	Medium	10-14	12 buc
	small	5-7	6 buc
Flower diameter		1.8-2.2	2 mm
Seed diameter		0.8-1	0.9 mm
Seed lenght		1.4-1.6	1.5 mm
Total weight of the green inflorescence		3.7-4.1	3.9 g
Total weight of dry inflorescence		0.9-1.2	1.1 g

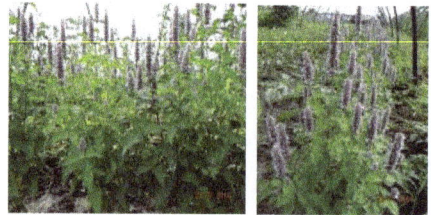

Figure. 1. Crop detail

Leaf sections have edges, sharp tip, slightly porous surface, similar to the *Urtica dioica*. (figure 1 and 2)

Figure 2. Leaves detail

Stems and shoots are grooved with square cross-section (figure 3)

Figure 3. Shoots detail

Nr. stamens in flower- 4 including two long and two short.

The blooming period is very long and is made in instalments from June until the coming of frost and plant specialists is ranked among the top bee plants in the world. Potential production per unit area is very high, of over 600 kg/honey per hectare. (figure 4).

Figure 4. Flower details

After completing the program of improvement, research has been channelled to developing specific technology culture. After research it was found that the species shows great flexibility in terms of culture technology but the best results were obtained from the culture seedling establishment.

It was found that a plant that is highly resistant to cold, frost resistant up to its limits and during periods of heat stress (cold) leaves change colour in green-purple due to anthocyanin pigments accumulation. (figure 5).

The best time for sowing seedlings is after February 20, it is recommended furrow sowing in hard or peat pots or directly in alveolar blades. If the bed is sown, it should work in palaces alveolar subculturing or pots and if is sown directly in pots palaces is recommended to rare the seedlings, leaving one plant in the alveola. Sowing is recommended close to the surface very carefully because the seeds are very small, less than 0.5 mm deep. Sunrise is

done in 10-15 days, if factors are insured optimum vegetation.

Planting seedling stage reached after 60 days of emergence (figure 6).

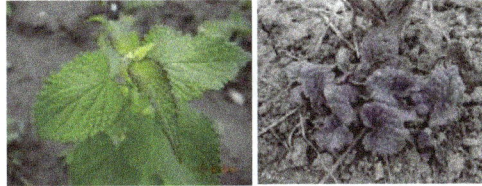

Figure 5. Termic stress, low temperatures

Figure 6. Seedlings details

The establishment of culture are taken into account several factors: the irrigation system used system used for maintenance vehicles that culture and space needed nutrition and development. It was found that the species supports several options for setting up culture technology but the best option checked in the research undertaken (figure 7 and 8) with 70 cm between rows and 30-35 cm between plants / row.

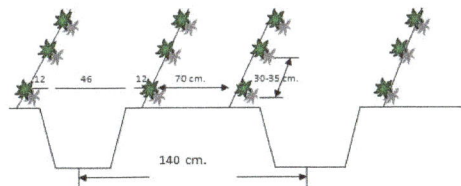

Figure. 7. Open field planting method-shaped soil

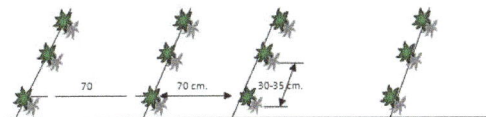

Figure 8. Open field planting method on plan soil

Establishment of culture can be done in a long time, depending on climatic conditions, starting April 15 and ending in late May.

It was found that culture are growing very well in the second and even third year. It can be said that it behaves as a perennial in our country conditions (figure 9).

Figure 9. Crop in the second year

Care works are common to all specific vegetable plant has outstanding technological requirements, no high demands from the ground, can be cultivated throughout the country, it is recommended to perform one or two mechanical hoeing and hand hoeing one or two depending on the physical condition of the land and its degree of weed.

Regarding water supply plant species fall into the group claims to moderate water. In the absence of rainfall is recommended to apply watering rules between 250-350 m3 of water / ha. The absence of water leads to the maturation of the plant and forced induction phenomenon dwarfs.

Research undertaken so far have shown that diseases and pests do not cause significant damage this species was reported only damaging nematode but found that even in areas heavily infested with nematodes, plants have survived, not significantly diminished production (figure 10).

Figure 10. Nematodes attack

Seed maturation is made in stages, starting from the base to the summit blossom. A 2520 gram seed = seed. A 17.5 g plant seeds with a total of over 44 100 units. A well developed and carefully harvested plant can produce one hectare of crop seed required.

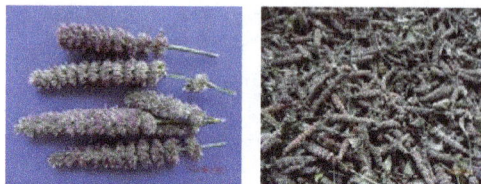

Figure 11. Dry inflorescences

Harvesting can be done only once or in stages. If the plants are harvested once mature, production per hectare is 18-22 t vegetative mass but if harvesting is done in stages, by cutting the shoots and regenerating bush, production can increase significantly.

The variety can be grown successfully in the ecological system or to protect other crops that are exposed to rodents (mice, rabbits, deer, etc.) strong flavor nook at the touch of all vegetative parts of the plant creates rodent repellent, protecting crops successfully .

CONCLUSIONS

The research was completed with success. The species can be grown successfully throughout Romania. In the process of breeding a new variety was obtained, which was signed at L3 SIVTR for approval, since 2015, as Aromat de Buzau. To develop specific culture technology has been a significant amount of seeds produced and performed by diffusion and culture.

REFERENCES

Zielińska, Sylwia, and Adam Matkowski, 2014 "Phytochemistry and bioactivity of aromatic and medicinal plants from the genus Agastache (Lamiaceae)."*Phytochemistry Reviews* 13.2:P. 391-416.

Fuentes-Granados, Roger Guillermo. 1997 "Genetic studies of Agastache."

Fuentes-Granados, Roger G., and Mark P. Widrlechner. "Diversity among and within populations of Agastache foeniculum." *Proceedings of the 14th North American Prairie Conference. Kansas State University, Manhattan, Kansas, July 12–16, 1994.* Ed. D. Hartnett. 1995.

11

ACTIVE FUNGAL ENDOPHYTES AGAINST PHYTOPATHOGENIC FUNGI- DWELLERS OF ROMANIAN AND CANARIAN *ARTEMISIA* SPP.

Andreea COŞOVEANU[1], Cristina Emilia NIŢĂ[2] Beatrice Michaela IACOMI[3]
Samuel RODRIGUEZ SABINA[1], Raimundo CABRERA[1]

[1] UDI Fitopatología, Facultad de Ciencias, Sección Biología, Universidad de La Laguna (ULL), ES-38206 La Laguna, Tenerife, Spain
[2] University of Agronomic Sciences and Veterinary Medicine of Bucharest, Faculty of Horticulture, 59 Marasti Blvd, District 1, Bucharest, Romania
[3] University of Agronomic Sciences and Veterinary Medicine of Bucharest, Faculty of Agriculture, 59 Marasti Blvd, District 1, Bucharest, Romania
Corresponding author email: andreeacosoveanu@gmail.com

Abstract

Endophytic fungi have been isolated from Artemisia austriaca (isolates HRO184, HRO183, HRO169 and HRO115), A. vulgaris (isolate HRO158) and A. thuscula (isolates HLP7, HLP26, HLP27, and HLP44) in Romania and La Palma, Canary Islands, Spain. The strains were studied for their bioactivity against phytopathgenic fungi (Alternaria alternata, Alternaria dauci, Alternaria brassicicola, Fusarium oxysporum, Fusarium solani, Fusarium moniliforme, Sclerotinia sclerotiorum, Botrytis cinerea, Phoma sp., Geotrichum sp. and Cladosporium sp.) using dual culture, agar dilution and volatile compounds (VOC) techniques. In dual culture assays, all Canarian isolates were strong antagonists of at least one pathogen and two Romanian isolates, HRO169 against S. sclerotiorum and HRO158 against A. brassicicola. In VOC assays, a difference was observed between slow growth isolates (1-2 cm/week) and normal growth (4-5 cm/week), therefore three moments of fungal pathogen inoculation were used: after three and five days from the endophyte inoculation. The highest inhibition gradually produced by the volatile compounds was observed with HLP44 isolate against Cladosporium sp. (%I = 31.5) at 7th day. Solvent extracts were obtained from HRO169, HRO158, HLP44, HLP27 and HLP7 isolates and further two of them (HRO158 and HLP44) were fractionated using vacuum-liquid chromatography eluted with n-hexane:EtAc:MeOH gradient to give seven fractions. Extract obtained from HLP44 isolate strongly inhibited A. brassicicola (%I = 44), F. moniliforme (%I =52.78) and S. sclerotiorum (%I = 50.4), at 0.1mg ml⁻¹. Ethyl acetate fraction was the most active against A. brassicicola, F. solani and B. cinerea (%I = 42.68, 59.17 and 49.36, respectively, at 0.1mg ml⁻¹) followed by the EtAc:MeOH (90:10) fraction which also inhibited A. brassicicola and F. solani (%I = 35.83 and 39.53, respectively).

Key words: Artemisia, bioactivity, endophytic fungi, phytopathogens

INTRODUCTION

Fungal endophytes are microorganisms that live in the intercellular spaces of stems, petioles, roots and leaves of plants without causing observable manifestation of their existence (Strobel and Long, 1998). Same substances can be produced by both endophytes and plants (Stierle et al, 1993; Strobel et al, 1997; Lee et al, 1995; Kusari et al, 2008). According to Strobel (2002), plants with medicinal value or unusual longevity, plants that survive under extreme conditions often harbour potential fungal endophytes that produce bioactive metabolites. Medicinal plants are revealed as host endophytes which in turn provide protection from infectious agents (Strobel et al, 2002). A caval-cade of endophytic species dwelling in medicinal plants revealed bioactivity features in various studies (Li et al, 2005; Raviraja et al, 2006; Chowdhary and Kaushika, 2015; Purwantini et al, 2015). The symbiotic relation between the endophyte and its host is considered 'defensive mutualism' id est the resistance of the host to pathogens, phytophagous insects and environmental conditions increases; secondary metabolites are involved in most cases (Gonzalez-Coloma et al, 2016). An array of compounds belonging to various chemical groups are metabolized by endophytic fungi such as phenols, steroids, flavonoids, quinones, terpenoids, xanthones, peptides, alkaloids, aliphatic compounds, phenylpropanoids, isocoumarins, benzopyranones, tetralones, cytochalasines and enniatines (Schulz et al. 2002, Rocha et al. 2011, Schulz and Boyle 2005, Aly et al. 2010, Santos et al, 2003). New compounds were isolated from endophytes inhabiting *Artemisia annua* (Ge et al. 2010; Lu et al, 2000) and *Erythrina crista-galii* (Weber et al, 2004). The production of antioxidant compounds by plants, like phenolic acids and their derivatives (Huang et al., 2007), isobenzofuranones (Strobel et al., 2002), isobenzofurans (Harper et al., 2003), as well as mannitol and other carbohydrates (Richardson et al., 1992), is attributed to the presence of reactive oxygen species (ROS) generated by endophytes (Gonzalez-Coloma et al, 2016). Volatile compounds (VOCs) are also produced by fungal endophytes among other microorganisms, but less is known about the pathways in which they are produced. Many of them are either metabolic transformation products of lipids, proteins, heterocyclic metabolites or other components of living tissues or degradation end-products ('waste products') of fungal catabolic pathways (Bennett et al, 2012). Species of endophytes produce antifungal and antimicrobial VOCs like *Muscodor albus* (Strobel et al, 2001), *Oxyporus latemarginatus* (Lee et al, 2009) and *Gliocladium* sp. (Stinson et al, 2003), respectively. *Artemisia* is a wide studied genus of plants for its medicinal and bioactive properties (Soylu et al, 2005; Brudea, 2008; Dancewicz and Gabrys, 2008; Garcia et al, 2015; Abad et al, 2012) and recently its endophytic fungi communities have been taken into observation looking for bioactivity features (Haniya et al, 2013; Qian et al, 2014; Purwantini et al, 2015, Cosoveanu et al, 2016). The present study selected several species of fungal endophytes previously isolated from *Artemisia austriaca*, *Artemisia vulgaris* and *Artemisia thuscula* to evaluate their bioactive potential. It is noteworthy to mention that is the first notification on fungal endophytes isolated from *A. austriaca* and *A. thuscula* and their bioactivity.

MATERIALS AND METHODS

Plant sampling and isolation techniques
Plant samples were collected and processed in 2012 in Romania: *A. austriaca* from Tuzla, Murighiol, Babadag Lake; *A. vulgaris* from Mahmudia; and in La Palma Island (Spain): *A.* *thuscula* from La Galga and Tigalate. A surface sterilization method was used in order to suppress epiphytic microorganisms from the plant samples. Briefly, plants were immersed first in sterile H2O, followed by 1min in EtOH 70%, 1min in sodium hypochlorite 15%, 1 min in EtOH at 70% and finally washed with sterile H2O (changed from Schulz et al., 1993). The isolation procedure was performed according to Cosoveanu et al. (2014). In order to analyse the fungal diversity, each replicate obtained from distinct stem fragments was registered. When an endophyte was acquired in pure culture it was preserved (Czapek, T=5°C and Glycerol 20% DI H2O, T= -30°C), bioactively tested and identified.

Dual culture assays
Dual culture technique was employed to find endophytic fungi that produce metabolites which inhibit *S. sclerotiorum*, *F. oxysporum*, *F. moniliforme*, *F. solani*, *A. brassicicola*, *A. dauci*, *A. alternata*, *Phoma* sp. *Geotrichum* sp. and *Cladosporium* sp. mycelial growth *in vitro*. PDA plates were incubated at 25°C in darkness for 7 days and observed daily; plates were left for a further week to check the stability of the interaction. The following criteria based on Kusari et al. (2013) were used to interpret the results:

0-Mycelia grow until making contact with each other

1-Mutual inhibition (both mycelia stop growing at a certain distance)

2-Mycelia grow until making contact with each other and in the area where the contact is produced morphological changes occur/ the growth is stopped in a convex form

3-Pathogen growth is detained at a certain distance from the endophyte (<2 mm)

4-Pathogen growth is detained at a certain distance from the endophyte (>2 mm)

RDP- Rapid development and parasitism of the endophyte

RD- Rapid development of the endophyte

RDL- Rapid development of the endophyte and lysed mycelia of the pathogen

L- Opponent fungus presents lysed mycelia

P- Endophyte displays parasitism on pathogen

It should be noted that also the pathogen may respond similarly to criteria 3, 4, RDP, RD, RDL, L and P (further, the results for this case are noted with *). However, this study focuses only on the endophyte response towards pathogen.

Fungal isolates

Fungal isolates (endophytic and pathogenic strains) were maintained on PDA, T=25°C, in darkness. Endophytes were selected based on their results in preliminary assays of antagonism. Pathogens were chosen due to their different interactions with the host and their high economic importance: *Alternaria alternata, A. dauci, A. brassicicola, Fusarium oxysporum, F. moniliforme, F. solani, Sclerotinia sclerotiorum, Botrytis cinerea, Phoma* sp, *Geotrichum* sp. and *Cladosporium* sp.

Biometric agar dilution assays

Tests were carried out to determine the biological activity of extracts using biometric agar dilution method. The extracts were incorporated into PDA as follows: 1, 0.5 and 0.1 mg ml^{-1}. The final percentage of ethanol in the media was adjusted to a concentration of 1% (v/v). Plates containing the solvent (ethanol) were used as negative control. Each pathogen was spot - inoculated at eight equidistant points to PDA media amended with the fungal extracts at tested concentrations. Three replicates were used per treatment. For each extract and concentration, inhibition of radial growth (%I) compared with the control was calculated after 72 hours of incubation at 25°C, in the dark. The radial growth was measured with an image - processing software Image J -Wayne Rasband (NIH).
Kruskal-Wallis Test and Mann Whitney U were performed using IBM SPSS Statistics 21.0.

Volatile compounds assays (VOC)

The VOC assays analyse the activity of volatile compounds produced by the endophytes on phytopathogens. The assays were done in Petri plates, using only the plate bases with nutritive media for the fungal inoculation as following: plate base with endophyte + plate base with pathogen and the controls: base without pathogen + base with endophyte and correspondingly for the pathogen. Plates were incubated for a week at T=25°C, in darkness. Assays were performed in triplicates. Measurements were made daily, calculating an average of two measurements of the same inoculum diameter. In the case of slow growth endophytic fungi three different moments for the pathogen inoculation were chosen allowing to the endophytic fungus to develop more as follows: inoculation of both fungi at the same time (Moment 0), inoculation of the pathogen after three days from the inoculation of the endophyte (Moment 1) and inoculation of the pathogen after five days from the endophyte's inoculation (Moment 2).

The following criteria were used to interpret the results:
A = E<CE: The pathogen inhibits the endophyte
B = P<CP: The endophyte inhibits the pathogen
C = CE<E: The pathogen enhances the endophyte's growth
D = CP<P: The endophyte enhances the pathogen's growth
* CE: Control of endophyte; CP: Control of pathogen; E: Endophyte in interaction; P: Pathogen in interaction

High scale cultivation of endophytic fungal isolates

Five of nine fungal isolates showing antagonistic activity were further explored for bioactivity of their crude extracts by multiplication on rice media. Rice medium was prepared in 500ml Erlenmeyer flasks containing 120g of rice grains with 40ml H$_2$O (autoclaved). The flasks were inoculated with 4-6 disks of endophytic fungus mycelium (25°C, darkness for three weeks).

Chemical solvent extraction of endophytic fungal isolates (crude extract and fractionation)

120ml of ethyl acetate (Sigma Aldrich) was poured on the rice grains covered with the fungus mycelium and kept aside for 24h. The content was filtered under vacuum using a Buchner funnel and the solvent extraction was repeated thrice. The collective extract was dried up with vacuum rotary evaporator under reduced pressure at 50°C and used as crude extract for further evaluation. Two of the crude extracts were selected to be fractionated. The extracts were previously subjected to partitioning between n-hexane and MeOH (V/V) to remove the fatty acids, twice. The MeOH fraction was chromatographed on a SiO$_2$ vacuum-liquid chromatography column (VLC) eluted with n-hexane:EtAC:MeOH gradient to give seven fractions (three times the volume of the column per solvent fraction). Fr-1 (n-hex. 100%), Fr-2 (n-hex:EtAc 90:10), Fr-3 (n-hex:EtAc 75:25), Fr-4 (n-hex:EtAc 50:50), Fr-5 (EtAc 100%), Fr-6 (EtAc:MeOH 90:10), Fr-7 (MeOH 100%). Fractions were further treated as the crude extracts.

RESULTS AND DISCUSSIONS

In dual culture assays, isolate HLP44 inhibited the development of six pathogens with a higher distance than 2mm (criterion 4) and three with a distance of 2mm (criterion 3) out of 11 pathogens (Table 4). Isolates HLP7 and HLP26 were also strong inhibitors of four and three pathogens, respectively. Also HLP27 and HRO169 isolates impeded the development of one pathogen each. In the agar dilution assays, HRO169 (extract code-1090) inhibited (%I= 47.70, at 0.1mg/ml) only *F. oxysporum* although the results from dual culture did not predict it (criterion 1). The crude extract of HLP7 (code- 1114) inhibited *A. brassicicola* (%I= 32.55, at 0.1mg/ml) but not *Cladosporium* sp. (%I= 9.7, at 1mg/ml) although in dual culture the observed reactions were similar for both pathogens (criterion 4). Despite the interesting activity in dual culture (criterion 3), the extract was not active against *A. alternata* and *A. dauci*. (%I= 15.5 and 5.02, respectively; at 1mg/ml). HLP44 extract (code- 1092) was the most active and therefore it was fractionated. 1092 inhibited with interesting values seven of the nine tested pathogens (Table 3). Although all three species of *Alternaria* reacted in a similar manner in dual culture assay, the response of the extract in dilution agar assays variated (vs. *A. alternata* %I= 38.6, vs. *A. brassicicola* %I= 41.7. vs. *A. dauci* %I= 19.5, at 1mg/ml). As for the three species of *Fusarium*, results of inhibition were similar (Table 3). The bioactivity of this extract was disjoined, having various active fractions with different pathogens (Table 3). Briefly, the fraction which most inhibited the mycelial growth of tested pathogens with interesting values was the one eluted with ethyl acetate (vs. *A. brassicicola* %I= 42.7, vs. *F.*

solani %I= 59.2, vs. *F. moniliforme* %I= 43.2 and vs. *B. cinerea* %I= 49.4, at 0.1mg/ml). Ethyl acetate and methanol extracts of an endophytic *Chaetomium globosum* isolate were more effective than hexane extract against *S. sclerotiorum* (Kumar et al, 2013). Yet, the non-polar fractions (L0 and F2, eluted with hexane and n:hexane-EtAc 90:10, respectively) strongly inhibited *B. cinerea* (L0- %I= 35.1; F2- %I= 43.4, at 0.1mg/ml). Previous reports on hexane extracts obtained from *Colletotrichum globosum* showed antifungal properties against *B. cinerea* (Nakashina et al, 1991). Fungal endophytes were tested for their VOCs bioactivity and separated in two groups due to their type of growth: regular growth and slow growth. It was hypothesized that the more mass of endophytic mycelia is produced, the more VOCs would be generated a posteriori in the interaction with the pathogen. In the slow growth group the endophytes were left to develop in the absence of the pathogen for three and five and five days. The highest percentage of inhibition, reached gradually, (%I= 31.5, at Moment 3) was calculated for the interaction between *Cladosporium* sp. and HLP44, confirmed by Kruskal Wallis test (p= 0.027). Overall there were no significant differences between the inoculation moments, but neither interaction with high percentages of inhibition. In the group of regular growth, the interactions between the same pathogen and two endophytes were compared. No significant difference resulted (p<0.05), therefore no difference in the sensibility of the pathogen exposed to more than one reputed bioactive endophytes, was found. The structural groups of VOCs detected in different individuals of *A. vulgaris* collected from various countries and regions show similarities as the main groups belong to monoterpenes followed by sesquiterpenes (Zhigzhitzhapovae et al, 2016). *Artemisia* spp. essential oils have antimicrobial properties (Baykan Erel et al, 2010), antiparasitic and cytotoxic activity (Martinez-Diaz et al, 2015) so the community of harboured endophytes seems likely to be also producing bioactive VOCs, as previously shown (Strobel et al, 2011).

Table 1. VOC interaction between pathogens and endophytes with regular growth: percentages of inhibition (expressed as average and Standard deviation STD) and criteria of interpretation

Interaction	%I pathogens- AVG (STD)			%I endophytes- AVG (STD)			CRITE-RIA
	Day 1	Day 4	Day 7	Day 1	Day 4	Day7	
A. alternata &HLP26	20.74 (3.03)	2.86 (6.92)	2.33 (5.58)	-3.33 (4.71)	10.56 (2.66)	17.53 (3.91)	
A. alternata & HRO158	19.16 (4.43)	9.65 (4.50)	9.32 (3.05)	40.99 (14.62)	44.18 (39.66)	6.07 44.92)	
A. brassicicola & HRO158	7.69 (10.88)	-8.11 (8.63)	19.21 (5.33)	6.78 (12.04)	0.86 (5.17)	-3.15 (6.07)	B
A. brassicicola & HLP26	-18.59 (4.53)	-18.62 (19.19)	2.61 (11.18)	0.00 (0.00)	21.07 (10.44)	16.96(4.67)	
A. dauci & HRO169	3.42 (2.42)	5.12 (5.13)	0.49 (6.85)	3.04 (2.15)	-7.91 (7.22)	-15.64 (6.22)	
A. dauci & HLP26	-2.41 (15.29)	-1.50 (8.77)	-2.07 (6.39)	0.00 (0.00)	5.25 (3.84)	11.36 (2.09)	
F. oxysporum & HLP26	-30.37 (35.09)	1.53 (1.24)	17.16 (3.74)	3.33 (4.71)	28.82 (3.09)	44.33 (4.99)	A
F. oxysporum & HRO184	-10.74 (34.91)	10.27 (1.98)	9.07 (5.64)	-0.67 (8.26)	9.60 (5.00)	27.69 (3.17)	A
F. moniliforme & HRO158	-4.22 (9.71)	-0.11 (3.04)	2.55 (8.48)	-7.77 (14.87)	-5.47 (18.68)	-18.49 (19.38)	
F. solani & HLP26	-3.92 (7.02)	0.02 (3.74)	5.94 (1.38)	-20.00 (0.00)	9.06 (1.69)	22.19 (4.93)	A
F. solani & HRO184	-1.96 (5.11)	-2.06 (4.03)	7.15 (2.90)	-8.08 (18.24)	-7.23 (8.23)	8.31(7.22)	
B. cinerea & HRO158	27.23 (15.87)	-16.94 (14.07)	-21.89 (14.24)	6.66 (24.97)	-11.05 (30.42)	-15.70 (17.65)	
B. cinerea & HRO183	10.65 (11.44)	-7.38(10.86)	-10.03 (12.71)	-2.04 (6.73)	10.04(1.41)	7.47 (4.82)	
S. sclerotiorum & HRO158	-39.91 (60.20)	-20.20(28.57)	-20.20 (28.57)	4.40 (14.92)	10.42(26.87)	17.22 (58.66)	
S. sclerotiorum & HRO169	-33.07 (65.05)	-18.18 (25.71)	-18.18 (25.71)	6.73 (11.59)	6.29 (1.45)	2.36 (3.61)	

Table 2. VOC interaction between pathogens and endophytes with slow growth: percentages of inhibition (expressed as average %I and standard deviation STD), moments of inoculation of pathogen and criteria of interpretation.

Interaction	Moment	%I pathogen			%I endophyte			CRITERIA
		Day 1	Day 4	Day 7	Day 1	Day 4	Day7	
A. brassicicola & HLP44	M0	15.0 (13.2)	17.1 (4.8)	6.8 (15.3)	0.0 (0.0)	15.4 (8.2)	-4.4 (16.1)	
	M1	-0.8 (8.9)	36.2 (1.5)	33.2 (4.4)	13.5 (8.2)	2.7 (9.9)	-0.3 (9.5)	B
	M2	-3.0 (5.2)	35.3 (11.8)	24.1 (4.1)	2.4 (14.8)	-0.3 (9.5)	-1.9 (12.9)	B
A. brassicicola & HLP27	M0	20.1 (1.3)	23.0 (7.6)	-1.2 (5.9)	0.0 (0.0)	4.5 (5.1)	5.8 (3.3) a	
	M1	-2.3 (26.6)	24.3 (9.5)	8.7 (24.8)	-1.2 (4.0)	3.9 (3.8)	15.2 (3.1)b	
	M2	-18.2 (18.2)	16.7 (3.7)	25.2 (1.5)	-8.6 (8.1)	-9.3 (5.4)	-6.6 (8.3)c	B
A. dauci & HLP44	M0	23.3 (2.9)	15.9 (9.6)	2.6 (20.9)	0.0 (0.0)	5.7 (5.6)	-0.7 (9.0)	
	M1	16.7 (3.6)	19.4 (3.6)	5.6 (6.3)	9.6 (8.9)	12.8 (4.8)	12.4 (5.0)	
A. dauci & HLP7	M0	6.7 (11.5)	1.5 (10.3)	6.5 (4.7)	0.0 (0.0)	-92.5 (62.3)	-77.8 (79.1)	
	M1	20.8 (7.2)	-4.7 (6.4)	-3.0 (3.9)	-92.3 (60.1)	-73.0 (68.4)	-68.7 (72.7)	
	M2	9.1 (0.0)	8.4 (1.8)	9.8 (4.2)	-104.5 (83.5)	-119.1 (94.2)	-131.1 (109.0)	
B. cinerea & HLP27	M0	18.5 (29.4)	2.5 (5.4)	2.5 (5.4)	-8.3 (14.4)	-4.7 (4.1)	-5.2 (1.2)	
	M1	2.2 (3.8)	6.5 (12.3)	4.4 (12.7)	-1.2 (4.0)	-2.0 (5.3)	13.5 (4.7)	
	M2	8.9 (10.2)	3.6 (6.4)	1.8 (6.4)	-3.3 (5.0)	-8.7 (6.2)	-11.2 (10.2)	
Cladosporium sp. & HLP44	M0	16.2 (16.7)	8.3 (15.7)	3.0 (10.5)a	0.0 (0.0)	4.2 (7.2)	-5.6 (20.6)	
	M1	2.5 (13.1)	17.7 (7.7)	17.1 (4.8)b	1.5 (12.4)	3.0 (5.2)	4.7 (4.6)	
	M2	-18.6 (25.2)	15.1 (2.0)	31.5 (3.2)c	10.3 (13.7)	13.8 (7.6)	15.0 (6.2)	B
Cladosporium sp. & HLP7	M0	-11.9 (33.0)	0.0 (23.8)	-5.5 (13.1)	0.0 (0.0)	-18.1 (28.5)	16.2 (22.3)	
	M1	14.1 (4.4)	11.0 (1.7)	6.6 (12.2)	-88.3 (80.6)	-103.5 (116.9)	-122.8 (137.8)	
	M2	-12.8 (23.4)	2.0 (10.1)	16.2 (21.4)	-30.2 (62.1)	-46.6 (85.5)	-45.0 (86.2)	

Values with different letter have statistical difference P<0,05, U Mann Whitney Test

Table 3. Dilution agar assays with crude extract and fractions versus fungal pathogens- %I (STD)

Extract	[C]	A. a.	A. b.	A. d.	F. o.	F. s.	F. m.	S. s.	B. c.	Clad.
972	1 mg/ml	20.8 (5.8)						12.2(9.4)		
1090	1 mg/ml	29.8(4.1)	12.8(5.2)		45.8(3.3)a	23.3(7.8)a	16.2(7.4)	14.8(5.8)		
	0.5 mg/ml				57.0(5.2)b	31.2(6.0)b				
	0.1 mg/ml				47.8(4.8)a	6.3(7.8)c				
1091	1 mg/ml		44.5(4.7)	-6.6(8.9)				2.6(9.9)	1.4(8.6)	
1114	1 mg/ml	15.5(2.6)a	42.3(8.0)a	5.0(4.6a						9.7(12.5)a
	0.5 mg/ml	18.7(2.8)a	41.2(5.7) b	11.7(4.6) b						13.1(17.1)a
	0.1 mg/ml	18.0(2.8)a	32.6(4.3) b	4.8(4.6)a						-16.4(29.5) b
1092	1 mg/ml	38.6(1.6)a	41.7(6.1)a	19.5(4.1)a	37.9(10.2)a	34.3(5.5)a	42.5(7.2)a	71.5(8.0)a	45.2(2.5)a	
	0.5 mg/ml	29.6(1.2) b	46.0(5.3)a	13.8(*3.0)b	26.6(6.5) b	18.4(4.2)b	28.7(6.9)b	67.9(8.8)a	51.1(5.0)b	13.4(10.9)a
	0.1 mg/ml	17.2(2.0)c	44.0(5.3)a	3.5(*2.1)c	16.9(7.5) c	-1.5(4.3) c	52.8(7.8)c	50.4(33.7)a	18.6(8.8)c	25.5(19.0a
1092LO	0.1 mg/ml	-1.0(11.6)	9.2(7.5) *			5.4(6.7)	6.8(7.7)		35.1(11.9) *	
1092F2	0.1 mg/ml	4.8(6.7)	29.3(4.9) *			15.3(5.1) *	2.7(10.2		43.4(5.4) *	
1092F3	0.1 mg/ml	-4.6(4.9)	7.6(4.8) *			7.5(9.2)	0.5(6.8)		14.4(7.1) *	
1092F4	0.1 mg/ml	4.5(5.2)	27.2(4.8) *			34.0(4.5) *	21.2(9.1) *		19.6(2.8) *	
1092F5	0.1 mg/ml	21.5(2.8) *	42.7(3.8) *			59.2(0.9) *	43.2(5.5) *		49.4(3.1) *	
1092F6	0.1 mg/ml	3.2(9.2)	35.8(5.2) *			39.5(5.4) *	10.8(10.7)		32.9(1.6) *	
1092F7	0.1 mg/ml	2.0(5.4)	19.3(3.6) *			24.7(5.3) *	-17.0(6.6)		19.1(5.7) *	

U Mann Whitney was applied in the case of the assays with fractions and pathogens to check the statistical difference between control and treatment (values marked with * have p<0.05) and between treatment at different concentrations (values with different letter have statistical diference p<0.05). A.a.= *A. alternata*, A.b.= *A. brassicicola*, A.d.= *A. dauci*, F.m. = *F. moniliforme*, F.s.= *F. solani*, F.o.= *F. oxysporum*, S.s.= *S.sclerotiorum*, B.c.= *B. cinerea*, Clad= *Cladosporium* sp.

Table 4. Dual culture assay: fungal endophytes and pathogens (interaction criteria)

	HLP7	HLP26	HLP27	HLP44	HRO115	HRO158	HRO169	HRO183	HRO184
F. o.	0	3, P*	2*	3, P*	0	2	1	0	4
F. m.	0	0, P*	2*	3, P*	2	0	0	0	P*
F. s.	0	3, P*	2*	3, P*	0	2	1	0	4
B. c.	4	4	4	4	0	0	2	3, P	2*
S. s.	4	2	2	4	2	2	4	RD, 1	RD*P*
A. a.	3	3	0	4	0	0	3	1	0
A. d.	3	4	P	4	P	3	1	P	2
A. b.	4	4	3	4	P	3, P	3*	P	0
Phoma	4*	4*	4*	1	RD, L	4*	4*	4*, RD	4*
Geot.	4*	4*	4*	1	4*	4*	4*	4*, RD	nt
Clad.	4	4	1	1	P		1	P, RD	4*

A.a.=*A. alternata*, A.b.= *A. brassicicola*, A.d.= *A. dauci*, F.m. = *F. moniliforme*, F.s.= *F. solani*, F.o.= *F. oxysporum*, S.s.= *S.sclerotiorum*, B.c.= *B. cinerea*, Clad= *Cladosporium* sp. , Geot= *Geotrichum* sp. 0-Mycelia grow until making contact with each other; 1- Mutual inhibition (both mycelia stop growing at a certain distance); 2- Mycelia grow until making contact with each other and in the area where the contact is produced morphological changes occur/ the growth is stopped in a convex form; 3- Pathogen growth is detained at a certain distance from the endophyte (<2 mm); 4- Pathogen growth is detained at a certain distance from the endophyte (>2 mm); RD- Rapid development of the endophyte; L- Opponent fungus presents lysed mycelia; P- Endophyte displays parasitism on pathogen. *= the pathogen is evaluated with the correspondent criteria; the rest of the cases are applied to the action of the endophyte on the pathogen.

CONCLUSIONS

Our study has shown the potential of two fungal endophytes against important plant pathogens. The most interesting fungal endophyte is HLP44 isolate (from *A. thuscula*), as its valences are multiple being a spring of active compounds against *A. brassicicola*, *F. solani* and *B. cinerea* found mainly in the ethyl acetate fraction but also in the hexane fraction. More, this isolate strongly inhibited *B. cinerea*, *S. sclerotiorum*, *A. alternata*, *A. dauci*, *A. brassicicola* and *Cladosporium* sp. in dual culture and *Cladosporium* sp. in VOC assay which converts it into a tool for biocontrol. Further studies will be carried out to identify the active compounds responsible for the inhibition of the mycelium growth of pathogens. One good candidate for *in vivo* further assays would be HRO158 isolate (from *A. vulgaris*) which inhibited the growth of the mycelium of *A. brassicicola* and *A. dauci* in dual culture assay.

ACKNOWLEDGEMENTS

This research work was carried out partially supported by grant of La Caixa- Fundacion Caja Canarias para Posgraduados (2014).

REFERENCES

Abad M.J., Bedoya L.M., Apaza L., Bermejo P., 2012. The *Artemisia* L. genus: A review of bioactive essential oils. Molecules 17, 2542-2566 doi:10.3390/molecules17032542.

Aly A.H., Debbab A, Kjer J., Proksch P., 2010. Fungal endophytes from higher plants: a prolific source of phytochemicals and other bioactive natural products. Fungal divers 41, 1–16.

Baykan Erel S., Reznicek G., Senol S.G., Karabay Yavasogulu N.U., Konyalioglu S., Zeybek A.U., 2012. Antimuctobial and antioxidant properties of *Artemisia* L. species from western Anatolia. Turk J Biol 36: 75-84

Bennett J.W., Hung R., Lee S., Padhi S., 2012. Fungal and bacterial volatile organic compounds: an overview and their role as ecological signaling agents. In: Fungal Associations, 2nd Edition. The Mycota IX DOI: 10.1007/978-3-642-30826-0_18.

Brudea V., 2008. The efficacy of some biopesticides and vegetal metabolites in the control of spireas aphid aphis *Spiraephaga* Muller (O. Homoptera – F. Aphididae). Lucrari Stiintifice, vol. 52, seria Agronomie, 611-616.

Chowdhary K., Kaushik N., 2015. Fungal endophyte diversity and bioactivity in the indian medicinal plant *Ocimum sanctum* Linn. PLOS ONE 10(11): e0141444. doi:10.1371/journal.pone.0141444.

Cosoveanu A., Hernandez M., Iacomi-Vasilescu B., Zhang X., Shu S., Wang M., Cabrera R. 2016. Fungi as endophytes in Chinese *Artemisia* spp.: juxtaposed elements of phylogeny, diversity and bioactivity. Mycosphere 7(2): 102–117.

Cosoveanu A., Gimenez-Marino C., Cabrera Y., Hernandez G., Cabrera R., 2014. Endophytic fungi from grapevine cultivars in Canary Islands and their activity against phytopathogenic fungi. International Journal of Agriculture and Crop Sciences. 7(15): 1497-1503.

Dancewicz K., Gabrys B., 2008. Effect of extracts of garlic (*Allium sativum* L.), wormwood (*Artemisia absinthium* L.) and tansy (*Tanaceum vulgare* L.) on the behaviour of the peach potato aphid *Myzus persicae* (Sulz.) during the settling on plants. Pestycydy/Pesticides, (3-4): 93-99

Garcia J.J., Andres M.F., Ibanez A., Gonzalez-Coloma A., 2015. Selective nematocidal effects of essential oils from two cultivated *Artemisia absinthium* populations. Zeitschriftung fur Naturforschung. 70(9) DOI: 10.1515/znc-2015-0109.

Ge H.M., Peng H., Guo Z.K., Cui J.T., Song Y.C., Tan R.X., 2010. Bioactive alkaloids from the plant endophytic fungus *Aspergillus terreus*. Planta Med 76, 822–82.

Gonzalez-Coloma A., Cosoveanu A., Cabrera R., Gimenez C., Kaushik N., 2016. Endophytic Fungi and their Bioprospection. In: Deshmukh S.K., Misra J.K., Tewari J.P., Papp T., (Eds.), Fungi: Applications and Management Strategies. CRC Press, 14-31

Haniya A.M.K., Padma P.R., 2013. Free radical scabenging activity of *Artemisia vulgaris*, L. leaf extract. World Journal of Pharmacy and Pharmaceutical Sciences 2(6): 6381-6390

Harper J.K., Arif A.M., Ford E.J., Strobel G.A., Porco J.A., Tomer D.P., Oneill K.L., Heider E.M., Grant D.M. 2003. Pestacin: a 1,3-dihydro isobenzofuran from *Pestalotiopsis microspora* possessing antioxidant and antimycotic activities. Tetrahedron, 59: 2471–2476.

Huang W.Y., Cai Y.Z., Xing J., Corke H., Sun M. 2007. A potential antioxidant resource: endophytic fungi from medicinal plants. Econ. Bot., 61, 14–30.

Kumar S., Kaushik N., Proksch P., 2013. Identification of antifungal principle in the solvent extract of an endophytic fungus *Chaetomium globosum* from *Withania somnifera*. Springerplus, 2:37.

Kusari S., Lamshoft M., Zuhlke S., Spiteller M., 2008. An endophytic fungus from *Hypericum perforatum* that produces hypericin, J. Nat. Prod. 71: 159-162.

Kusari P., Kusari S., Spiteller M., Kayser O., 2013. Endophytic fungi harbored in *Cannabis sativa* L.: diversity and potenyial as biocontrol agents against host plat-specific phytopathogens. Fungal Diversity 60:137-151

Lee J.C., Lobkovsky E., Pliam N.B., Stroble G.A, Clardy J., 1995. Subglutinol A and B: immunosuppressive compounds from the endophytic fungus *Fusarium cubglutinans*, J. Org. Chem. 60, 7076-7077.

Lee S.O., Kim H.Y., Choi G.J., Lee H.B., Jang K.S., Choi Y.H., Kim J.C., 2009. Mycofumigation with *Oxysporus latemarginatus* EF069 for control of postharvest apple decay and *Rhizoctonia* root rot on moth orchid. J. Appl Microbiol 106: 1213-1219.

Li H., Qing C., Zhang Y., Zhao Z., 2005.Screening for endophytic fungi with antitumour and antifungal activities from Chinese medicinal plants, World J. Microbiol. Biotechnol. 21 (2005) 1515-1519.

Lu H., Zou W.X., Meng J.C., Hu J., Tan R.X., 2000. New bioactive metabolites produced by *Colletotrichum* sp., an endophytic fungus in *Artemisia annua*, Plant Sci. 151, 67-73.

Martinez-Diaz R.A., Ibanez-Escribani A., Burillo J., Heras L., Prado G., Argullo-Ortuno M.T., Julio L.F.,

Gonzalez-Coloma A., 2015. Trypanocidal, trichomonacidal and cytotoxic components of cultivated *Artemisia absinthium* Linnaeus (Asteraceae) essential oil. Mem Inst Oswaldo Cruz, 110(5): 693-699.

Nakashina N., Moromizato Z., Matsuyama N., 1991. The antifungal substance produced by *Chaetomium trilaterale* var. diporum RC-5 isolated from sclerotia of *Sclerotinia sclerotiorum*. Ann Phytopath Soc Japan, 57:657-662.

Purwantini I., Mustofa W., Asmah R., 2015. Isolation of endophytic fungi from *Artemisia annua* l, and indentification of their antimicrobial compounds using bioautography method. International Journal of Pharmacy and Pharmaceutical Sciences. 7(12): 95-99.

Qian Y., Kang J., Geng K., Wang L., Lei B., 2014. Endophytic fungi from *Artemisia argyi* Levl. et Vant. and their Bioactivity. Chiang Mai J. Sci. 41(4): 910-921.

Raviraja N.S., G.L. Maria, Sridhar K.R., 2006. Antimicrobial evaluation of endophytic fungi inhabiting medicinal plants of the western ghats of India, Eng. Life Sci. 6: 515-520.

Richardson, M.D., Chapman, G.W., Hoveland, C.S., Bacon, C.W., 1992. Sugar alcohols in endophyte-infected tall fescue. Crop Sci., 32: 1060–1061.

Rocha A.C.S., Garcia D., Uetanabaro A.P.T., Carneiro R.T.O., Araújo I.S., Mattos C.R.R., Góes-Neto A., 2011. Foliar endophytic fungi from *Hevea brasiliensis* and their antagonism on *Microcyclus ulei*. Fungal Divers 47:75–84.

Rude M.A., Schirmer A., 2009. New microbial fuels: a biotech perspective. Curr Opin Microbiol 12: 274-281.

Santos R.M., Rodrigues G., Fo E., Rocha W.C., Teixeira M.F.S., 2003. Endophytic fungi from *Melia azedarach*, World J. Microbiol. Biotechnol. 19: 767-770.

Schultz B., Wanke S., Draeger S., Aust H.J., 1993. Endophytes from herbaceous plants and shrubs: effectiveness of surface sterilization methods. Mycol. Res. 97 (12): 1447-1450.

Schulz B., Boyle C., Draeger S., Aust H.J., Römmert A.K., Krohn K., 2002. Endophytic fungi: a source of novel biologically active secondary metabolites. Mycol Res, 106:996-04.

Schulz, B., Boyle, C., 2005. The endophytic continuum. Mycological Research 109:661–686.

Singh S.K., Strobel G.A., Knighton B., Geary B., Sears J., Ezra D., 2011. An endophytic *Phomopsis* sp.

possessing bioactivity and fuel potential with its volatile organic compounds. Microb Ecol 61:729-739.

Soylu E.M., Yigitbas H., Tok F.M., Soylu S., Kurt S., Batsal O., Kaya A.D., 2005. Chemical composition and antifungal activity of the essential oil of *Artemisia annua* L. against foliar and soil-borne fungal pathogens. J. of Plant Diseases and Protection, 112 (3): 229–239.

Stierle A., Strobel G.A., Stierle D., 1993. Taxol and taxane production by *Taxomyces andreanae*, an endophytic fungus of Pacific yew, Science 260(5105):214-6.

Stinson E., Exra D., Hess W.M., Sears J., Strobel G., 2003. An endophytic *Gliocladium* sp. of *Eucryphia cordifolia* producing selective volatile antimicrobial compounds. Plant Sci. 165: 913-922

Strobel G.A., 2002. Microbial gifts from rain forests. Can J Plant Pathol 24:14-20.

Strobel G.A., Dirkse E., Sears J., Markworth C., 2001. Volatile antimicrobials from *Muscodor albus* a novel endophytic fungus. Microbiology 147: 2943-2950.

Strobel G.A., Hess W.M., 1997. Glucosylation of the peptide leucinostatinA, produced by an endophytic fungus of European yew, may protect the host from leucinostatin toxicity, Chem. Biol. 4: 529-536.

Strobel G.A., Long D.M., 1998. Endophytic microbes embody pharmaceutical potential, ASM News 64 (1998) 263-268.

Strobel G.A., Singh S.K., Riyaz-U-Hassan S., Mitchell A.M., Geary B., Sears J., 2011. An endophytic/pathogenic *Phoma* sp. from creosote bush producing biologically active volatile compounds having fuel potential. FEMS Microbiol Lett 320: 87-94.

Strobel, G.A., 2002. Rainforest endophytes and bioactive products. Criti. Rev. Biotechnol., 22(4):315–333.

Weber D., Sterner O., Anke T., Gorzalczancy S., Martino V., Acevedo C., 2004. Phomol, a new anti-inflammatory metabolite from an endophyte of the medicinal plant *Erythrina crista-galli*, J. Antibiot. 579, 559-563.

Zhigzhitzhapova S.V., Radnaeva L.D., Gao Q., Chen S., Zhang F., 2016. Chemical composition of volatile organic compounds of *Artemisia vulgaris* L. (Asteraceae) from the Qinghai-Tibet Plateau. Industrial Crops and Products, http://dx.doi.org/10.1016/j.indcrop.2015.12.083

12

ANALYSIS OF ANATOMICAL AND MORPHOLOGICAL CHARACTERS OF THE *SILENE CAPPADOCICA* BOISS. & HELDR. AND *SILENE SPERGULIFOLIA* BIEB. (CARYOPHYLLACEAE) SPECIES

Yavuz BAĞCI, Hüseyin BİÇER

Department of Biology, Faculty of Science, Selçuk University, Konya, Turkey
Ardıçlı Mh., Alaaddin Keykubat Kampüsü, Diş Hekimliği Fakültesi Kampüs, Merkez/Konya, Turkey
Corresponding author email: ybagci@selcuk.edu.tr

Abstract

In this study, analysis of anatomical and morphological characters of the Silene cappadocica Boiss & Heldr and Silene spergulifolia Bieb. specie which belong to family of Caryophyllaceae were determined. In morphological studies of these species, parts of stem, leaves, flower and fruit were measured and given as tables. In anatomical investigations of these two species were taken section from root, stem, and leaves by microtom and hand. These sections were painted and were made constant slide. After that, it was taken photograph of these slides with assist of microscope camera. Stomatal characteristics were examined by section taken superficial from these plants leaves and stomatal index was calculated.

Key words: Anatomy, Caryophyllaceae, Morphology, Silene, Endemic.

INTRODUCTION

Silene is one of the largest genera of flowering plants in the world consisting of about 750 species with the generality of them distributed in Mediterranean region (Greuter, 1995). In Turkey, the genus is represented by 148 species (Coode and Cullen 1967; Davis et al. 1988, Greuter 1995; Tan and Vural 2000; Özhatay and Kültür, 2006; Özhatayet al. 2009; Bağcı et al. 2007; Aksoy et al., 2008; Bağcı 2008; Tugay and Ertuğrul 2008; Kandemiret al. 2009; Yıldız and Dadandı 2009; Hamzaoğlu et al. 2010; Yıldız and Erik 2010; Yıldız et al. 2010; Hamzaoğlu et al. 2011, Hamzaoğlu, 2012).

Little work appears to have been done on the anatomy of vegetative organs of *Silene* species especially halophytic ones. Anatomical fluctuations about the plants structure are related with the habitats of plants. Millner (2006) reported that the anatomical structures of two *Silene* species is correlated with a wide range of environmental conditions. The high salinity of soils and the soil's moisture has a major impact on halophytes' anatomical structures and has formative effects. Their cumulated action has accompanied the halophytes evolution through time, as an active and dynamic component of the evolutionary "adventure" (Grigore and Toma, 2007). From this point of view, the present study the anatomical and ecological properties of two local endemic species of *Silene* (glikophytic and halophytic ones) have been investigated.

Silene genus is named various names as locally. Usually it is named "nakil çiçeği" in Turkish. The other names are used as *"salkım çiçeği, gıvışgan otu, gıcı gıcı, acı gıcı, gicime, cıvrıncık, çığıstak, gıvırsık, ecibücü, ibiş gıbış, kıvırşık, kıvışgan, kıvışık, kıvışkan, kıvrışık, kıvşıyık, tavuk yastığı"* for *Silene* genus in Turkish *(Baytop, 1997).*

MATERIALS AND METHODS

Materials

The investigated species have been collected from their natural habitats when they are mature. The localities of species are below:

Silene cappadocica: C4: Konya; Cihanbeyli; Tuz Gölü, Gölyazı, 2 June 2012, 38° 45.672 K, 33°06.801 D, 925 m °28.546 K, 32°43.904 D, 1750-1770 m, Bağcı 4145.

Silene spergulifolia: C4: Konya; Hadim-Taşkent between, 2 June 2012, 38° 45.672 K, 33°06.801 D, 1300 m, Bağcı 4156.

Methods

Morphological method

The species have been diagnosed by Davis (1967) and our observations have been stored in KNYA herbarium. Morphological researches for plant samples were done according to Flora of Turkey and the related articles. The diagnostic parts of *Silene* species such as plant length, basal and cauline leaves, petal, sepal, capsule (fruit) dimensions were calculated on 20 plant samples.

Anatomical method

For anatomical studies, roots, stems and leaves were used in paraffin method. And also some parts of plants were taken by hand with the aid of a razor blade. Paraffin sections were stained with safranin-fast green and hand sections were investigated directly. Canon EOS 450D cameras which attached to Leica DM 1000 light microscope were used for photograph and were calculated the cells dimensions with Cameram 21 programme.

RESULTS AND DISCUSSIONS

S. cappadocica
Morphological characteristics

Perennial. Stems ascending to erect, retroversely puberulent, 10-50 cm. Leaves elliptic to oblanceolate, usually without sterile shoots in their axils, puberulent, less than 5 mm broad. Inflorescence a rather loose though strict panicle. Calyx 3-5 mm in functionally female flowers, 5-11 mm in hermephrodite flowers, puberulent, often glandular. Petals white to grennish yellow, deeply bifid into ± linear lobes. Anthophore (in hermaphrodite flowers) 3-4 mm. Capsule ovoid, trigonous or 3-sulcate, long acuminate, included in the calyx. *Fl.* 5-7. *Steppe, slopes, etc. 800-2300 m* (Davis 1967), (Figure 1).

Figure 1. *S.cappadocica*, (A): general view (B): Fruit, (C): Basal leaves

Table 1. Morphological data of *S. cappadocica*

Plant Parts	wide		length	
	MİN	MAX	MİN	MAX
Basal leaf	0.2 cm	0.3 cm	1.2 cm	2.2 cm
Stem leaf	0.3 cm	0.2 cm	1.7 cm	2.3 cm
Calyx	0.3 cm	0.5 cm	0.6 cm	0.8 cm
Corolla	0.1 cm	0.15 cm	0.6 cm	1.2 cm
Fruit	0.2 cm	0.4 cm	0.6 cm	0.8 cm
Calyx teeth length	-	-	0.1 cm	0.1 cm
Bracts	-	-	0.1 cm	0.3 cm
Anthophore length	-	-	0.3 cm	0.5 cm
Plant length	-	-	30 cm	44 cm
Stem diameter	-	-	0.15 cm	0.4 cm

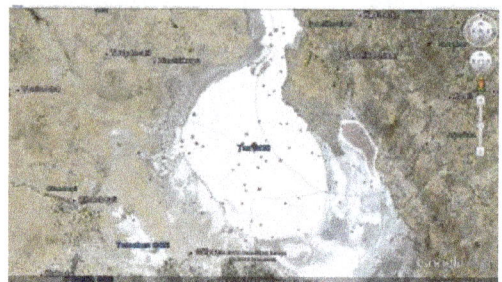

Figure 2. Location of *S.cappadocica* species

Figure 3. Location of *S. spergulifolia* species

Anatomical results

ROOT ANATOMY

Root usuaally comprises of four parts as anatomical; peridermis, cortex, vascular bunds and pith (1, 2 A, B). Peridermis encloses the outermost surface of *S. cappadocica* root cross sections. Below the peridermis, there is 8-10 layered cortex. Vascular system is well developed; there is 2-3 rowed cambium between phloem and xylem. Trachea diameter is 50-100 μm. Pith region is covered with xylem and the pith rays are 1-2 rowed.

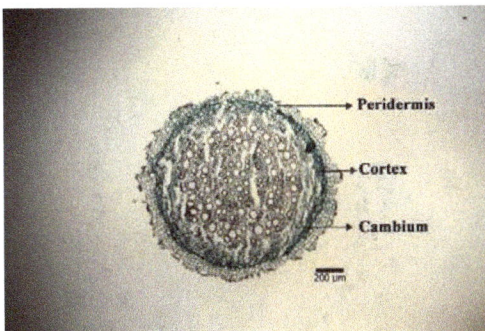

Figure 4. The cross sections of *S.cappadocica* root

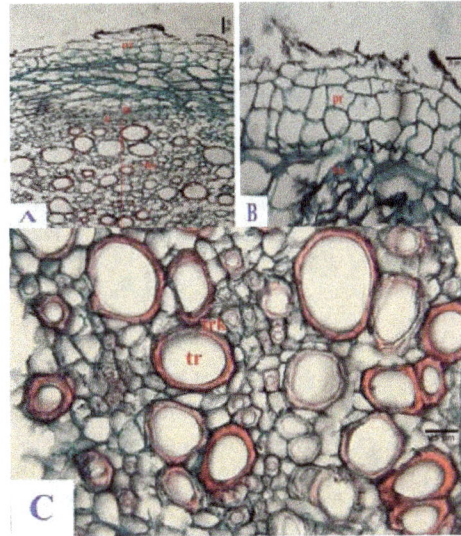

Figure 5. The cross sections of *S.cappadocica* root (A): **pr:** peridermis; ph:phloem; ca:cambium; xy: xylem; (B): pr: peridermis; (C): **tr:** trachea; trk: traceid

STEM ANATOMY

The cross sections of herbaceous stems of *S. cappadocica* have rectangular-oval shaped epidermis on the outermost surface. Under epidermis the 3-5 rowed cortex with chloroplast is located and they have druses in some cells. Endodermis which pentagon shaped and single line lies below the cortex. Sclerenchyma occupies large area (10-12 rowed) under the endodermis as uninterrupted parallel to peripheral. Vascular bundles type is open collateral definitely. Cambium splits up the phloem and xylem. The pith region is composed of parenchymatic cells and contains druses or usually it is empty Figure 3, 4 A-D).

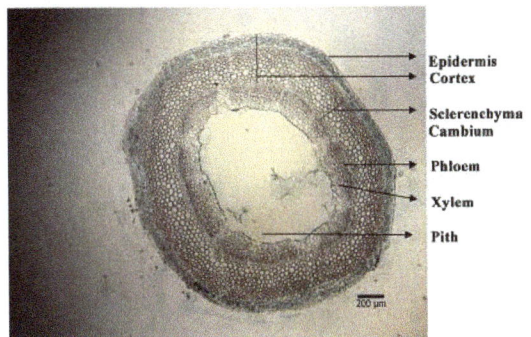

Figure 6. General view of the cross sections of stem parts of *S.cappadocica*

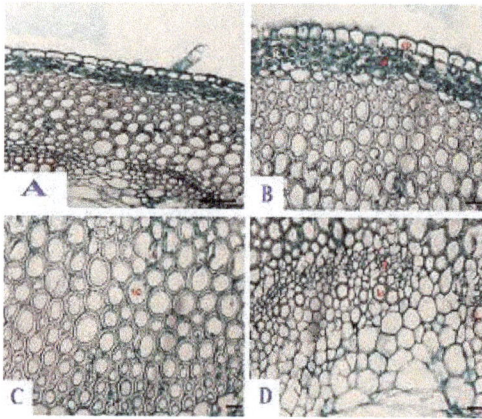

Figure 7. The cross sections of stem parts of
S.cappadocica, (A): General view of whole parts; (B):
ep: epidermis, **kl:** klorenkima, (C): **sc:** sclerenchyma
rings; (D): **xy:** xylem, ph: phloem

LEAF ANATOMY

In cross sections of stem leaves, on the outer surface, there is single row rectangular-oval shaped epidermis. Epidermis is covered by cuticle. Mesophyll is composed of palisade and spongy parenchyma cells (equifacial type). Mesophyll has rarely druses in its large space. Vascular bundles are collateral type and single row bundle-sheath cells are around them. Stem leaves are amphistomatic and stomata of species are diasitic type. The secretory trichomes are abundant in superficial sections of stem leaves (Figure 8, 9, 10).

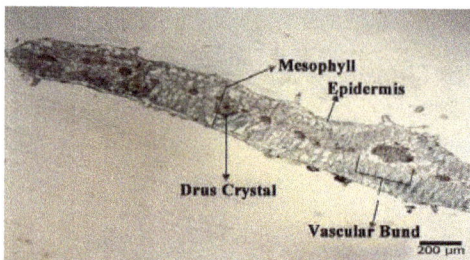

Figure 8. The cross sections of leaves of *S.cappadocica*

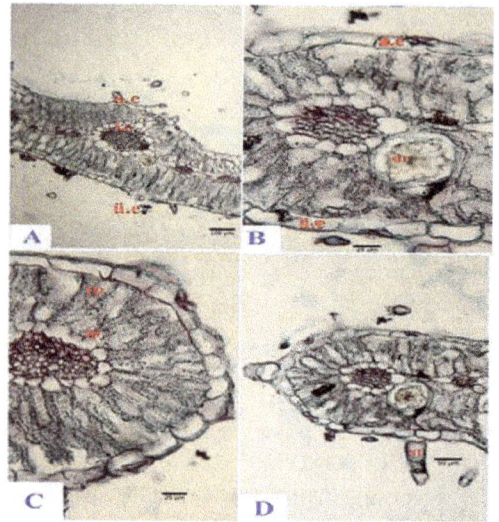

Figure 9. The cross sections of leaves of *S. cappadocica*
(A): ie: vascular bund, u.e:upper epidermis, l.e: lower
epidermis; (B): u.e: upper eridermis, l.e: lower
epidermis, dr: druses crystal; (C): pp: palisade
parenchyma, sp: spongy parenchymas; (D): st: secretory
trichomes

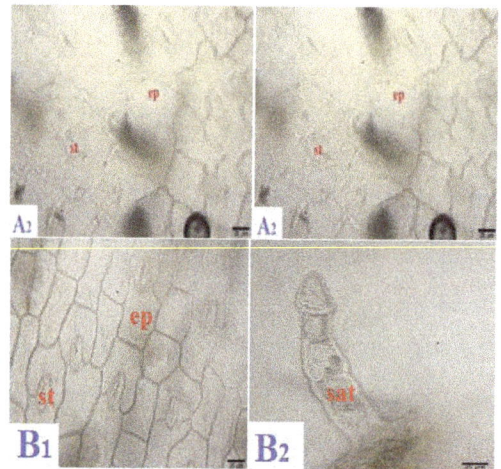

Figure 10. Surface sections of leaves of *S.cappadocica*
(A1-A2) lower surface, (A1): Lower surface, sat:
secrotery trichome (A2): **ep:** epidermis, **st:** stomata; (B1-
B2): upper surface, (B1): upper surface, **ep:** epidermis,
st: stomata; (B2): sat: secrotery trichome

Table 2. Anatomical data of *S. cappadocica*

Plant part	Item	Silene cappadocica Boiss. & Heldr.									Number of measurement
		Wide (µM)			Length Boy (µM)			Diameter / thick (µM)			
		Min	Max	average	Min	Max	average	Min	Max	average	
Root	Peridermis	0.223	0.669	0.37	0.09	0.489	0.24	7413	132.1	1175	110
	Cortex	-	-	-	-	-	-	25.65	75.36	±200	110
	Trachea	-	-	-	-	-	-	50	100	±87.5	110
Stem	Epidermis	9.54	33.07	20.77	9.03	24.58	14.87	-	-	-	110
cortex	cortex	-	-	-	-	-	-	-	-	±225	110
	Sclerenchyma	85.355	256.7	117.9	95.385	260.05	170.15	98.655	353.4	181.25	110
	Trachea	-	-	-	-	-	-	7.96	23.5	21.3±	110
Leaf	Lower Epidermis	12.44	28.53	16.76	14.30	39.73	20.54	-	-	-	50
	Mesophyll	-	-	-	-	-	-	-	-	2.312	50
Upper	Upper Epidermis	14.193	30.105	35.69	13.165	32.22	19.18	-	-	-	50

Table 3. Numerical data of *S.cappadocica* leaves

Leaf	Leaf		
	Min	Max	Ort.
Number of stomata of lower surface / mm²	122	204	162
Number of stomata of upper surface / mm2	82	183	144
Number of lower surface / mm2	367	693	482
Number of upper surface / mm2	408	672	591
Stomata index of lower surface	7.95		
Stomata index of upper surface	7.05		
Stoma index ratio	0.886		

Silene spergulifolia Bieb.

Morfologic Results

Perennial. Stems ascending to erect, retroversely puberulent, 30-44 cm. Leaves lineare to oblange, usually witht sterile shoots in their axils, puberulent, less than 5 mm broad. Basal leaves 8-16 mm×0.5-1.5 mm. Bract 1-1.5 mm, Inflorescence a rather loose though strict panicle. Calyx 10-14 mm in flowers, puberulent, often glandular. Petals white to grennish yellow, deeply bifid into ± linear lobes. Anthophore (in hermaphrodite flowers) 4-6 mm. Capsule roundade, trigonous or not 3-sulcate, long acuminate, included in the calyx. *Fl. 5-7. Screes and slopes, steppe, 800-3100 m*, (Figure 11).

Table 4. Morphological data of *S.spergulifolia*

Plant parts	Measurement data			
	wide		length	
	MİN	MAX	MİN	MAX
Basal leaf	0.05 cm	0.15 cm	0.8 cm	1.6 cm
Stem leaf	0.1 cm	0.2 cm	1 cm	1.5 cm
Calyx	0.2 cm	0.4 cm	1 cm	1.4 cm
Corolla	0.1 cm	0.1 cm	0.8 cm	1.5 cm
Fruit	0.2 cm	0.4 cm	0.6 cm	0.8 cm
Calix teeth length	-	-	0.1 cm	0.2 mm
Bracts	-	-	0.1 cm	0.15 cm
Anthophor length	-	-	0.4 cm	0.6 cm
Plant length	-	-	30 cm	40 cm
Stem diameter	-	-	0.1 cm	0.3 cm

Figure 11. *S.spergulifolia* (A): General view, (B): Mature fruit (C): Young flower; (D) Semi-mature fruit

Anatomical Results
ROOT ANATOMY

Root usually comprises of four parts as anatomical; peridermis, cortex, vascular bunds and pith (Figure 12-15). Peridermis encloses the outermost surface of *S. spergulifolia* root cross sections. Below the peridermis, there is 10-12 layered cortex. Vascular system is well developed; there is 2-3 rowed cambium between phloem and xylem. Trachea diameter is 35-70 µm. Pith region is covered with xylem and the pith rays are 1-2 rowed.

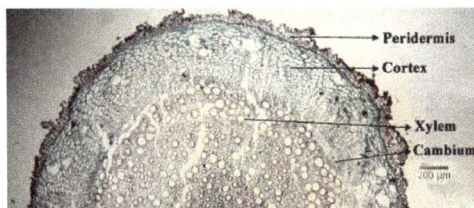

Figure 12. General view; The cross sections of *S.spergulifolia* root

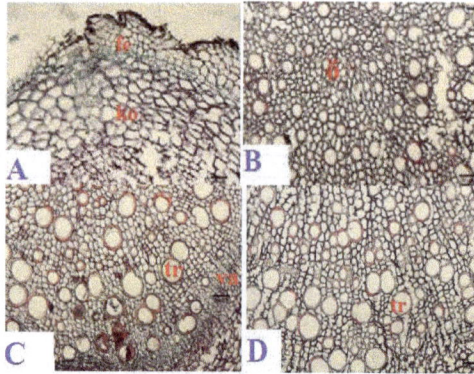

Figure 13. The cross sections of *S. spergulifolia* root,
(A): **ko:** cortex, **pe:** peridermis; (B): p:pith region
parenchyma (C): **tr:** trachea, **va:** vasculer region (D):
tr: trachea

STEM ANATOMY

The cross sections of herbaceous stems of *S. spergulifolia* have rectangular-oval shaped epidermis on the outermost surface. Under epidermis the 3-5 rowed cortex with chloroplast is located and they have druses in some cells. Endodermis which pentagon shaped and single line lies below the cortex. Sclerenchyma occupies large area (9-10 rowed) under the endodermis as uninterrupted parallel to peripheral. Vascular bundles type is open collateral definitely. Cambium splits up the phloem and xylem. The pith region is composed of parenchymatic cells and contains druses or usually it is empty. Probably pith regioan may be decomposed while it is mature (Figure 14-15).

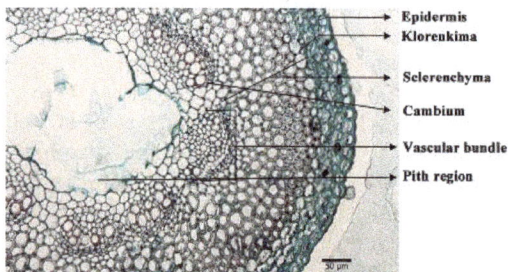

Figure 14. The cross sections of
S.spergulifolia stem

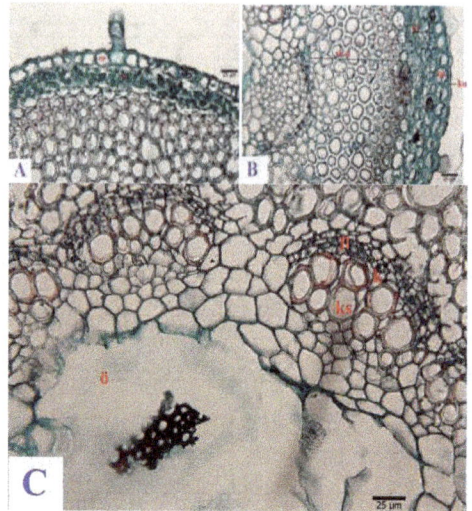

Figure 15. The cross sections of *S.spergulifolia* stem,
(A): **ep:** epidermis, **kl:** klorenkima; (B): cr:
cycleranchima ring; **kl:** klorenkima; **ep:** epiderma; **ku:**
cuticle (C): xy: xylem, **ph.:** phloem, p: pith region, **k:**
cambium

LEAF ANATOMY

In cross sections of stem leaves, on the outer surface, there is single row rectangular shaped epidermis. Epidermis is covered by cuticle. Mesophyll is composed of palisade and spongy parenchyma cells (equifacial type). Mesophyll has druses in its large space. Vascular bundles are collateral type and single row bundle-sheath cells are around them. Stem leaves are *amphistomatic* and stomata of species are *diasitic* type. The secretory trichomes are abundant in superficial sections of stem leaves (Figure 16-18).

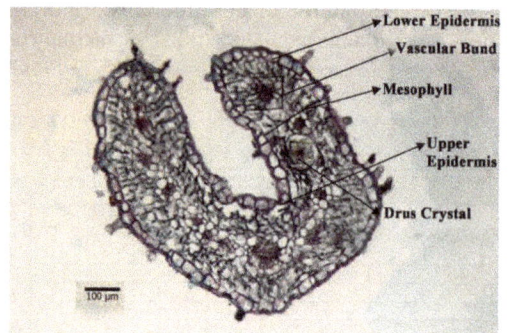

Figure 16. The cross sections
of leaf of *S.spergulifolia*

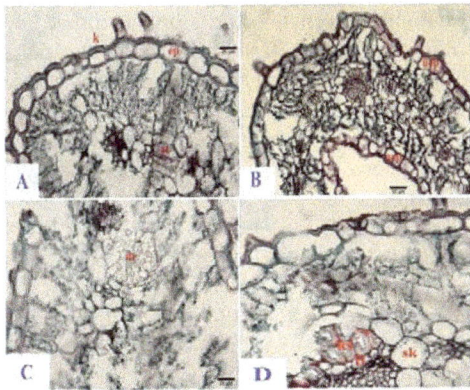

Figure 17. The cross sections of leaf of *S.spergulifolia*, (A): ep: epidermis, k: cuticle, **st**: stomata; (B): uep: upper epidermis, lep; lower epidermis; (C): dr: druse crystal; (D): sk: sclerenchyma, xy: xylem, ph: pholem

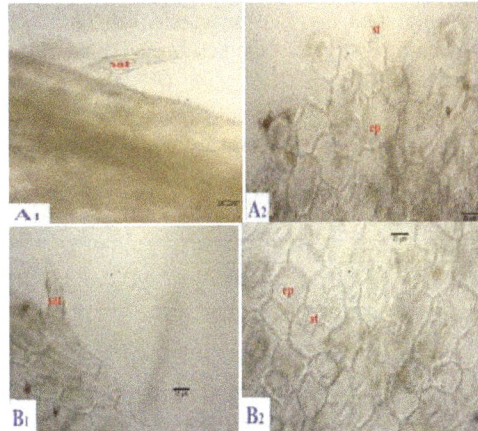

Figure 18. Surface sections of leaves of *S.spergulifolia*, (A1-A2) Lower surface, (A1): Lower surface, **sat:** secrotery trichome (A2): **ep:** epidermis, **st:** stomata; (B1-B2): upper surface, (B1): upper surface, **ep:** epidermis, **st:** stomata; (B2): **sat:** secrotery trichome

Table 5. Anatomical measure data of *S.spergulifolia*

Plant part	Item	Silene spergulifolia Bieb.									Number of measurement
		Wide (µM)			Lenght (µM)			Diameter/ Thick (µM)			
		Min	Max	Average	Min	Max	Average	Min	Max	Average	
ROOT	Peridermis	8.612	53.871	22.891	8.824	32.261	16.182	27.848	95.07	175±	110
	cortex	-	-	-	-	-	-	-	-	50±	110
	Trachea	-	-	-	-	-	-	35	70	50	110
STEM	Epidermis	6.07	7.024	22.55	23.87	36.51	20.04	-	-	-	110
	cortex	-	-	-	-	-	-	27.59	50.44	37.45	110
	Sclerenchyma	7.858	28.307	13.53	10.19	29.055	16.845	-	-	-	50
	Trachea	-	-	-	-	-	-	-	-	14.67	110
	Pith	-	-	-	-	-	-	-	-	-	-
LEAF	Lower epidermis	13.54	42.90	21.45	10.53	29.02	20.44	-	-	-	50
	Mesophyll	-	-	-	-	-	-	36.249	202.3	135.81	110
	Upper epidermis	14.56	40.943	24.38	11.92	27.11	19.67	-	-	-	50

Table 6. Numerical datas of leaves of *S.spergulifolia*

Leaf	Leaf		
	Min	Max	Ort.
Number of stomata of lower surface / mm^2	122	306	204
Number of stomata of upper surface / mm2	122	428	329
Number of lower surface / mm2	346	611	491
Number of upper surface / mm2	387	713	532
Stomata index of lower surface	10		
Stomata index of upper surface	15.35		
Stoma index ratio	1.535		

Table 7. The morphological differences between investigated species

Characters	Silene cappadocica		Silene spergulifolia	
	According to Results of our	According to Flora of Turkey (Davis, 1967)	According to Results of our	According to Flora of Turkey (Davis, 1967)
plant length	30-44 cm	10-50 cm	30-40 cm	30 - 50 cm
basal leaves	12-22×2-3mm eliptice – oblanceolate	Less than 5 mm eliptice – oblanceolate	8-16×0.5-1.5 mm Eliptic-oblanceolate	eliptic– oblanceolate
stem leaves	17-23x2-3 mm eliptic– oblanceolate	30-60x1-3 mm eliptic–oblanceolate	10-15 x 1-2 mm Linear -oblong	Linear - oblong
calyx	6-8×3-5 mm	3-5mm	10-14×2-4 mm	-
anthophore	3-5 mm	3-4mm	4-6 mm	-
stem diameter	1.5-4 mm	-	1-3 mm	-
flowering period	June-July	June-July	June-July	June-July
location	Konya/Tuzgölü	Konya/Hadim	Konya/Hadim	
bract	1-3 mm	-	1-1.5 mm	-

The root, stem and leaf anatomical differences between investigated species are given below in Table 8-10.

Table 8. The root anatomical differences between investigated species

Species tissue	Silene cappadocica	Silene spergulifolia
peridermis	Average ± 117.5 μm	average ± 175 μm
cortex	150-200 μm / 8–10 layer	400-600 μm/ 10-12 layer
trachea	average : 87.5 μm	average : 50 μm

Table 9. The stem anatomical differences between investigated species

Tissue-species	S. cappadocica	S.spergulifolia
epidermis	9.03-24.58 x 9.54-33.07	6.07-23.87 x 7.024x 36.51
cortex	±225 μm	27.59-50.44 μm
trachea	average 21.3 ± μm	average 14.7 ± μm

Table 10. The leaf anatomical differences between investigated species

Species tissue		S. cappadocica	S. spergulifolia
Present of stomata on leaf surface		Amfistomatik	Amfistomatik
According to epidermis position of stomata		Kseromorf	Kseromorf
Type of stomata		Diasitik	Diasitik
Place in druse		Mesophyle layer	Mesophyle layer
Stomata index rate	Upper surface	7.05	15.35
	Lower surface	7.95	10
	Stomata index ratio	0.886	1.535

CONCLUSIONS

Although S.cappadocica and S.spergulifolia are related species, there are anatomical and morphological different between the two species.

From morphological point of view, Silene spergulifolia has leaf narrow than Silene cappadocica. Silene spergulifolia has different fruit shapes according to fruit of Silene cappadocica.

S. cappadocica has 30-44 cm length while S.spergulifolia has 30-40 cm length. S. cappadocica is taller than S. spergulifolia. Calyx is 6-8 mm length in S.cappadocica but calyx of S.spergulifolia is 10-14 mm length. Anthofor length is 3-5 mm at S. cappadocica but calyx length of S. Spergulifolia is 4-6 mm

Fruits of S.cappadocica are smaller than S. spergulifolia. Also, S. cappadocica has a split on fruit (capsule) and this feature is a distinct character for this two species.

In anatomical terms, while stomata index rate of S.cappadocica is 0.886, iken stomata index rate is 1.535 at S. spergulifolia. Leaf of S. cappadocica has thinner cuticle layer than leaf of S. sperguliolia. While S.cappadocica has 4-5 layer peridermis, S. spergulifolia has 8-9 layer peridermis.

ACKNOWLEDGEMENTS

We would like to thank to Selcuk University (BAP Project no: **13201051**) for financial support during this study.

REFERENCES

Akman, Y., 1990, İklim ve Biyoiklim (Biyoiklim Metodları ve Türkiye İklimleri), *Palme Yayınları*, Yayın No:103, Ankara.

Aktaş, K., 2006, Türkiye'nin *Petrorhagıa* (Ser.) Link (Caryophyllaceae) cinsi türleri üzerinde taksonomik bir araştırma, Doktora Tezi, *Celal Bayar Üniversitesi Fen Bilimleri Enstitüsü*, 230 sayfa, Manisa.

Algan, G., 1981, Bitkisel dokular için mikroteknik, İstanbul: Fırat Üniversitesi Fen Edebiyat Fakültesi Yay Bot No: 1

Bağcı, Y., Uysal, T., Ertuğrul, K. and Demirelma, H., 2007, *Silene kucukodukii* sp. nov. (Caryophyllaceae) from South Anatolia, Turkey. Nordic Journal of Botany 25, 306-310.

Bağcı, Y., 2008, A new species of *Silene* L. (Caryophyllaceae) from South Anatolia, Turk J Bot 32,11-15.

Baytop, T., 1992, Trakya ve Türkiye Florasına ilave Kayıtlar, *Doga T. Journal Bot.* 16: 15-17, Ankara.

Baytop, T., 1997, Türkiye bitki adları sözlüğü, *TDK yayınları*, Ankara.

Bolat, N., 1989, Edirne ve yöresi *Silene* L. (Caryophyllaceae) cinsinin G grubu türlerinin sistematiği ve morfolojisi, Yüksek Lisans Tezi, *Trakya Üniversitesi Fen Bilimleri Enstitüsü*, 38 sayfa, Edirne.

Bouyoucos, G.J., 1955. Hydrometer method improved for making particle size analysis of soil. *Agr. Jour.*, Vol 54: 3.

Budak, Ü., Koç, M., 2010, *Silene hamzaoglui* (Caryophyllaceae), A New Species from Çekerek (Yozgat, Turkey), Department of Biology, Faculty of Science and Arts, Bozok University, 66200, Yozgat – TURKEY.

Carlquist, S., 1995, Wood Anatomy of Caryophyllaceae: Ecologıcal, Habıtal, Systematıc, and Phlogenetıc Implıcatıons from Claremont. CA 91711-3157.

Chapman, H. D., Pratt, F. P., 1961. Methods of analysis for soil, plants and waters. *Priced Publication 4034, University of California*, California.

Davis, P.H. (ed.),1967, Flora of Turkey and the East Aegean Island, *Edinburgh Univ. Press.* Vol. 2, Edinb. Üniv. Press., London.

Deniz, İ. G. and Düşen, O. D., 2004. *Silene sumbuliana* (Caryophyllacea), a new species from SW Anatolia, Turkey. Ann. Bot. Fennici, 41, 293-296.

Ekim, T., Koyuncu, M., Vural, M., Duman, H., Aytaç, Z., Adıgüzel, N., 2000, Türkiye Bitkileri Kırmızı Kitabı, Türkiye Tabiatını Koruma Derneği, *Van Yüzüncü Yıl Üniversitesi*, Ankara.

Erdir Erten, M., 2009, Türkiye *Saponaria* L. (Caryophyllaceae) cinsi üzerinde taksonomik, morfolojik ve anatomik çalışmalar, Doktora Tezi, *Eskişehir Osmangazi Üniversitesi Fen Bilimleri Enstitüsü*, 339 sayfa, Eskişehir.

Ersöz Poyraz, İ., 2008, Türkiye *Velezia* L. (Caryophyllaceae) cinsi revizyonu, Doktora Tezi, *Osmangazi Üniversitesi Fen Bilimleri Enstitüsü*, 219 sayfa, Eskişehir.

Fidan, M., 2011, Türkiye *Gypsophila* L. (Caryophyllaceae) cinsine ait Hagenia A. Braun. seksiyonunun revizyonu, Yüksek Lisans Tezi, *Yüzüncü Yıl Üniversitesi Fen Bilimleri Enstitüsü,* 11 sayfa, Van.

Gümüştaş, A., 2005, Erciyes nakilli (*Silene argaea* fisch. ve C.A. mey.)'nın anatomik, morfolojik ve ekolojik özellikleri, Yüksek Lisans Tezi, *Erciyes Üniversitesi Fen Bilimleri Enstitüsü*, 100 sayfa, Kayseri.

Güner, A., Özhatay, N., Ekim, T., Başer, K.H.C. (eds). 2000. Flora of Turkey and the East Aegean Islands, Vol. 11 (Supplement), *Edinburgh Üniv. Press,* Edinburgh.

Hamzaoğlu, E., 2011, A new species of *Gypsophila* and a new name for *Silene* (Caryophyllaceae) from Turkey, *Turkish Journal of Botany*, 36, 135-139.

Hamzaoğlu, E., Aksoy, A. and Budak, Ü., 2010, A new species of *Silene* (Caryophyllaceae) from Turkey. Turk J Bot, 34, 47-50.

Jackson, M. L. 1962. Soil Chemical Analysis. Prentice Hall. Inc. 183 New York.

Kandemir, A., Genç, G. E. and Genç, İ., 2009, *Silene dumanii* (Caryophyllaceae), A new species from East Anatolia, Turkey. Ann. Bot. Fennici, 46, 71-74.

Kepek, M., 2003, İstanbul Üniversitesi Fen Fakültesi Herbaryumu'nda (ISTF) bulunan *Silene* L. (Caryophyllaceae) örneklerinin revizyonu, Yüksek Lisans Tezi, *İstanbul Üniversitesi Fen Bilimleri Enstitüsü,* 69 sayfa, İstanbul.

Kılıç, S., 2007, Türkiye'nin *Silene* L. (Caryophyllaceae) cinsi Brachypodeae Boiss. ve Auriculatae Boiss. seksiyonları üzerinde biyosistematik çalışmalar, Doktora Tezi, *Süleyman Demirel Üniversitesi Fen Bilimleri Enstitüsü,* 267 sayfa, Isparta.

Kılıç, S., 2009, Anatomical and Pollen characters in the genus *Silene* L. (Caryophyllaceae) from Turkey, *Botany Resarch Journal* 2 (2-4), 34-44.

Kılıç, S., Özçelik H., 2009, Comparative Morphological and Anatomical Studies on the Genus *Silene* L. Sect. Auriculatae Boiss. (Caryophyllaceae) Species in Turkey, *Journal of Plant & Environmental Sciences,* 5-15.

Korkmaz, M., 2007, Türkiye 'de yetişen tek yıllık Gypsophila L. (Caryophyllaceae) taksonları üzerinde biyosistematik çalışmalar, Doktora Tezi, *Süleyman Demirel Üniversitesi Fen Bilimleri Enstitüsü,* 247 sayfa, Isparta.

Marie, R., 1963. Flore de l'Afrique du Nord (Maroc, Algerie, Tunisie, Tripolitaine, Cyrenaıque et Sahara). 10, Encycl. Biol. 62.

Muca, B., 2009, Türkiye *Ankyropetalum* Fenzl (Caryophyllaceae) cins taksonları üzerinde anatomik, palinolojik, taksonomik ve morfolojik araştırmalar, Yüksek Lisans Tezi, *Süleyman Demirel Üniversitesi Fen Bilimleri Enstitüsü,* 49 sayfa, Isparta.

Özgökçe, F., Tan, K., Stevanovic, V., A New Subspecies of *Silene acaulis* (Caryophyllaceae) from East Anatolia, Turkey, Helsinki 27 April 2005.

Öztürk, M, Pirdal, M. vd., 1997, Bitki ekolojisi uygulamaları, Ege Üniversitesi Fen Fakültesi Kitapları Serisi No: 157, Bornova-İzmir.

Öztürk, M.A. ve Seçmen, Ö., 1999, Bitki Ekolojisi, Ege Üniversitesi Fen Fakültesi Yayınları, Yayın No: 141, Ege Üniversitesi Basımevi, Bornova-İzmir.

Rechinger, K.H. (ed.), 1988. Flora Iranica, Flora des Iranischen Hochlandes und der Umrahmenden Gebirge, 163. Graz.

Richard, L. A. 1954. Diagnosis and Improvement of Saline and Alkaline Soils. Handbook: 60, U.S. Dept. Of Agriculture.

Sahreen, S., Khan, M., Khan, R., 2009, Leaf epidermal anatomy of the genus *Silene* (Caryophyllaceae) from Pakistan, Pakistan.

Sarıoğlu, A., 2006, Samsun ve çevresinde yayılış gösteren bazı *Silene* L. (Caryophyllaceae) türleri üzerinde anatomik, morfolojik ve taksonomik bir araştırma, Yüksek Lisans Tezi, *Ondokuz Mayıs Üniversitesi Fen Bilimleri Enstitüsü,* 82 sayfa, Samsun.

Schoeder, D. 1972. Bodenkunde in Sticworten, Verlag Ferdinant Hirt, Kiel

Tugay, E., 2005, Konya ilindeki bazı *Silene* L. (Caryophyllaceae) taksonları üzerine karyolojik bir araştırma, Yüksek Lisans Tezi, *Selçuk Üniversitesi Fen Bilimleri Enstitüsü,* 46 sayfa, Konya.

Tugay, O. and Ertuğrul, K., 2008, A new species of *Silene* (Caryophyllaceae) from east Anatolia, Turkey. Botanical Journal of the Linnean Society, 156, 463–466.

Tunalı, H., 2004, İzmir ili *Silene* L. türleri üzerinde sistematik, morfolojik ve anatomik çalışmalar, Yüksek Lisans Tezi, *Eskişehir Osmangazi Üniversitesi Fen Bilimleri Enstitüsü,* 120 sayfa, Eskişehir.

Tutin, T. G., Burges,N. A., Chater, A. O., Edmondson, J. R., Heywood, V. H., Moore, D.M., Valentine, D. H., Walters, S. M. & Webb, D. A., (ed) 1993, Flora Europaea. ed. 2, 1. Cambridge.

Tüzüner, A. 1990. Toprak ve Su Analiz Laboratuvarı el kitabı, Tarım, Orman ve Köy İşleri Bakanlığı, Köy Hizmetleri Genel Müdürlüğü, Ankara

Yalçınkaya, Z., 2006, Ankara Üniversitesi Fen Fakültesi Herbaryum'undaki (ANK) Caryophyllaceae familyasının revizyonu, Yüksek Lisans Tezi, *Ankara Üniversitesi Fen Bilimleri Enstitüsü,* 181 sayfa, Ankara.

Yarcı, C., 1987, Trakya bölgesi *Minuartia L.* (Caryophyllaceae) türleri üzerinde, morfolojik ve sistematik çalışmalar, Yüksek Lisans Tezi, *Trakya Üniversitesi Fen Bilimleri Enstitüsü*, 53 sayfa, Edirne.

Yıldız, K., 1990, Tokat çevresinin *Silene* L. türleri üzerinde morfolojik araştırmalar, Yüksek Lisans Tezi, *Marmara Üniversitesi Fen Bilimleri Enstitüsü*, 44 sayfa, İstanbul.

Yıldız, K., 2005, A palynological investigation on *Silene* L. (Caryophyllaceae) species distributed in North Cyprus and West Anatolia. C.B.U. Journal of Science, 1 (2), 61-71.

Yıldız, K., 2006, A Morphological İnvestigasion on *Silene* L. (Caryophyllaceae), Species Distributed in West Anatolia and North Cyprus, *Pak. J. Bot.*, 38(1): 67-83,

Yıldız, K. and Dadandı, M. Y., 2009, *Silene cirpicii*, a new species from Turkey. Ann. Bot. Fennici, 46, 464-468.

Yıldız, K., Minareci, E. and Çırpıcı, A., 2009, Karyotypic study on *Silene*, section Lasiostemones section from Turkey. Caryologia 62 (2), 134-141.

Yıldız, K. and Erik, S., 2010, *Silene aydosensis* (Caryophyllaceae), a new species from Anatolia, Turkey. Ann. Bot. Fennici, 47, 151-155.

13

ASSESSMENT OF LANDSCAPE COMPONENTS IN COMANA NATURAL PARK

Vladimir Ionuț BOC, Robert Mihai IONESCU

University of Agronomic Sciences and Veterinary Medicine of Bucharest, Department of
Landscape Architecture, Biodiversity and Ornamental Horticulture. 59, Marasti Bd., 011464,
Bucharest, Romania, email: vladimirboc@gmail.com

Corresponding author email: vladimirboc@gmail.com

Abstract

The approach consists in proposing an assessment method regarding landscape typologies and components, applied in Comana Natural Park, located 20 km south of Bucharest, in Giurgiu County. The research methodology is focused on assessing the share of the individual landscape components in each type of landscape identified. Thus, the components of the landscape are grouped in a tabular analysis into two broad categories: natural and anthropogenic, with specific subcategories: relief, soil, water, vegetation and sunlight for the natural category; infrastructure and architecture for the anthropogenic one. Each subcategory varies depending on the type of landscape, e.g. vegetation varies from forest to agricultural or palustrine. Following the assessment of the components, the anthropic impact level and the diversity index result for each type of landscape. Within the analysis, six types of landscape have been identified: forest, palustrine, agricultural, fallow, old rural, recent rural. After assessing the frequency of the landscape components two categories have been identified: common components (frequently found) and specific components (found only in particular cases). Both groups include natural, anthropic or mixed elements. The study brings in new approaches in identifying and assessing the determining factors in terms of landscape identity and typology, deepening the relations between the different components of the landscape.

Key words: Landscape Assessment, Landscape Typologies, Landscape Identity, Natural Heritage, Comana Natural Park.

INTRODUCTION

According to the European Landscape Convention launched in Florence in 2000 and ratified by Romania in 2002, the signatory states are committed to: "a) (i) to identify its own landscapes throughout its territory; (ii) to analyze their characteristics and the forces and pressures transforming them; (iii) to take note of changes; b) to assess the landscapes thus identified, taking into account the particular values assigned to them by the interested parties and the population concerned" (CE, 2000). Unfortunately, the implementation of provisions at national level is deficient and the landscape assessment studies in Romania are isolated, being realized mostly in the academic environment.

In this context, a priority in initiating steps for identifying, mapping and assessing landscapes is researching sites which belong to the natural and cultural heritage. One such case is Comana Natural Park, the largest wetland in Bucharest metropolitan area, which includes valuable landscape both at natural and anthropogenic level.

Comana protected area was declared a natural park in 2004, on an area of 24,963 ha. The park includes 8 communes with a range of cultural and historical heritage objectives, Comana and Neajlov River - a Ramsar and Natura 2000 site, Comana Forest - including 2 floral reserves (Ruscus aculeatus and Paeonia peregrina).

MATERIALS AND METHODS

The methodology is focused on the assessment of the relationship between typologies and the specific components of the landscape. Every landscape typology is assessed in terms of the characteristic components within a tabular analysis.

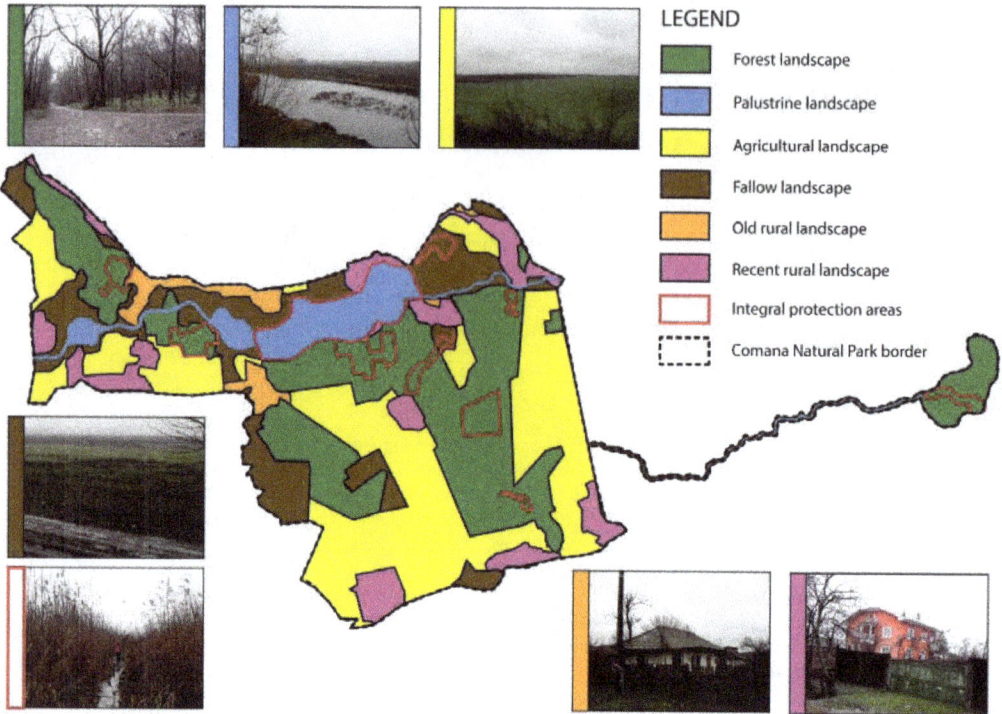

Figure 1. Landscape Typologies in Comana Natural Park

The individual components of the landscape are divided into two broad categories: natural and anthropogenic, with specific subcategories: relief, soil, water, vegetation and sunlight level for the natural category; infrastructure and architecture for the anthropogenic one. Each subcategory contains different components depending on the type of landscape.

Following the assessment of the components for each type of landscape, the anthropic impact level and the diversity index result. The anthropic impact level is determined through the ratio between natural and anthropogenic elements of each landscape. The diversity index results based on the proportion of the components encountered. The last part of the study includes an analysis of the frequency of each component found in Comana Natural Park depending on landscape typologies and on the total area.

The final results concluded two main categories of landscape components: common factors and specific factors. The first category comprises the most frequent landscape elements, while the second one includes the rarest landscape components in Comana Park.

RESULTS AND DISCUSSIONS

Following the site assessment and mapping, 6 landscape typologies resulted in Comana Natural Park: forest, palustrine, agriculture, fallow, old rural, recent rural. Of the 6 landscape types, the forest and palustrine landscape include integral protection areas (flora and fauna reserves within the forest landscape and avifauna reserve within the palustrine landscape) (Figure 1).

After analyzing the landscapes distribution, the agricultural (34%) and forest (28%) typologies resulted as the dominants. The fallow (15%), the new rural (11%) and the palustrine (8.5%) landscapes presented an average share. The lowest share resulted for old rural landscape (3.5%) (Figure 2).

The first step in landscape typologies assessment was to identify the presence of landscape components for each of the 6 cases. In the tabular analysis, for each type of landscape natural components (relief, water, soil, light) and anthropogenic characteristics (architecture and infrastructure) have been identified (Figure 1). Depending on the natural

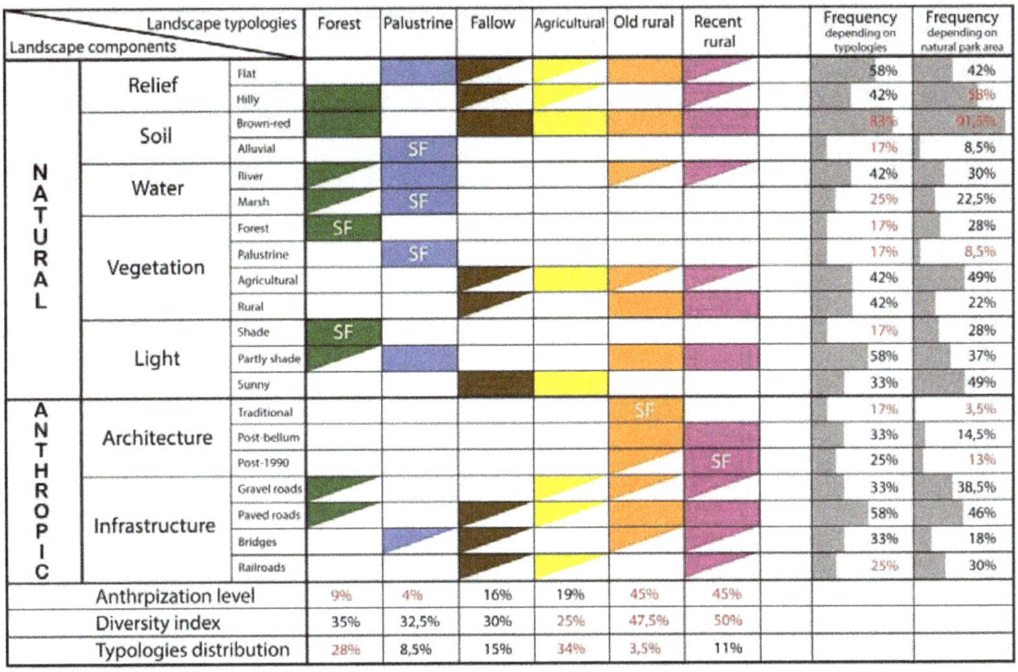

Landscape components		Landscape typologies	Forest	Palustrine	Fallow	Agricultural	Old rural	Recent rural		Frequency depending on typologies	Frequency depending on natural park area
N A T U R A L	Relief	Flat								58%	42%
		Hilly								42%	58%
	Soil	Brown-red								83%	91,5%
		Alluvial		SF						17%	8,5%
	Water	River								42%	30%
		Marsh		SF						25%	22,5%
	Vegetation	Forest	SF							17%	28%
		Palustrine		SF						17%	8,5%
		Agricultural								42%	49%
		Rural								42%	22%
	Light	Shade	SF							17%	28%
		Partly shade								58%	37%
		Sunny								33%	49%
A N T H R O P I C	Architecture	Traditional					SF			17%	3,5%
		Post-bellum								33%	14,5%
		Post-1990						SF		25%	13%
	Infrastructure	Gravel roads								33%	38,5%
		Paved roads								58%	46%
		Bridges								33%	18%
		Railroads								25%	30%
	Anthrpization level		9%	4%	16%	19%	45%	45%			
	Diversity index		35%	32,5%	30%	25%	47,5%	50%			
	Typologies distribution		28%	8,5%	15%	34%	3,5%	11%			

SF - Specific factor (specific landscape components for certain typologies)

Figure 2. Landscape components assessment depending on landscape typologies

- anthropogenic ratio and on the variety of landscape elements, the diversity index and the anthropic level impact resulted. Thus, the dominant components revealed the following indicators (Figure 2):
- Forest landscape: hilly relief, red-brown soil, forest vegetation, shaded areas, gravel roads; Anthropization: 9%; Diversity: 35%;
- Palustrine landscape: flat relief, alluvial soil, Neajlov River, Comana Marsh, palustrine vegetation, partly shaded areas, wooden decks; Anthropization: 4%; Diversity: 32.5%;
- Fallow landscape: flat and hilly relief, red-brown soil, herbaceous vegetation, sunny areas, paved roads, bridges, railways; Anthropization: 16%; Diversity: 30%;
- Agricultural landscape: flat and hilly relief, red-brown soil, agricultural vegetation, sunny areas, paved roads, railways; Anthropization: 19%; Diversity: 25%;
- Old rural landscape: flat relief, red-brown soil, rural vegetation (orchards, vegetable gardens, etc.), sunny and shaded areas, traditional and post-bellum architecture, paved and gravel roads; Anthropization: 45%; Diversity: 47.5%;

- New rural landscape: hilly and flat relief, red-brown soil, rural vegetation (orchards, vegetable gardens, etc.), sunny and shaded areas, post-bellum and post-communist architecture, paved and gravel roads, railways, bridges; Anthropization: 45%; Diversity: 50%;
The last phase of the study is based on assessing the frequency of components within the 6 types of landscape. Thus, depending on the results, two broad categories of anthropogenic and natural components of the landscape have been identified: common factors – frequently encountered in Comana Natural Park in general and specific factors – rare ones, representative only for certain landscape types within the studied area (Figure 2). The first category includes: red-brown soil, flat and hilly relief, partly shaded areas, paved roads, hilly relief, the Neajlov River, agricultural vegetation, rural vegetation. The second category includes traditional architecture, post-1990 architecture, shaded areas, palustrine vegetation, forest vegetation, alluvial soil, Comana Marsh (Figure 3).

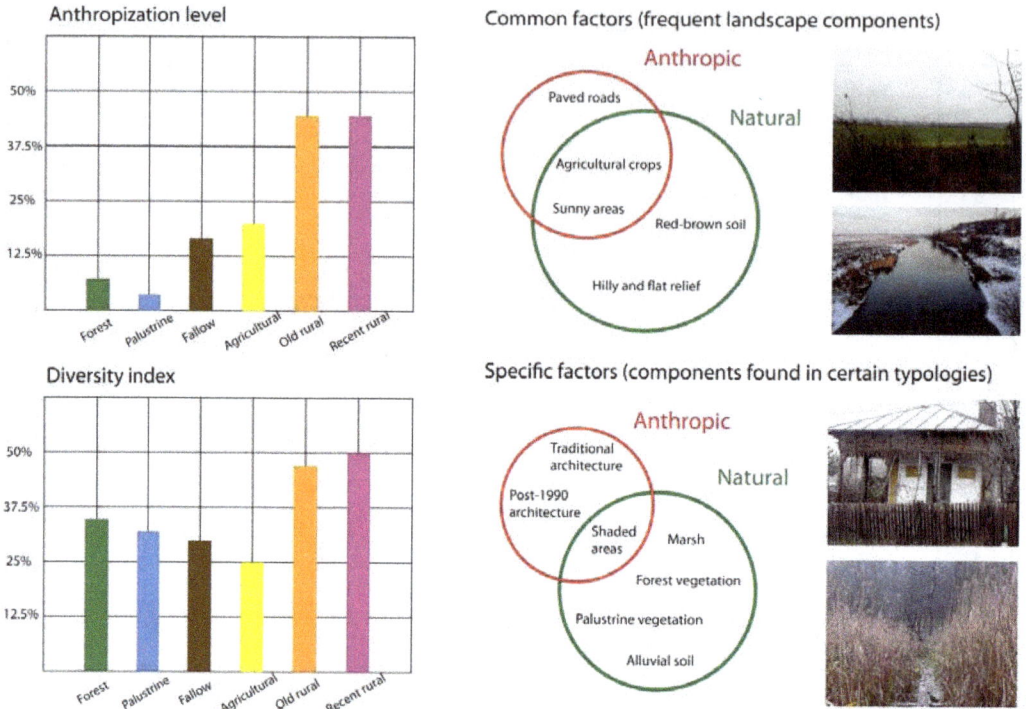

Figure 3 – Common and specific factors in landscape typologies in Comana Natural Park

CONCLUSIONS

The study aims to identify common and specific components of the landscape in Comana Natural Park and proposes a new approach in the assessment of landscape typologies. The importance and originality of the research consist of proposing a comprehensive and integrated analysis method addressing the complex dimension of the landscape. The method complements the sphere of knowledge in the field of landscape assessment by introducing a new approach in quantitative analysis. Its results highlight the qualitative dimension of landscape components. The innovative character of the analysis consists in identifying common and specific factors to certain types of landscapes depending on their frequency within the site, correlated with the spatial distribution of the typologies, the anthropization level and the diversity of landscapes. The results of the proposed method highlight the common elements which define the general character of the landscape in the studied area and determine

the particularizing character of typologies. Thus, the results can be integrated into management and local development strategies, including tourism promotion through place branding initiatives. Both specific landscape factors and common ones should be considered within landscape conservation and protection measures for Comana protected area. The presented landscape assessment method can be applied at different scales, from local to territorial level, in both anthropogenic and natural environments. Also, this approach can be developed in order to be integrated in various landscape studies and strategies.

REFERENCES

Comana Natural Park Administration, 2013. Integrated Management Plan for Comana Natural Park
Council of Europe (CE), 2000. European Landscape Convention, Florence
Criveanu I., 2010. Protected Areas – Course Notes, "Ion Mincu" University of Architecture and Urban Planning, Bucharest
LaGro J., 2008.Site Analysis: A Contextual Approach in Sustainable Land Planning and Site Design, Ed. Wiley and Sons, Hoboken, New Jersey

14

CURRENT APPROACHES IN METROPOLITAN GREEN INFRASTRUCTURE STRATEGIES

Vladimir Ionuţ BOC

University of Agronomical Sciences and Veterinary Medicine of Bucharest, Department of
Landscape Architecture, Biodiversity and Ornamental Horticulture. 59, Marasti Bd., 011464,
Bucharest, Romania, email: vladimirboc@gmail.com
Corresponding author email: vladimirboc@gmail.com

Abstract

In the last decade, in developed countries the awareness of green infrastructure and its impact on quality of life has increased considerably. Thus, increasingly more cities have initiated development and conservation plans for metropolitan green infrastructure. The research consists of a comparative analysis of a number of green infrastructure strategies from different cities around the world, including major cities such as New York, Sydney, London, and smaller metropolitan areas like Milwaukee (USA) and Cambridge (UK). Within the study, green infrastructure plans are analyzed in terms of structure, underlying studies, visions, objectives, approached themes, complexity, relating to national and international directives, etc. The results reveal the complexity and interdisciplinary character of green infrastructure development plans. The strategies contains various current global issues approached at local level, such as climate change, energy efficiency, pollution reduction, storm water management, biodiversity conservation, public health, etc. The study shows different green infrastructure planning approaches, highlighting an increasingly interest to integrate green areas in urban development strategies and policies.

Key words: Development Strategies, Green Infrastructure Plan, Landscape Planning, Metropolitan Areas, Sustainable Development.

INTRODUCTION

The continuous development of the concept of green infrastructure in the last two decades has led international organizations, central and local authorities to develop specific policies and strategies at international, national and local level. Such measures have been initiated especially in the last 10 years, mostly in developed countries such as USA, Canada, Australia or Western Europe. Since 2008 the European Union introduced green infrastructure into its institutional discourse through the European Environmental Bureau (EEB, 2008). Subsequently the concept was taken over by the European Commission (EC, 2012; EC, 2013), which intends to develop a general strategy at EU level on GI (green infrastructure) till 2020. The European documents presents the importance of green infrastructure benefits, particularly for urban areas and their role in combating threats to human security and to the environment (Boc,

2014). Thus, metropolitan green infrastructure strategies can be used to propose natural solutions to various global challenges such as climate change, energy efficiency, urban microclimate conditions, food security, carbon footprint, water management, etc.

MATERIALS AND METHODS

The following comparative analysis illustrates different approaches to green infrastructure strategies for metropolitan areas from Western Europe, North America and Australia, developed in the last 5 years. In the analysis are studied both large cities such as London, New York, Sydney and smaller metropolitan areas such as Milwaukee (USA) and Cambridge (UK). The results of the research are listed within a table, which contains a synthesis of the analysis, and also in a descriptive manner by presenting each criterion gradually according to which the metropolitan green infrastructure strategies have been analyzed (Table 1).

Table 1. Comparative analysis of green infrastructure strategies

	GI Strategies for metropolitan areas				
	Large areas (over 4 million inhabitants)			Medium areas (1,5 mil. inhab.)	Small areas (0,3 mil. inhab.)
	London	New York	Sydney	Milwaukee	Cambridge
Title of the project	Green Infrastructure and Open Environments: the all London Green Grid (ALGG), 2012	NYC Green Infrastructure Plan - 2010 (updated yearly)	Metropolitan Strategy for Sydney, 2012 (Chapters 3.6 Infrastructure, 3.8. Environment)	Regional Green Infrastructure Plan, 2013	Cambridgeshire Green Infrastructure Strategy, 2013 (a review of the 2006 strategy)
Prepared for	Greater London Authority	NYC Depart. of Environmental Protection	NSW Department of Planning and Infrastructure	Milwaukee Metropolitan Sewerage District	Cambridge City Council, Cambridge County Council
Major structure (contents)	1. Introduction 2. Vision 3. Delivery 4. Functions (Benefits) 5. Green grid areas	1. Build cost-effective grey infrastructure 2. Optimize the wastewater system 3. Control 10% of water runoff 4. Management, modeling impact, monitoring 5. Stakeholders	Infrastructure: 1. Planning 2. Funding 3. Social infr. 4. Green infr. Environment: 1. Environment 2. Natural hazard 3. Climate change 4. Waste 5. Sustainability	1. GI in Milwaukee 2. Regional GI Plan Goals 3. Analysis and results 4. GI watershed priorities 5. GI benefits and costs	1. Background 2. Developing the GI Strategy 3. The strategic network 4. GI priorities 5. Delivery of the strategic network
Visions and objectives	A GI network of interlinked, multi-purpose open and green spaces with good connections, the Green Belt and the Blue Ribbon Network, especially the Thames.	Improving water quality that integrates green infrastructure, such as swales, rain gardens and green roofs, with investments to optimize the existing system.	Open space should be treated in a holistic and integrated way as a GI system including parks, reserves, protected lands, landscapes, trails, foreshores, national parks and waterways.	To capture more storm water, harvest more rainwater for reuse, and to provide social economic and environmental benefits for all.	Objectives: 1. Revise the decline in biodiversity 2. Mitigate and adapt to climate change 3. Sustainable economic development 4. Healthy living
Implementation	20 years	20 years	20 years	22 years	15 years (2007-2021)
Approached themes	Climate change, energy efficiency, food security, biodiversity, air quality, water management healthy living, accessibility, sustainable tourism	Storm water management, climate change, air quality, energy efficiency, green roofs, bio-swales	Climate change, energy efficiency, food security, biodiversity, water management, sustainable tourism, healthy living, landscape	Storm water management, climate change, air quality, energy efficiency, green roofs, bio-swales	Climate change, energy efficiency, food security, landscape, biodiversity, air quality, water management healthy living, accessibility, tourism, heritage
Relation to international directives	INTEREG Climate-Change Project	-	-	-	- European Landscape Convention - RAMSAR - SPA, SAC

The criteria included in the analysis are: the general structure of the strategy, vision and objectives, the expected period to implement strategies, main themes, relating to international guidelines and the number of inhabitants of each metropolitan area.

The conclusions show the common elements of the strategies and the main factors which generates different approaches in metropolitan green infrastructure planning.

RESULTS AND DISCUSSIONS

Structure of strategies. The approached green infrastructure strategies have different structures depending on the main themes. In general, the first stage includes a presentation of the current situation of green infrastructure in the overall context of local development strategies and policies. The second part comprises the vision and the main objectives of the GI strategy. Afterwards, the priorities in the development of green infrastructure and the specific benefits are mentioned.

The implementation phase is presented either after vision, in the case of London or at the end of the strategy in the case of Cambridge. The common element encountered in all strategic plans is the development vision, which is designed generally for a period of 20 years (Table 1). The exception is the metropolitan area of Cambridge, where the strategy was developed as a review of the strategic plan from 2006, aimed to be implemented during 15 years (2007-2021). In this case, the main goal was to update the strategic objectives for 2021 and to present programs and projects already implemented or ongoing (CCC, 2013).

Visions. In the UK and Australia the vision and objectives of the GI strategies are approached from an integrated perspective, with a strong interdisciplinary character. The green infrastructure development means to create a complex network of interconnected green areas with ecological, economic, social and cultural role. In the case of the American strategies, the focus is primarily on solving storm-water management issues through sustainable methods in environmental and economic terms.

Approached themes. The strategies from Australia and the UK approach numerous

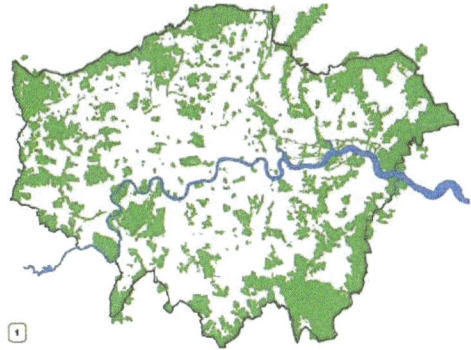

Figure 1. Green infrastructure in London (up), Managing climate change flooding (down), (source: Yurisic, 2014)

topics mentioned in general within the EU directives, such as climate change, energy efficiency, food security, biodiversity conservation, air quality, high accesibility in green areas, encouraging an environment and style healthy living, sustainable tourism development. Regarding the multiplicity of topics addressed, the most complex is the strategy of Cambridge metropolitan area. In this case, in addition to the above mentioned themes, the cultural dimension of green infrastructure is introduced by integrating the concepts of landscape and heritage within the strategic objectives (GLA, 2012). In contrast, US strategies propose an approach based on the role of green infrastructure in storm-water management. Thus, issues such as climate change, energy efficiency and air quality are addressed in the background, especially in relation to sustainable water management. In New York and Milwaukee, strategic plans aimed at developing green roofs, bio-swales, rain gardens, wetlands and green corridors in order to reintegrate the rainwater into the natural biogeochemical circuit (MMSD, 2013). Relation to international directives. In general, the studied strategies are not related, at least not directly, to international conventions, policies or programs. Such directives are only specified in the strategies from the UK. In the case of London, the INTERREG trans-boundary program, launched by the European Union, is integrated to combat climate change (Figure 1). The objectives of the program include: urban heat island management, flood prevention, reducing CO2 emissions and improving the quality of life through a range of practical activities (GLA, 2012). The GI Strategy for Cambridge metropolitan area has the widest coverage and is strongly related to international directives. The strategic plan integrates principles of the European Landscape Convention. Specific issues as the erosion of the character of cultural and natural landscapes are mentioned. The objectives related to the European Landscape Convention include landscape restoration and creation of new development projects involving the local community. Beside this, the GI strategy for Cambridge highlights the importance of managing natural areas which are protected through international conventions involved in biodiversity conservation, such as RAMSAR – worldwide and SAC and SPA - at European level (CCC, 2013).

The size of metropolitan areas. The main aspect which varies in this regard is the scale and the level of detail in spatial zoning of green infrastructure.

CONCLUSIONS

The comparative analysis is noted that the American GI strategies present a sectorial character, geared mainly towards sustainable storm-water management problem. In the case of strategic plans from UK and Australia, the vision is more comprehensive, addressing numerous issues of contemporary global human security sphere - climate security, energy and food, public health, sustainable tourism, etc. The main common elements in all strategies include an implementation period lasting about 20 years, presenting a general view of the priorities and highlighting the benefits of green infrastructure. At the same time, all GI strategies highlights the importance of green infrastructure to ensure a sustainable future for metropolitan areas. Therefore, the metropolitan authorities foresee increasingly significant investments in GI programs and projects in the coming decades.

REFERENCES

Boc, V., 2014. Green Infrastructure from the Perspective of European Institutions, University of Agronomic Sciences and Veterinary Medicine - Scientific Papers. Series B. Horticulture, Vol LVIII, p.297-300
Cambridge City Council (CCC), Cambridgeshire, 2013. Green Infrastructure Strategy, Cambridge, UK
Council of Europe, 2000. European Landscape Convention, Florence
European Commission (EC), 2012. The Multifunctionality of Green Infrastructure, Brussels
European Commission (EC), 2013. Communication from the Commission to the European Parliament, the Council, the European Economic and Social Committee and the Committee of the Regions; Green Infrastructure (GI) — Enhancing Europe's Natural Capital, Brussels
European Environmental Bureau (EEB), 2008. Building Green Infrastructure for Europe, Brussels.
Greater London Authority (GLA), 2012. Green Infrastructure and Open Environments: the all London Green Grid (ALGG), London
Milwaukee Metropolitan Sewerage District (MMSD), 2013. Regional Green Infrastructure Plan, Milwaukee, Wisconsin, USA
NSW Department of Planning and Infrastructure (NSW), 2012. Metropolitan Strategy for Sydney, (Chapters 3.6 Infrastructure, 3.8. Environment), Sydney
NYC Depart. of Environmental Protection (NYC-DEP), 2010. New York City Green Infrastructure Plan, New York
Yurisic, M. J., 2014. Westbourne Urban Valley – Hydro-Logical Urban Infrastructure Recovery of the Hydro-Logical Network Of London As Strategy Against Flood, Master Project, Kingston University, London

15

AGRICULTURE AS A PROVIDER OF LANDSCAPES, TICVANIU MARE VILLAGE CASE STUDY

Alexandru CIOBOTĂ, Smaranda BICA

Politehnica University of Timişoara, 2 Vasile Pârvan Blvd, 300223, Timişoara, Romania,
Email: alciobota@yahoo.com smaranda.bica@arh.upt.ro
Corresponding author email: alciobota@yahoo.com

Abstract

Considering agriculture as both a large-scale user of land and a provider of landscapes, this paper aims to present the evolution of landscape in Ticvaniu Mare, Caraş Severin County, located in the Romanian Banat Region due to different agriculture policies. The paper debates on different historical periods: Habsburg Empire (beginning with the 18th century), between the wars (1918-1939) with The Romanian Agricultural Reform (1921), communism and post communism period (1990) until present day. We are interested in how agriculture modified the landscape over time and if there still are landscape elements bearing witness to such changes in the present. The research focuses on different scales: a small scale, the village, its tissue, plots' structure, homestead, specific architecture and a large scale, outside the village, agriculture fields, orchards, meadows/pastures and agriculture infrastructure and buildings. The research data has been obtained through different research methods: archive research (The Romanian National Archives in Timisoara and Caransebeş, Municipality of Ticvaniu Mare archives), map comparing and several field observation. The research on agriculture policies during different periods and landscape changes reveal that the two are well interconnected and that landscape should be taken into consideration by the local/national/European agricultural policy.

Key words: agriculture, landscape, landscape change, agriculture policy.

INTRODUCTION

Considering agriculture as both a land user and a provider of landscapes (Lefebvre et. al., 2012) this paper intends to analyse different agriculture policies and the way they changed the landscape in Ticvaniu Mare village. Landscape means an area whose character is the result of the action and interaction of natural and human factors (Council of Europe, 2000) and agricultural landscape is the result of the land use and management system in an area (Kizos et. al., 2006). Also, landscape is considered to have memory, some characteristics we see today come from the past, representing different historical periods and management systems (Haines-Young, 2005). The research on landscape changes in Ticvaniu Mare due to different agriculture policies is part of a wider research dealing with agriculture landscape in different villages in Banat, the historical perspective being the central focus. Agricultural policies applied different on Banat's area during historical periods, in accordance with natural and social conditions. Therefore in some villages the

historical periods with agriculture policies are more present than in other and determined different landscape changes. There are typical archaic Romanian villages in Banat, still practicing traditional agriculture where the landscape systematization during Habsburg Empire did not apply, nor communist collectivization (Crivina de Sus in Timiş county or Cornereva in Caraş Severin). On the other hand, Banat plains suffered great landscape changes during Habsburg Empire when a great part of the marshes were drained out with the help of Dutch engineers (Griselini, 1984) and transformed in agricultural land. During this period new villages were settled, following new predefined typology (Buβhoff, 1938) the most common typology being the cess table (Biled and Hatzfeld in Timis County) and the cross shape typology (Bogarosch). This period is also associated with the modernization of agriculture in Banat. Later, probably one of the most important agriculture reforms in Romania is the Agriculture Reform in 1921, that applied a set of laws for each region in Romania, in Banat applied the Law For Agriculture Reform in Transilvania, Banat,

Crişana and Maramureş from the 30th of June 1921. Few of the mutations generated by the agricultural reform were:

-the reduction of big properties and the raise of the percentage of medium and small properties;

-the raising of medium surface of medium and small properties and the reduction of it in case of the big properties;

-the reduction of the bipolarity of the agriculture property in Romania; (Otiman, 2007).

Not in the end, period of the collectivization process during communism starting with the law 187/1945 for the agrarian reform (Otiman, 2006) is probably one of the most intensive reforms with a great impact on rural society and landscape:

-the private property was transferred to the public/governmental property;

-there were organized local centres for agriculture machines (the future SMT);

-the most efficient farms (between 50-100 Hectares) were totally or partially turned into governmental property;

-with the Decree 133 in April 1949 agriculture cooperatives (first named GAC) began to be founded.

Ticavniu Mare the village is mentioned in 1699 to have a majority Romanian population, the property of Petru Macskási and later it appears in official Empire documents (Lotreanu, 1935) and detailed maps. In time the landscape in Ticvaniu Mare suffered radical changes due to agriculture policies and agriculture land management as it was subject of all agricultural reforms.

MATERIALS AND METHODS

The research proposes a historical approach on the village of Ticvaniu Mare (Agnolleti, 2007) and the methodology focuses on two different scales:

- A village scale with its tissue, plots' structure, homestead, circulation and architecture
- A large scale analyse, outside the village (Ticvaniu Mare's territory) with agriculture field, orchards, meadows/pastures agriculture infrastructure and buildings.

The results will be correlated with short references on social impact and changes in social structure.

Research materials are obtained from four main sources:

- Archives study: National Archives in Caransebeş, Municipality of Ticvaniu Mare Archives, National Agency of Land Improvement, Timiş- Mureş inferior Teritorial Branch Archives (ANIF) and personal archives;
- Historical map and plans comparing;
- Studies and research on Banat;
- Field observations and landscape analyse (James et. al., 2008);
- Photographic documentation;

RESULTS AND DISCUSSIONS

1. The village scale analyse

Being a majority Romanian village, Ticvaniu Mare has a typical organic tissue. Usually, Romanian villages structures before Habsburg Empire colonisation are organic, in relation to the landscape and natural conditions, the houses are grouped and the circulation system is not very well developed (Ciobotă et. al., 2014) (Figure 1). Generally, the house is surrounded by a garden or there is a small orchard nearby. In Romanian villages, usually young couples leave their family's home and move out building their own, phenomenon named "roire". (Gheorghiu, 2008). During this period, the houses in Romanian villages were made out of clay, straw and reed or the walls were made out of knitted birches and clay. Usually a house had two rooms, one used as a kitchen and one as a dormitory.

Figure 1 Ticvaniu Mare. The First Military Survey 1763-1787 (http://mapire.eu/)

After Kempelen visited Banat, to observe the colonization process, the Empire ordered a set of regulations for the colonization named "Impopulations-Haupt-Instruktion" (Roth, 1988). This regulation affected not only the new colonized villages but also Romanian and Serbian typical villages. The house was to be placed perpendicular to the street so that the access inside was protected. Not only the house and its relation to the street was affected by the new regulations, but also the village structure, circulations became more coherent and houses got in better relation to them (Figure 2).

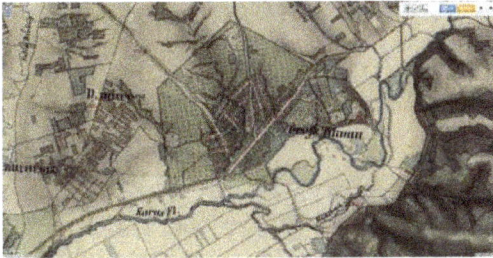

Figure 2 Ticvaniu Mare. The Second Military Survey 1806-1869 (http://mapire.eu/)

Village architecture is well related to community occupations and the level of welfare. A quick analyse on Ticvaniu Mare architecture reveals different periods in the village history, most of them connected to agriculture. The analyze of Ticvaniu Mare's Second Military Survey map, reveals a quite dense compact village, with the plots orientated perpendicular to the street (Figure 3), with a very strong axis (the road from Resita to Gradinari). All the houses have small cultivated gardens. Romanian peasant's agriculture techniques are described to be poor, and they cultivate only some roots and few plants for eating (Griselini, 1984).

Figure 3 Typical clay house of a 19th century peasant with later extension at street side (original)

In time, the house typology changes in Ticvaniu Mare. Houses perpendicular to the street are less built and the house parallel to the street begins to be typical. The new house typologies are bigger, more compact and together with its annexes define an interior yard

well protected. The crop garden is outside this ensemble (Figure 4).

Figure 4 House No. 342. Typical compact house of a rich peasant 20th century (original)

All these coherent structure and architecture was brutally completed during communism period with technological/urban architecture such as the Machine and Tractors Station (SMT), the Veterinary Centre, animal stables or the engineer's block of flat. In 1971, People's Council of Caraş-Severin County orders a research for architectural solutions in rural areas in Caraş-Severin County. The project No. 255 *Research about housing in rural areas* is accomplished, offering 10 architectural solutions in accordance with different cultural landscapes (Unit for Planning in Caraş-Severin County, 1972). Nevertheless, few years later the project No. 1645 orders the construction of a six apartments block of flat that should be finished until 1979 (Figure 5).

Figure 5 Block of flat for agriculture engineers, 1979 (original)

Part of the communist buildings and farms were put down after December 1989, but few still remained as marks of the period.
Even if the structure typology of Ticvaniu Mare was with the houses facades close to each other, describing a continuous front, street side, the situation has changed in the late years when part of the Romanians, because of economic problems and weak agriculture reform, left the village, selling their houses to gipsy community that built their own typical houses (Figure 6).

Figure 6 Typical continuous house front interrupted by a new out of the scale house, 21st century (original)

2. The large scale analyse (Ticvaniu-Mare territory level)

On the First Military Survey map, agriculture landscape in Ticvaniu Mare is quite modest. There are few gardens or orchards near houses and few arable terrains along Caraş river. For this period, cultivating the land is very well connected to nature and it has a small impact on the landscape. A family produces only for its own needs. Another reason could be the lack of agriculture knowledge, as Romanian peasants don't plough the land or they do it very bad and they don't fertilise the land with compost (Griselini, 1984). Situation changes, the map of The Second Military Survey reveals that the whole village of Ticvaniu Mare is surrounded by orchards, and first lots are noticed. Another sign that the agriculture developed are the two mills along Caraş river, one of them still in the same location, still functioning until modernization and later closing after miller's death. During Habsburg Empire there were several regulations for beautification of villages, but also for planting fruit trees and especially Morus alba for silk production (Griselini, 1984). Therefore, alignments of trees along roads already appear on the Second Military Survey map of Ticvaniu Mare.

The situation of Romanian peasants is not very good during this period. A statistical survey in Transylvania, Banat, Crişana and Maramureş conducted in 1914 reveals the fact that a very important part of the agricultural terrain, 11.283.818 jugăre (1 jugăr= 0,5775 hectares) belonged to minorities (6 jugăre/peasant), meanwhile Romanians had only 1 jugăr/peasant (Georgescu, 1943). This situation, but also agriculture situation in other Romanian regions, was the fundament for an agricultural reform, process started in 1918 and finished in Banat in 1921 with the Law For Agriculture Reform in Transilvania, Banat, Crişana and Maramureş.

As an immediate result, there was an excessive fragmentation of land generating a diverse landscape mosaic (a complete plot was of 7 jugăre and a colonisation plot had 16 jugăre) (Law For Agriculture Reform in Transilvania, Banat, Crişana and Maramureş, 1921) and a decrease in production due to poor endowment. Later, in April 1949, the process of agriculture organising into a system of collective farms, following the Soviet model of kolkhozes, started also in Romania. In Ticvaniu Mare the Collective Agricultural Farm (GAC) was founded later in 1952. At first, GAC was founded by 42 families but statistics show that only in 7 years there were already 215 families as GAC members (Collective Agriculture Farm report, 1960). Still GAC Ticvaniu Mare wasn't a very stable organization, dealing with great variations of land use from one year to another as it is revealed in The Annual Production and Financial Plans 1962-1969. The situation turned to be different starting with 1966 (Production and Financial Plan of GAC Ticvaniu Mare, 1962-1969) when GAC was transformed in Agriculture Co-operative for Production (CAP).

In 1976, the Agriculture Co-operative for Production (CAP) Ticvaniu Mare plants the first plot with orchards (P1) (Figure 7) having a total surface of 28 hectares (13.852 trees) and during 1977-1978 plantings in P1 continue and the second plot (P2) is planted with a total surface of 93 hectares (76.500 trees). In 1977 The Economic Inter-cooperative Fruit-growing Association is founded and the orchard surfaces are turned into farms (P1 turns into Farm No.1 and P2 turns into Farm No.5.).

Figure 7 Land preparing for tree planting. Plot P1, 1976 (eng. Costescu G., director of AEIP Ticvaniu Mare personal archive)

In 1978, a plot belonging to Farm No.2 is planted (P3) and later in 1979 and 1981 trees are planted on the last plot of the farm (P4). In 1984 and 1986 the last two small areas are planted in Farm No.1, one with cherry trees and another with raspberry (Figure 8). Even if

AEIP was founded in Ticvaniu Mare, only three of the five fruit tree farms owned by the association, were situated on Ticvaniu Mare's territory. Farms No.3 and No.4 were located on Cârnecea and Secăşeni territory.

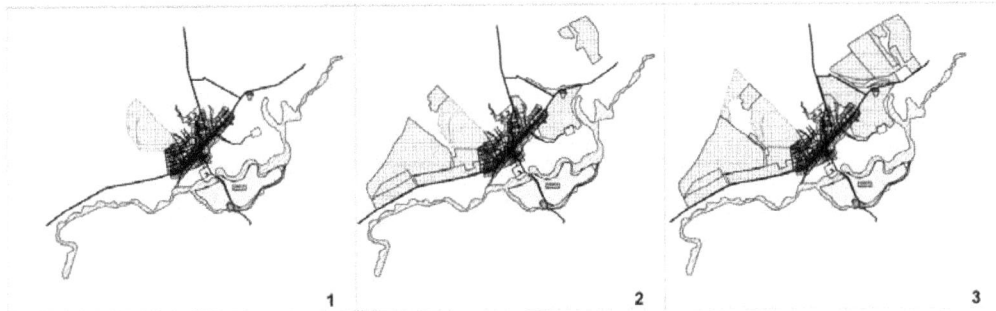

Figure 8. Fruit trees planting situation at AEIP Ticvaniu Mare between 1976-1986 1. Tree planting in 1976;
2. Tree planting 1977 and 1978; 3. Tree planting in 1979, 1981, 1984 and 1986
(original, after an eng. Costescu G., AEIP Ticvaniu Mare's director situation, dated 1983)

Between 1976 and 1986 a total surface of 406,78 hectares of land were transferred from the Agriculture Co-operative for Production (CAP) Ticvaniu Mare to AEIP Ticvaniu Mare and they were converted into orchards. An analyse on the land use situation before being converted, proves that for parcel P1, a total surface of 100,64 hectares was converted (97 hectares of natural pastures and 3 hectares with other usage). For parcel P2, a total surface of 139,38 hectares was converted (95,64 hectares of arable land, 24,53 hectares of pastures, 7,65 hectares of meadows and 11,18 hectares with other usage). For parcel P3, a total surface of 40,22 hectares was converted (37,92 hectares of pastures and 2,3 hectares with of other usage). For parcel P4, a total surface of 126 hectares was converted (70,59 hectares of arable land, 40,27 hectares of pastures, 9,65 hectares of meadows, 1, 08 hectares of orchards and 4,95 hectares of other usage). (AEIP Ticvaniu Mare, 1980). This land conversions together with land drainage and land erosion control works along Caraş river are maybe the greatest changes in Ticvaniu Mare's landscape since Habsburg Empire. In 1985 the National

Agency of Land Improvement (ANIF) starts a large project of erosion control and agriculture land drainage. For better management, the area around Ticvaniu Mare is divided in two subzones, Vărădia-Secăşeni with an impact area of 3.734 hectares and Greoni-Ticvani subzone with an impact area of 7.855 hectares. The channels' total length for Vărădia-Secăşeni subzone has 121.235 meters (73.520 drainage channels and 25.680 meters erosion control channels) and for Greoni-Ticvani subzone the channels' total length is 138.451 meters (122.601 drainage channels and 15.850 erosion control channels). These works had also architectural elements such as bridges, concrete tubes and abrupt discharges (ANIF archives, 1985-1989). Even if all those quantities were not implemented only on Ticvaniu Mares's territory, the impact on the landscape and especially in Caraş flooding valley was quite significant. The marshes here were drained out and their specific vegetation (Salix alba, Salix fragilis, Populus tremula, Populus alba, Sambucus nigra, Rubus idaeus) were cut down. All this natural landscape was transformed in agricultural landscape (Figure 9).

Figure 9. Drainage channels system and flooding protection dike (original, after an ANIF map in archives)

Figure 10 In the foreground abandoned apple trees orchard and in the background forested plum trees orchard at Farm No.2 Ticvaniu Mare (original)

After 1989, together with the transition from communism to capitalism the agriculture situation was not very good. Even if dissolving both AEIP and CAP in Ticvaniu Mare was very well welcomed, the peasants and agriculture situation wasn't getting better. The new Law 18/1991 was applied. Even if the law had two objectives: to give back property to people and to make the agriculture reform, it is proved to be very weak and its impact wasn't the expected one. There were two immediate effects on the situation of Ticvaniu Mare:

-peasants received small properties, maximum 10 hectares (even if they had more before collectivization) and the new farms were very bad endowed.

-the migration of the village population to the city and the aging of the remained population (Otiman, 2007).

As a short term effect on Ticvaniu Mare's agricultural landscape, there was a great fragmentation of the arable landscape and a great diversity of land mosaic with a growing trend in land abandonment because of a weak agricultural reform. As for the great potential of fruit farms (they were in full economic production), even if the land was returned to peasants property, they didn't have the specific technology and knowledge to continue production. As a result, on a medium term, the fruit farms declined and turned into almost forested areas (Figure 10).

From the large surface of fruit trees farms (708 hectares belonging to AEIP Ticvaniu Mare) only a small surface was still maintained (2 hectares of apple trees farm) until 2003.

Because of the Law No.7/1996 of Land Cadastre articles, of the already poor rural communities and of the land low price compared to other EU countries, a phenomenon of large personal farms and large agriculture land property appeared in Romania. In Timiş and Arad counties, in 2007, based on an unofficial data-base, one third of the agricultural surface was already part of this kind of large farms property of foreign companies (Otiman, 2007). In Ticvaniu Mare this phenomenon determined a great landscape change, agriculture landscape mosaic being less diverse (Figure 11).

Figure 11. Large farms on Ticvaniu Mare's territory (Agency of Agriculture Payments and Interventions in Agriculture Caraş-Severin)

Starting with the year 1880 there was a constant decrease in population and in ethnic diversity due to different reasons (Table 1). Between the periods with the greatest impact, besides the two world wars are the communism period and the period after December 1989 until now. The communist period was one of the most tragic periods for Romanian peasants culminating with HCM 308/1953 for the expropriation of land. This was the moment when lots of the peasants in Ticvaniu Mare left their village and moved to industrial places around (Reşiţa, Oraviţa, Ciudanoviţa, Anina) to work in industry or mining. After December 1989, the Ticvaniu Mare's community identity loss process continued because of economic problems and lack of a real agriculture reform with a great impact on landscape and village architecture.

Table 1 Population in Ticvaniu Mare village between 1880 and 2011 (Varga E. Árpád, 2000 (updated in 2008). Nationality and confessional census in Transylvania, III. Arad, Caraş Severin and Timiş Counties between 1880 and 2002, Pro-Print Publishing, Miercurea Ciuc. For 2011 census data from the Municipality of Ticavniu Mare)

Year	Total	Romanian	Hungarian	German	Gipsy	Other	Possible causes of the decrease in population
1880	1844	1775	2	24	-	1	
1890	1871	1834	2	31	-	1	
1900	1832	1761	13	36	-	3	
1910	1684	1589	6	46	-	1	World War I
1920	1422	1348	4	21	-	-	
1930	1452	1388	5	43	13	1	World War II
1941	1262	1095	3	36	-	-	
1956	1054	-	-	-	-	-	
1966	880	698	1	3	175	3	Collectivization process culminating with HCM 308/1953 for the expropriation in industry interest and later Decree No. 115/1959. The moving from the village to industrial areas.
1977	832	652	1	5	173	-	Economic problems, weak agriculture reform and migration to EU countries
1992	782	482	7	10	280	3	
2002	831	470	6	8	342	-	
2011	862	363	3	3	483	-	

CONCLUSIONS

Different policies applied during different historical periods produced dramatic changes into Ticvaniu Mare's landscape along the past centuries.

The application of this historical analyse method shows that there are landscape elements transmitted from the past to the present time. This kind of research should build the basis of a future strategy of management, protection and restoration of such heritage as part of the local identity.

Another conclusion is that agriculture policy did not change only the landscape but it also had a very strong impact on the community. During Ticvaniu Mare's history there are big changes in the community structure.

As a main conclusion, the historical approach studying agriculture policy in Banat and particularly the way they applied and influenced Ticvaniu Mare, reveals a very strong interconnection between policies and landscape dynamic with a great impact on community identity.

ACKNOWLEDGEMENTS

This work was partially supported by the strategic grant POSDRU/159/1.5/S/137070 (2014) of the Ministry of National Education, Romania, co-financed by the European Social Fund – Investing in People, within the Sectorial Operational Programme Human Resources Development 2007-2013.

REFERENCES

AEIP Ticvaniu Mare, 1980. Situation of land transferred from CAP Ticvaniu Mare for fruit tree planting, Municipality of Tcvaniu mare Archives, Ticvaniu Mare.

Agricultural Reform in Transylvania, Banat, Crişana and Maramureş, 30th of July 1921.

ANIF- National Agency of Land Improvement, Timiş-inferior Mureş Teritorial Branch Archives, 1985-1989.

Anoletti M., 2007. The degradation of traditional landscape in a mountain area of Toscany during the 19th and 20th centuries: Implications for biodiversity and sustainable management. Forest Ecology and Management 249 (2007) 5-17, Elsevier.

Bußhoff L., 1938. The Landscape changes and the Schwab colonisation in Banat. Max Schick Publishing, Munchen.

Ciobotă A., Rusu R., Obradovici V., 2014. The cemetery as an element in the evolution of the cultural landscape. German communities in Banat. West University Publishing, Timişoara.

Council of Europe, 2000. European Landscape Convention. Adopted by the Committee of Ministers of the Council of Europe on 19 July 2000 and opened for signature by its Member States in Florence on 20 October 2000.

Collective Agriculture Farm report, Oravitas' District Peoples' Council 1960. Caransebeş State Archives Found 376, Inventory 740, No. 30/1960.

Decree No. 133, 2nd of April 1949. The Romanian Council of Ministries, published in the Official Bulletin No.15 the 2nd of April 1949.

Georgescu M., 1943. Principles and methods for Romanian laws for the agriculture reform. The Romanian Institute for Agriculture Laws and Agriculture Economy. Bucovina I. E. Toroutiu Publishing

Gheorghiu T. O., 2008. Traditional rural habitation in Banat- Crişana area. Artect Publishing.

Griselini F., 1984. Attempt at a political and natural history of Timişoara's Banat. Facla Publishing, Timişoara.

Haines-Young R., 2005. Landscape pattern: context and process. Issues and Perspectives in Landscape Ecology. Cambridge University Press, Cambridge, p. 103-111.

James A., LaGro Jr., 2008. Site analysis. A contextual approach to sustainable land planning and site design. John Wiley & Sons Inc. Publishers, New Jersey.

Kizos T., Koulouri M., 2006.Agricultural landscape dynamics in the Mediterranean: Lesvos (Greece) case study using evidence from the last three centuries. Environmental Science & Policy 9 (2006) 330-342, Elsevier.

Law No.18, of Land Property, 19th of February1991. The Romanian Parliament, published in Romanian Official Monitor No. 37 from 20th of February 1991.

Law No.7, of Land Cadastre and Real Estate, 13th of April 1996. The Romanian Parliament.

Leferbe M, Espinosa M., Paloma S. G., 2012. The influence of the Common Agricultural Policy on agricultural landscapes. European Union Joint Research Centre, European Union.

Lotreanu I., 1935. Banat monography. The institute of graphic arts "Ţara", Timişoara, p. 402.

Otiman I. P. 2006. Rural sustainable development in Romania, Romanian Academy Publishing, Bucharest p.49.

Otiman I. P., 2007. The Romanian rural life on its long road between hunger and European Union or the drama of the Romanian village and peasant in a century of illusion, disappointment and hope. Romanian Academy Publishing, Bucharest.

Roth E., 1988. Large scale plan of the communities from Banat's military border 1765-1821. R. Oldenberg Publishing, Munchen.

Unit for Planning in Caraş-Severin County, 1972. Project No.255- Research about housing in rural areas. People's Council of Caraş-Severin, Systematization, Architecture and Control Direction Reşiţa. In Municiplaity of Ticvaniu Mare Archives.

16

CHARACTERISTICS OF GROWTH AND DEVELOPMENT OF THE SPECIES *SPARTIUM JUNCEUM* L. IN THE REPUBLIC OF MOLDOVA

Ion ROȘCA, Elisabeta ONICA, Alexei PALANCEAN

Botanical Garden (Institute) of the Academy of Sciences of Moldova,
18 Padurii Street, Chisinau, Republic of Moldova
Corresponding author email:onicaelisaveta@yahoo.com

Abstract

This article describes the characteristics of growth and development of the species Spartium junceum L. (weaver's broom, Spanish broom), which can be used in various branches of the national economy. The most effective method of propagation was sowing the seeds, which had been previously treated hydrothermally with hot water of 70°C for 16 hours and with 0.01% gibberellin solution for 24 hours, in trenches, in spring – in April or May.

Key words: generative propagation, Spartium junceum L.

INTRODUCTION

In relation to climate change, dangerous processes of vegetation degradation and worsening conditions for the development of woody plant species occur, making the continuous mobilization, conservation and rational use of biodiversity more necessary than ever.

The family *Fabaceae* L. includes trees, shrubs, subshrubs, perennial and annual grasses. Over 700 genera and 17,000 species, spread almost all over the globe, are part of this family.

The genus *Spartium* L. includes only one species. *Spartium junceum* L. (common names: weaver's broom, Spanish broom) was introduced in the Republic of Moldova in 1867 [1, 2]. Bast fibres extracted from this plant may be used as raw material for making thick fabrics and weaving baskets. This ornamental, melliferous shrub has been ranked "A" for its decorative qualities and can be cultivated in central and southern districts of Moldova, but, for unknown reasons, it isn't widely used. In the north of the country, this species is affected by low temperatures.

It is native to the Canary Islands and the Mediterranean Basin. That is why we decided to study the peculiarities of growth, development and reproduction of this species, in order to use it in various sectors of national economy.

MATERIALS AND METHODS

Plants of 5-6 years-old that grew and developed in the collection of the Botanical Garden (I) of the ASM from Chisinau served as research materials.

Phenological observations were carried out according to methodical indications [5], in 2013-2016. The seeds, collected in August and September, were stored differently, cleaned by various methods and treated according to the methodology [3]. Prior to the incorporation in fine loose soil, the seeds were treated with hot water of 70°C for 30, 60 minutes and 16 hours, until the water chilled, and with 0.01% and 0.03% gibberellin solution - for 24 hours. The seeds were sown in two periods, in February-March, in boxes, in greenhouses, in a mixture of soil, sand and peat in a ratio of 2:1:1, and in April-May, depending on climatic conditions. The fruits and seeds collected during the years of research were analyzed by several morphological parameters, in the Dendrology Laboratory of the Botanical Garden (Institute) of the ASM.

RESULTS AND DISCUSSIONS

Spartium junceum **L.** is an evergreen shrub, which grows up to 2-3 m tall. Its stems are erect, cylindrical, light-green, glabrous and leafless or with few leaves. The leaves are simple, narrow-lanceolate or lanceolate, glabrous or sparsely hairy, ephemeral, blue-green, 1-2.5 cm long. The growing season started, in 2014-2016, 10-15 days earlier than the multiannual average. The leaves developed in April-May (Table 1.). The stems remained green throughout the growing season until next spring. In the years of the research, the leaves fell in October. The flowers are papilionaceous, yellow, fragrant, 2-2.5 cm across, in erect terminal racemes. The calyx is bilabiate, the upper labium is double-toothed and the lower – undivided during flowering. The wings are longer than the keel (carina), the keel – acuminate or curved, the stamens – monadelphous, the anthers – hairy, the ovary – sessile, multiovulate, with linear style, slightly curved at the tip, elongated stigma. The fruits mature in August-September (Table 1).

Table 1. Phenological stages of the species *Spartium junceum* L.

Phenological stages	2013	2014	2015	2016
Bud swelling	18 April	1 April	6 April	2 April
Bud opening	20 April	9 April	25 April	10 April
Beginning of flowering	18 May	16 May	20 May	9 May
End of flowering	18 June	24 June	20 June	15 June
Beginning of fruit maturation	10 August	5 August	11 August	8 August
End of fruit maturation	10 September	3 September	6 September	5 September

The abundance of flowering and fruiting is in close correlation with the air temperature and the amount of rainfall during the respective stages. The flowers develop on annual shoots.
The fruits are polyspermous, linear pods, 4-8 (10) cm long and 0.5-0.7 cm wide, at first villous, then glabrescent. Depending on the climatic conditions in the years of research, the percentage of fruiting ranged between 48 and 65 %. Temperature fluctuations had a negative impact on the process of ontomorphogenesis of seeds and only half of them were viable, the rest - sterile. The viable seeds differed from the sterile ones in colour, size and weight.
Spartium junceum L. is an evergreen shrub, very decorative in the flowering stage, with fragrant flowers, which contain essential oil. It flowers and bears fruit from the age of 3-4 years. *Spartium junceum* L. grows fast in the first 3-4 growing seasons and reaches a height of 2 m, is light tolerant, drought resistant, grows on arid soils and develops well on slopes exposed to the sun. In cold winters, a part of the plant may be affected, but it recovers by producing root sprouts. After pruning, the number of shoots increases and the flowering stage starts later than usual. Hydrothermally treated seeds, sown in February-March in the greenhouse, germinated evenly in 15 days, while those sown in April-May germinated unevenly in 20-25 days. The germination of the seeds, treated with 0.01% gibberellin solution for 24 hours and sown in trenches, was more even and by 10-15% higher as compared to the untreated seeds. The 0.01% gibberellin solution influenced positively the germination capacity of seeds and the germination percentage, in the years of research, constituted 85-90%.
Before sowing, the seeds had been soaked in hot water for 16 hours, until the water cooled. Transplanting the 5-7 cm tall seedlings, obtained from the seeds sown in February-March, did not give the expected results. The seedlings obtained from seeds treated with hot water for 16 hours had a less developed root system as compared with those obtained from seeds treated with gibberellin. The seedlings obtained from seeds treated with 0.01% gibberellin solution for 24 hours, sown in trenches, in April-May, reached a height of 30-35 cm, by the end of the growing season, and had a well-developed root system.
These seedlings were planted in loose soil. The survival rate of seedlings was about 80-90%, depending on climatic conditions and compliance with the technology in the years of research. The weight of 1000 fruits was 350-420 g and the weight of 1000 seeds ranged between 14 and 30 g.

Analysing the data from Table 2, we can conclude that the fluctuations in temperature and the amount of rainfall in the years of research, 2013-2014, had a negative impact on the morphological parameters of fruits: the percentage of fruiting, the share of seeds in the mass of fruits, the number of viable seeds in a fruit. The lowest percentage of fruiting was recorded in 2015.

Table 2. Morphological characteristics of fruits of *Spartium junceum* L.

Morpho-biological characteristics	2013	2014	2015	2016
Weight of 1000 fruits, g	350	420	400	380
Weight of 1000 seeds, g	14	30	18	22
Share of seeds in the mass of fruits, %	4.0	7.14	4.5	5.8
Seed diameter, mm	4.0	3.0	4,0	4.0
Number of seeds in a fruit, units	11	15	12	16
Fruiting percentage, %	60	65	48	50
Peduncle length, mm	6.0	8.0	7.0	8.0
Fruit length, mm	75	73	65	81

The seeds were sown about 2 cm deep. The norm of pure and fertile seeds was 3-4 g per linear meter. Transplanting can be carried out in the first 2-3 growing seasons because the survival rate decreases by 40-50% as the plants get older. The well developed and deep root system hinders transplanting of mature plants in the field.

CONCLUSIONS

1. *Spartium junceum* L. is a heliophile, melliferous species, decorative in the flowering stage. It grows fast and is undemanding to soil, can be transplanted in the first 2-3 growing seasons.
2. The duration of the flowering stage of this shrub, the abundance of flowering and fruiting is in close correlation with the climatic conditions at the time.
3. The optimal method of propagation of *Spartium junceum* L. was sowing in spring, in April or May, the seeds that had been previously treated hydrothermally with hot water of 70°C for 16 hours and with 0.01% gibberellin solution for 24 hours.
4. *Spartium junceum* L. can be used as an ornamental plant in landscape architectture, planted in groups with other, taller species or along small alleys.

REFERENCES

Palancean A., 2015. Dendroflora cultivată din Republica Moldova. Autoref. tezei de doctor abilitat în științe biologice. Chișinău, 46 p.

Palancean A., Comanici I., 2009. Dendrologie (Asortimentul de arbori, arbuști și liane pentru împăduri și spații verzi). Chișinău: F.E.-P: „Tipografia Centrală", 519 p.

Palancean A.I., 2013.Reproducerea speciilor lemnoase (lucrare metodică). Chișinău.48 p.

Деревья и кустарники СССР,1958, М.-Л, Т. IV, с. 68-70.

Методика фенологических наблюдений в ботанических садах СССР,1972, V 113,с.3-8

17

RESTORATION STUDY IN ORDER TO INTEGRATE NEW FUNCTIONS IN THE ACTUAL STRUCTURE OF OROMOLU MANOR

Elisabeta DOBRESCU, Mihaela Ioana GEORGESCU

University of Agronomic Sciences and Veterinary Medicine of Bucharest, 59 Mărăşti Avenue, District 1, 011464, Bucharest, Romania

Email: veradobrescu@yahoo.com

Corresponding author email: veradobrescu@yahoo.com

Abstract

In time mansions and palaces that belonged to nobles and representatives of the Romanian monarchy underwent transformations that were due to changes in the architectural ensemble of dominant functions. Many of them have lost the function of dwelling and turned in office buildings or public interest. Mansion at Pausesti Maglasi Valcea County, which belonged to family can be a model for transforming ancient architectural structures with a single family dwelling leading role in buildings for training and leisure. The methodology to integrate new functions in the context of restoration of built architecture targeting and assembly of garden landscaping related Oromolu Mansion. To conduct the study of landscape restoration has outlined a set of specific analysis of complex landscape that provided information absolutely needed to successfully perform the task of adapting the current functioning of the whole history, atmosphere and ambience of the reference time fall monument, so restoring the whole area is not foreign to the values heads and authentic memoirs. After synthesizing this information it has been generated a series of major strategic directions that underpins the concept development and restoration solution / redevelopment landscaping.

Key words: restoration, adaptation, refunctionalisation.

INTRODUCTION

If the restoration of works of art or architecture, the study and intervention aims mainly accurate reconstruction of the original shape and appearance of that object, in the restoration of the historic landscape, the main objective is to reconstruct the atmosphere of that landscape and not necessarily to restore the original state of its component entities.

Garden, as part of the landscape, generates by itself a certain kind of atmosphere and implicitly refers to the archetypal patterns that generated the history of European garden art garden classic or romantic gardens.

Classic garden style praised supremacy of reason, order, geometry and human power to master nature, while romantic garden style rediscovered the beauty, perfection and harmony generated by human relationship with nature, gardens of that time were greatly reflecting natural model as unaltered by man. (Iliescu, 2008)

"The Romanian Garden "takes these features, the two styles of gardens and combines them according to the measure of local understanding, however, beyond simple imitation or use of formal models and principles characteristic of European gardens (Toma, 2001)

The spirit and atmosphere that gives life to Romanian garden is rather a natural space, abundant with fruits, which provides shelter from the scorching sun, path that you refresh in cold and clear water, shining of people cheerfulness, which resounded with music and noise of games and parties.

These are the characteristic features found in most Romanian gardens, which established them authenticity and pragmatic character, inclined to practical use without neglecting the harmonious relationship with European models taken as reference in the way of building the gardens.

MATERIALS AND METHODS

Since the World Heritage Convention drafted in 1972 in Paris, are considered part of the cultural heritage those sites that constitute the work of man or man and nature, serving as a valuable historically, aesthetic, ethnological and anthropological landmarks. Expression of the relationship between civilization and nature, place of leisure and reverie, the garden captures an idealized image of the world carrying over time, culture, style, age and originality of the creator.

In the absence of relevant historical documents and other sources of information today in an effort to restore the garden, were conducted historical research on similar buildings in the same space-time framework – XX century.

The set of analyzes carried out on the site aimed to identify the historical reference period and items of special historical value, assessment and quantification of the existing plant fund value, defining the existing functions, facilities, their current status and the way they relate with each other, the geoclimatic environment that houses the garden and its relationship with the surroundings.

Research has revealed a major empirical and spontaneous landscape design intended to ennoble the buildings constructed at that time.

Regarding the garden of Oromolu there were identified pieces of furniture having a special historical and aesthetic value (two benches carved in natural stone, but placed indiscriminately); plant specimens whose size and approximate age suggests that they are part of the composition of vegetable garden during the early twentieth century (linden - *Tilia platiphillos*, Maple - *Acer platanoides*, Ash - *Fraxinus excelsior*, Oak - *Qercus pedunculiflora*, pine - *Pinus nigra*, spruce - *Picea abies*, yew - *Taxus baccata*. fir - *Abies alba*) and at the same time revealed a spatial zoning that may suggest how the garden was organized (Figure 1.).

Figure 1. Spatial zoning of the garden (own source)

Certain geometric formations made of vegetal and elements and the vertical systematization of the land, suggested the existence of models using the principles of space organization in landscape architecture: garden's space of honor for the main façade of the manor consists of a sequence of two terraces whose difference is taken over by a slope with two sets of symmetrically placed stairs (Figure 2), on the left and right along the perspective that opens up the windows of the mansion.

Figure 2. Stairs – constructed elements that are part of the garden initial planning (own source).

On the upper terrace there is a spatial formation of circular invoicem made of *Buxus sempervirens* specimens (Figure 3). Specimens phytopathology status is precarious, the entities being aged, unstructured and aesthetically unpleasing. Spatial organization and the use of vegetation in thi area gives the mansion a representation framework that includes overall perspective to and from the edifice (the garden), geometric compositions, symmetrical to the central perpendicular axis for the main façade of the manor.

Figure 3. Circular space shaped by a group of shrubs of Buxus sempervirens the species - the initial arrangement of the garden (own source).

At the top of the forest garden was identified plant specimens arrangement according to principles of landscape architecture, trees and *Tilia* species *Tilia cordata platiphillos* being arranged in an inconsistent alignment, extending throughout the northern side of the garden. The alignment keeps a relatively constant distance to the north fence of the property, suggesting the possible presence of a utility road in this area of the garden (Figure. 4).

Figure 4. Alignment of trees and the utility road, part of the initial planning of the garden (own source).

In the "forest garden type" the spatial arrangement of vegetation and its diversity is an undeniable reference to the characteristics romantic landscape gardening, seemingly natural way through the bush and trees are composed in this area (Figure 5).

Figure 5. Forest garden (own source).

The vegetal spectrum, in this area covers both kind of species: from the local ecosystem, and exotic species such as Platanus acerifolia, Paulownia tomentosa, Acer saccharinum, Ginkgo biloba, Maclura aurantiaca, Sophora japonica, etc. The dominant species are represented by specimens of *Carpinus betulus, Juglans regia, Acer platanoides* and *Tilia platiphillos*.

The significant percentage, the favorable pedo-climatic regime and good status of exotic species in this ecosystem supports the idea of using some species thet are very similar with the existing ones. Also, the spontaneously herbaceous layer identified in the site indicate, in addition to invasive plant groups, a number of perennial grasses category (*Phleum, Festuca,*

Dactilis Lolium etc.), possibly favored by soil composition, which is an excellent environment for developing this type of perennial vegetation.

The existing plant fund and the evolutive state in which it is located makes a good relationship with the neighbouring landscape as vegetation having reached maturity, is part of a massive plant extending on several ha. in the north of the property. In contrast, in the southern part of the site passes the National Road 64A, which brings a transport station (a visual conflict with the character and atmosphere of the manor) and noisy traffic. In the distance, a series of hills bring back to sight an image and an atmosphere that is suitable to the dominant character of this historic monument.

In time, however, there is a risk that the image and character of these picturesque surroundings are altered by various elements built that does not fit in a landscape of such invoice. This involves taking measures limiting factor inappropriate assault on the historical monument. Unsightly constructions, noise, visual and air pollution are harmful factors that can be annihilated by legal measures of protection of historical monuments outside localities. This approach requires legal protection zone delineation and establishment of the historical monument (500 m. - Measured from the outer limit of the property), which had already limit accompany the historical and plans should provide for landscaping (Law nr.422 / 2001, art. 8). At the same time, planting trees and shrubs with dense contour perimeter of the property will be a visual and sound barrier that will allow isolation to the historic area in relation with national road.

Accessibility of motor vehicles on the property is made with great difficulty because there is no space to allow a stationary waiting outside the perimeter of national road 64A until the opening of doors by an authorized person (Figure 6).

This issue raises the necessity of establishing functional withdrawals for the purposes of traffic on National Road and the flow conditions in the roadway accesses the property (or a band's speed reduction or an outlet practiced within the property, the input roadway).

Figure 6. Site access (own source).

RESULTS AND DISCUSSIONS

In terms of the concept of landscaping, the solution is focused on highlighting all of the "strong" aspects that characterizes the historic mansion (studied in complex analysis of landscape) and integration of facilities and functions to meet the high requirements of new functional purpose of the building. In this regard were designed multifunctional spaces, areas of interest, objectives and diversified atmospheres which provided perfect relaxation, both active and passive activities that invite and encourage playful, and meditative, providing abundant color, tactile, olfactory and characteristic charm and harmony of the Romanian garden type and to give benefit of the murmur and the cool water, gentle warmth of a clearing or occurring in forest meadows.

At stylistic approach is proposed the mixed style, style that characterizes gardens established during the nineteenth and twentieth centuries, bringing gardens that harmonize the two styles that preceded it; classic and romantic. The area in the vicinity of the manor is treated in a conventional manner using geometric rigor and order to enhance the architectural value of the building.

The main facade is accompanied by a reminiscent circular lawn makink an imaginary circle of buxus species, but the overall appearance, and also in size, obstruct the affirmation of the value of construction in relation to land. The pedestrian walkways stresses the lawn circularity, then backing off two branches that lead to natural stone stairs that go down to the lower terrace. At the end of the path is a decorative pond, also circular in form, a keystone closing the semicircular design of the alleys on the lower terrace. In this space there is the possibility of organizing

outdoor group sessions or tea can provide a delightful atmosphere (Figure 7).

Figure 7. The representative area of the garden – virtual simulation (own source).

The transition between the levels of the two terraces is taken by a succession of stairs that are designed to emphasize the geometric nature of this area. The end point of the shaft which is drawn perpendicular to the main facade of the manor, is supported by an end view defined by a group of five pieces of pyramidal oak (*Quercus rubra* "Fastigiata"). This plant is a focus group color and a volume that will serve to structure and prioritize the entire space of representation in the vicinity of the manor.

To the right of the mansion, near the proposed park, a group of three specimens of the *Liquidambar styraciflua* (which turns bright red in autumn) founded another accent color point, making a very interesting contrast, is profiled against a background consisting of pieces of the species *Acer saccharinum* (colored in golden yellow in the same period). All proposed plant composition in the representativeness of the manor offers a varied but unified solution, based on a volumetric decor and color spread and sustained throughout the growing season.

Both species, existing and proposed, with persistent foliage or obsolete offers a harmony of shapes and colors through all the decorative elements: flowers, foliage texture and color coating, architecture and texture stem and canopy.

Moving in the "forest garden type" it has been proposed a number of elements to reinforce the potential landscape of manor garden. After studying topographical plan and following the requirements imposed by the theme launched (so to arrange the existing cave), were able to identify the key points that could be a spring area and a lake area in the garden. The stylistic approach for this romantic area encouraged us to introduce a watercourse in this proposal, the route meandered with stone or earth bank covered with grassland. The role of water in this landscaping proposal is to give life to the space, to reflect and double through reflection the trees verticality, increasing the character, sometimes mystical of the garden, but also to enrich through movement and sound the atmosphere of the place. The river route arise small islands formed almost natural way water finds its way through the forest, river captured by specific vegetation. Aquatic plants and perennial grasses accompanying riverside bring more atmosphere to the texture and color range of the whole arrangement. In the upper course of the river forms a waterfall, such as water around us would find the natural drainage outlet, on its way upstream (north side domain) to downstream, where it meets Olanesti River (south side). Fortunately, the natural existing terrain allows, without major land movements, the potentiation of the garden's space garden and the introduction, almost naturally, an element so much important for a romantic atmosphere in Historic Gardens, namely water (Figure 8).

Figure 8 Waterfall area – virtual simulation (own source).

In the vicinity of this area, with absolute power recharging energy, there could not be another area for passive recreational activities (contemplation, rest, relaxation). At the same time, this space can be a dining area outdoors, near the main access to the historical manor building, this being one of the main arguments underlying the establishment in that area of

functions that include relaxation and water. The platform covered with natural stone slabs carved with lawn between joints will create a special atmosphere and the decor will be absolutely delightful. For the furnishing of this platform there will be two options: either use easily removable furniture, allowing its transport inside when the weather conditions are unfavorable, or the option to mount a fixed body of natural stone table, with similar banquets, to provide strength and stability, but in a way that fits perfectly and harmoniously into the special atmosphere of this area. In this case, the decision will be subject to further analysis, together with the client space. The waterfall will become a spectacular element that will transform this dining area into one of the three major areas of interest in the proposed arrangement. The characteristic landscape of the "forest garden" lays the dining platform at the highest point, thus providing an overview of the watercourse of the river. In the forest, the lowest topographical land turns into a special place, by joining three other very important elements - lake, gazebo and lawn (Figure 9).

Figure 9 Lake and gazebo area – virtual simulation (own source).

During the vegetation analysis was identified in this area, one of the most spectacular plant specimens from the entire site. A superb specimen of the species *Platanus acerifolia*, a spatial structuring element is in this area of the garden. The conclusions of the analysis performed on the vegetation proposed to eliminate several common species around the dish to give it a major importance and landscape value.

To emphasize the romantic landscape value and character of the area, the presence of a gazebo situated partly in the console above the lake, and inspired by the traditional romanian architecture (pillars and arches) can complete a picture of romantic garden.

The pedestrian walkways and routes of natural stone thrown into the grass, walk you through this space by passing successively through the brook, forest, and even water, reminiscent of natural bridges, created naturally in natural streams.

At the left and the right access point in kiosk are placed two weeping willows (*Salix alba* 'vitellina") whose role is to enhance the romantic atmosphere and the presence of water in the area.

There were also put into value, view points that open from the enytrance into the property, by marking perspective endpoints with vertical forms like columnar oak (*Quercus rubra* "Fastigiata"), which emphasizes and highlight adjacent items (waterfall, mansion).

The proposal for landscaping will provide solutions to preserve and protect the image and atmosphere of the manor against harmful external factors that might alter the character of the historic and picturesque garden (Figure 10).

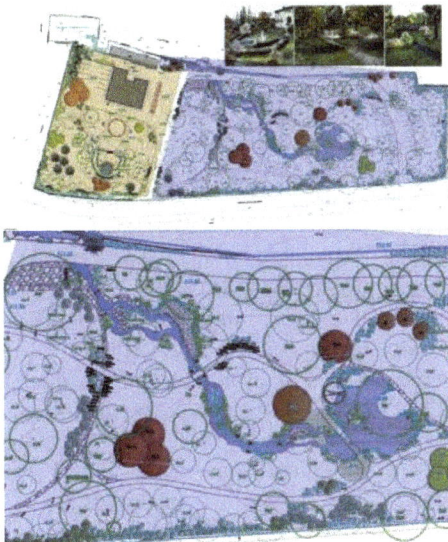

Fig. 10 Plan with the proposed design and detail over the forest garden area (own source)

CONCLUSIONS

Following the synthesis of analyzes it was revealed that the historical reference period to be taken into account in the garden landscape restoration approach is between nineteenth and twentieth centuries, a period in art history that is characterized by a combination of the two gardens reference styles (classic and romantic) and creating mixed style (or composite).

At the same time it is necessary to answer by landscaping restoration to all functional requirements of the new destination of the manor, adapted to current needs of modern society without altering the actual atmosphere, substance and character of valuable historical monument.

Pedo-climatic context favorable and presence in a state of emergency exotic plant specimens encourages the use of decorative species to create an authentic romantic atmosphere.

Auto and pedestrian circulation shall provide a maximum degree of accessibility and to serve the full potential of the area Oromolu landscape.

Landscaping proposal will solve all requirements and problems encountered so successfully to reconstruct the atmosphere of the early twentieth century gardens, and to fall in the current XXI century.

ACKNOWLEDGEMENTS

This research was conducted with support from the central bank, through a pilot project functional restructuring of historic properties.

REFERENCES

Toma Dolores, 2001. *Despre gradini si modurile lor de folosire,* Ed. Polirom
Iliescu, Ana-Felicia, *Arhitectura peisagera,* Ed. Ceres, Bucuresti, 2008.
***Legea nr.422/2001

18

SOME ASPECTS REGARDING FLOWER'S MORPHOLOGY ON SEVERAL LOCAL POPULATION OF *PRUNUS DOMESTICA* L. FROM PATÂRLAGELE (BUZĂU COUNTY)

Daniel Constantin POTOR, Mihaela Ioana GEORGESCU, Dorel HOZA

University of Agronomic Sciences and Veterinary Medicine of Bucharest,
59 Marasti Blvd., District 1, Bucharest, Romania
Corresponding author email: danutpotor@yahoo.com

Abstract

According to the literature, a number of morphometric values of flower from varieties of Prunus domestica may be correlated with their degree of fertility. We chose four local populations of plums from Pătârlagele area, Buzau County, to measure the floral components - sepals (length), petals (length), stamens (number and length), carpel (length) and to compare them with the literature. The results of our observations shows the presence of a local variability regarding the perianth, the average number of stamens, the average length of the carpel and the ratio between the number of stamens and the length of the carpel; for the T3 and T4 populations there are slight differences from literature on the average length of petals. The ratio of average number of stamens and carpel length shows low values for the four populations (range 1.35 to 1.63) compared with the data from literature, indicating the possible presence of self-incompatibility among the four local populations of Prunus.

Key words: *Prunus domestica* L., *flower components, self-incompatibility, local populations.*

INTRODUCTION

Research on the variability of infraspecific characters of the *Prunus domestica* was aimed primarily towards fruit morphology (Buia, 1956), biology of flowering (Branişte, 1994), morphology, viability and pollen germination (Gilani et al., 2010; Calicut et al., 2013) or degree of fertility of varieties (Ionita, 1956; Boredianu et al., 1965; Cociu, Bumbac, 1985). The differences in the morphology of the flower are fairly low compared to the general characterization of higher taxa; Flora RPR, 1956 shows the following characters for *P. domestica* flower: sepals' length 2.5-5 mm; petals' length 7-12 mm.

In a comparative analysis of variation of the morphological characters of the flower to 23 plum varieties, Mădălina Butac (2003) shows that they have been grouped into 12 classes according to the diameter of the flower, in 9 groups depending on the average length of the pistil, respectively in 9 groups after ratio between the number stamens and pistil length. Analysing these characters in relation to the fertility of the varieties it has been observed that the self-fertile and partly self-fertile varieties have the flower's diameter, the length of the pistil, the number of stamens and the ratio of the number of stamens and length pistil have much higher values and significantly different from those with self-incompatibility.

MATERIALS AND METHODS

Pătârlagele city is located in the north - west of Buzau County, at 45 ° 19 ' north latitude and 26 ° 21' east longitude, at a distance of 56 km from the city of Buzau. The city is located in the Pătâragele Basin from the Curvature Subcarpathians, on the Buzau River, at an altitude of about 400 meters, dominant landscape of the area consisting of hills.

It were selected 4 local populations, denoted T1, T2, T3 and T4, located in plantations set in the village. Populations of T1, T3, and T4 are in independent plantations, while T2 population is planted alongside other local varieties. The land with plantations made by T2, T3, T4 populations has north exposure while the T1 population is on a land with south exposure.

For the morphological characterization of the flower (Figure 1) the following indicators were chosen: the average length of the sepals, the average length of the petals and the ratio between them, the average number and length of the stamens, the average length of the carpel, the ratio between the length of the carpel and of the petals and the ratio between the number of stamens and length of carpel. From each population were taken twenty five flowers into flowering phenophase.

Results were used for comparative analysis of the four populations.

Figure 1. Flower of *Prunus domestica* population T3

RESULTS AND DISCUSSIONS

The morphometric measurements on T1 flowers variety (Table 1) shows that the petals have an average length of 9.16 mm and are about 2.14 times longer than sepals; average number of stamens is 19 and the carpel's average length is 12.90 mm; it exceeds the flower perianth of 1.41 times; the ratio between the average number of stamens and carpel average length is equal to 1.41.

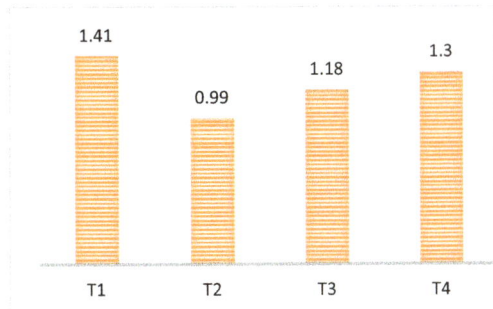

Figure 2. The ratio between the average length of the carpel and of the petals

The values of the floral components of the T2 population (Table 1) are close to the followings floral indicators of T1 population: sepals' average length (4.13 mm), average number of stamens (19.5) and average carpel length (12 mm); the size of the petals is obviously higher (12.16 mm) while the ratio between them and the average size of the carpel is lower (0.99).

The ratio between the average number of stamens and the average length of the carpel is 1.63 (Figure 3) is the highest value recorded for the 4 populations.

In the group of the four populations, the flowers from the T3 population (Table 1) shows the highest values to the average length of sepals (4.79 mm), petals (12.80 mm), carpels (15.10 mm) and average number of stamens (22); the value of the ratio between the average length of petals and carpel (Figure 2) is the smallest of the observed populations series; also, the ratio between the number of stamens and the average length of carpels - 1.46, is at the middle of the series of values for the 4 populations (Figure 3).

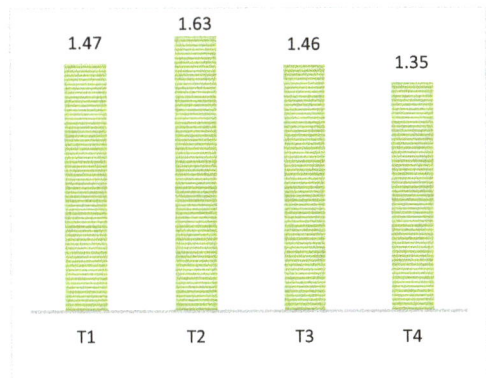

Figure 3. The ratio between the average number of stamens and the average length of carpel

The lowest values for the used indicators are found to the morphometric measurements of floral components of T4 population (Table 1), except the average length of the carpel (12.20 mm) and the ratio between the average length of carpel and the petals (Figure 2).

Comparing the results obtained with literature data it is observed that:

- The length floral perianth components fall within the values indicated in the description of the *Prunus domestica* flower (Romanian Flora, 1956), except populations T2 and T3 which the average length of petals slightly exceeds data from literature.

- For the T3 population there are values of floral components that exceeding the data in the literature like the average length of the carpel, while the remaining variants fall between the average values of the literature (Butac M., 2003).

- The ratio of average number of stamens and carpel length set shows low values for the four populations (range 1.35 to 1.63) compared with the data from literature (range 1.42 to 2.33) (Butac M., 2003).

Table 1. Morphometric characters of the floral components of *Prunus domestica* populations from the area Pătârlagele (Buzau County)

Population	Sepals - average length (mm)	Petals - average length (mm)	Petals/sepals - average length ratio	Stamens - average number	Stamens - average length (mm)	Carpel - average length (mm)	Carpel/petals - average length ratio	Stamens average number / carpel average length
T1 (Prun roşu)	4.28	9.16	2.14	19	7.99	12.90	1.41	1.47
T2 (Gras)	4.13	12.16	2.94	19.5	9.95	12	0.99	1.63
T3 (Gras CB)	4.79	12.80	2.67	22	8.96	15.10	1.18	1.46
T4 (Ciorăsc)	3.63	9.38	2.58	16.50	6.59	12.20	1.30	1.35

CONCLUSIONS

The four local populations of *Prunus domestica* examined in terms of the morphology of the floral components shows some differences with the data in the literature, and that indicated the existence of local variability on the flower's level.

The size of the flower's perianth and the ratio between the average number of stamens and length of carpel indicate that in the analysed populations, self-incompatibility and self-sterility may be present.

REFERENCES

Branişte N., 1994. Polenizarea şi fecundarea la prun. Mapă documentară, nr 39.

Buia Al., 1956. Familia Rosaceae. In Flora RPR vol. IV, Ed. Acad. RPR, Bucureşti, 835-834.

Butac Mădălina Maria, 2003. Biologia înfloritului şi fructificării soiurilor de prun din sortimentul naţional în scopul stabilirii celor mai buni polenizatori. Teză de doctorat. Bucureşti, 145.

Ćalić, D., Devrnja, N., Kostić, I., & Kostić, M., 2013. Pollen morphology, viability, and germination of *Prunus domestica* cv. Požegača. Scientia Horticulturae, 155, 118-122

Gilani, S. A., Qureshi, R. A., Khan, A. M., & Potter, D., 2010. Morphological characterization of the pollens of the selected species of genus *Prunus* Linn. from Northern Pakistan. African Journal of Biotechnology, 9 (20), 2872.

Ioniţă C., 1956, Contribuţii la studiul fertilităţii soiului de prun Tuleu Gras.Grădina, via şi livada, nr. 5.

Săvulescu T., Beldie AL., Buia AL., Guşuleag M., Nyarady E., Prodan I., Răvăruţ M., 1956. Flora Republicii Populare Române IV, Editura Academiei Republicii Populare Române.

19

RESEARCH ON QUALITATIVE AND QUANTITATIVE PERFORMANCE OF GERMAN ORIGIN VARIETIES IN ECOPEDOCLIMATIC CONDITIONS OF THE EXPERIMENTAL FIELD U.S.A.M.V. BUCHAREST

Marinela Vicuţa STROE, Cristinel IOANA

University of Agronomical Sciences and Veterinary Medicine of Bucharest, 59 Mărăşti,
011464, Bucharest, Romania
Corresponding author email: marinelastroe@yahoo.com

Abstract

It is well known that the area of culture defines fundamentally the phenotypic, agrobiological and technological manifestations, also the quantity and quality of varieties of grape-vines. They practically, by the values of the main elements that define their degree of adaptation, can achieve better results compared with the area in which they formed or were naturalized. In this study we analyzed six varieties of German white wine: 'Rhine Riesling', 'Müller Thurgau' 'Silvaner', 'Bacchus', 'Ortega' and 'Phoenix', which are found in the experimental, teaching and research field within U.S.A.M.V. Bucharest. The research was conducted in wine-year 2012-2013 and target tracking these sorts in terms of quantitative and qualitative performance in the production and also the veg-productive balance. The results obtained highlight the fact that the varieties had a good rapport between production and the wood removed at pruning, as evidenced by values 4,04-8.82 (Index of Ravaz) and 11,58 – 24,38% (Vegetative and productive balance index - VPBI).

Key words: balance, grapevine, index of Ravaz, maturity, varieties.

INTRODUCTION

The varieties of grape-vines, regardless of origin, force, production capacity, production direction, are characterized by morphological variability and enhanced technology, given by their genetics, but also that they are influenced by a large climatic factors and agrotechnical factors, thus manifesting differently depending on the area of culture. Therefore, in studies that aims adaptation of the varieties of grape-vines in areas different from their place of origin, increasingly more attention is given to outstanding research aimed at showing how they manifest veg-productive, as direct result of physiological processes and cultural practices applied (Belea, 2008). The study focused on tracking the behavior of six varieties of German origin, Rhine Riesling, Silvaner, Müller-Thurgau, Bacchus, Phoenix and Ortega, in south area of Romania. The varieties come from the same vineyard where they emerged as basic varieties, but in Romania, except the first two, were rarely investigated and the less cultivated. The varieties are distinguished by a high degree of similarity between them, having

in common a certain genetic lineage, as follows: the first three are found as genitors variety of Bacchus and Müller-Thurgau variety is a result of crossing between Rhine Riesling and Silvaner. In addition, Phoenix is a hybrid variety obtained between Bacchus and Villard Blanc variety and the Ortega variety is a cross between Müller-Thurgau and Siegerrebe (Table 1). Data were extracted from Vitis International Variety Catalogue (www.vivc.de). The study was discussed based on two reasons: to determine and assess the quantitative and qualitative performance of varieties in an area different from home and veg-productive balance assessment using indicators (Index of Ravaz, Vegetative and productive balance index - VPBI). To calculate these indicators, balanced loads of buds are left behind cuts, in order to link the photosynthetic capacity of the plant to the number and weight of the grapes, which regulates the two activities (vegetative and productive) and thus improve production quality. By properly sizing the number of buds and appropriate allocation of production elements, the ratio of the processes of growth and fructification are effectively adjusted in favor of the latter, it is increased or maintained the longevity of plantation and is obtained large crops of grapes, economic and relatively stable. Watching the relationship between the influence of fruit load on the quantity and quality of crop at Müller-Thurgau, Rhine Riesling and Silvaner varieties during the years 1976-1981 in conditions of Germany (Kiefer and Crusius 1984 quoted by Belea, 2008) have obtained variable production as follows: Müller-Thurgau variety, if awarded a load of

15 buds /m^2 was obtained an average of 20.61 t/ha, compared to 12.56 t/ha at a load of six buds/m^2.
The Rhine Riesling variety, production increased from 7.65 t/ha at a load of six buds/m^2 to 12.55 t/ha in 15 buds/m^2, and the variety Silvaner, production ranged from 9.26 t/ha and 14.99 t/ha. Basically, the allocation of large loads of buds/vine, increased, grape production without experiencing loss of quality. Were found, however, significant reductions on organoleptic and analytical quality during the years when productions were recorded over 15 to 23 t/ha.

Table 1. Genetic origin of studied varieties

Prime name	Rhine Riesling	Silvaner	Müller-Thurgau	Bacchus	Phoenix	Ortega
Variety number *VIVC*	10077	3865	8141	851	9224	8811
Country of origin of the variety	Germany	Germany	Germany	Germany	Germany	Germany
Species	*Vitis vinifera* L.	*Vitis vinifera* L.	*Vitis vinifera* L.	*Vitis vinifera* L.	*Vitis vinifera* L.	*Vitis vinifera* L.
Pedigree as given by breeder/bibliography	-	-	Riesling Weiss x Silvaner Gruen	(Silvaner Riesling) x Müller Thurgau	Bacchus x S.V. 12-375	Müller Thurgau x Siegerrebe
Pedigree confirmed by markers	- x Heunisch Weiss	- x Heunisch Weiss	Riesling x Madeleine Royale	(Silvaner x Riesling) x Müller Thurgau	-	Müller Thurgau x Siegerrebe
Prime name of pedigree parent 1	-	-	Riesling Weiss	Silvaner x Riesling	Bacchus Weiss	Müller Thurgau Weiss
Prime name of pedigree parent 2	-	Heunisch Weiss	Madeleine Royale	Müller Thurgau	Villard Blanc	Siegerrebe
Year of crossing	-	-	1882	1933	1964	1948
Last update	11.02.2015	11.02.2015	30.01.2015	29.09.2014	29.09.2014	29.09.2014

MATERIALS AND METHODS

The research was conducted in 2012-2013 wine year, in the experimental field of U.S.A.M.V. Bucharest and the varieties that are object of these research were applied the same technology culture: Guyot on semi-stem type cutting, planting distance of 2.2/1.2 m, with a load of 32 buds/vine (12 buds/m^2) considered optimal for obtaining quality white wines.
Climatic data focused on daily observations regarding the evolution of parameters - temperature, precipitation, insolation, which helped calculating climatic indexes that define the level of favorability of an area: - real heliothermic index (IHr), hydro-thermic coefficient (CH), vine plant bioclimatic index (Ibcv), oenoclimatic aptitude index (IAOe), and also Huglin index calculus. Huglin index

(HI) is calculated from April the 1st to September the 30th in the northern hemisphere and is defined as follows:

$$IH = \sum_{01.04}^{30.09} \frac{[(Tm-10)+(Tx-10)]}{2} \times k$$

Tm = Medium air temperature (°C)
Tx = Maximum air temperature (°C)
k = Day length coefficient in relation with latitude, with values between 1,02-1,06 for latitudes of 40-50° and for Romania ($44,1^0$ – $46,0^0$) this has the value of 1,04.

The benchmark index in viticulture is widely used in France because it provides information about the potential heat in the vineyard, showing importance in appropriate choice of product, on the one hand and is positively correlated with the amount of sugars accumulated in grapes, on the other hand. The values of this indicator in different wine regions causes a general classification of these areas and establishing minimum temperature necessary to conduct the vegetative cycle of varieties of grape-vines in that area, (Huglin, 1978, Tonietto and Carbonneau, 2004). From this perspective, recent research conducted at the Bavarian Research Center for Viticulture and Horticulture state established a minimum standard Huglin's index for fructification of investigated varieties, as follows: IH is 1300^0C for Ortega variety, 1400^0C for Müller-Thurgau varieties, Phoenix and Bacchus, 1600^0C Silvaner variety, Rhine Riesling variety 1700^0C (Ulrike and Schwab, 2011), 1700 ^{0}C for Chardonnay and Syrah variety almost 2100^0C. The minimum limit for grape-vine is considered by some authors to an IH = 1600^0C (Laget et al., 2008).

During the experience, observations and determinations were made used in determining the elements of fertility and productivity, with special focus on those who have shown interest in calculating the vegeto-productive index balance covered by this study: average weight of a grape, 100 berries weight, production/vine, sugars (g/l), total acidity (g/l tartaric acid).

To assess the balance between production of grapes and vine growth, in practice, is used Ravaz index in formula: IR = Production/removed wood.

In general, values of this indicator varies within wide limits from 1.2 to 27.7, values between 5-7 are being considered ideal; for varieties with medium vigor, the IR is ideal between 4-6; varieties with reduced vigor take the value 8; values lower than 3 and bigger than 10 should be avoided, since it causes big vigor or delays in maturation and reduced quality, as appropriate. (Celotti et al., 2001).

The relation between growth and fruiting was established using vegeto-productive balance index (VPBI). It highlights the percentage share of the vegetative part, expressed by weight of removed wood at pruning to achieve total production and it represents the ration between „weight of wood removed at pruning x 100/grapes production + removed wood" expressed in kg/vine (Maccarone and Scienza,

1996). If grape varieties are for quality wines, the result must be within the 22.1 to 33.5% (Celotti et al., 2000) at Cabernet Sauvignon and 18 to 23% (Dejeu et al., 2003) at Feteasca regala.

RESULTS AND DISCUSSIONS

The analysis of climatic elements for the wine year 2012-2013 was performed by comparing the defining climatic elements of this year with the annual average of the last 10 years (2001-2011), due to the frequency of extreme weather events and the lack of constancy of the values recorded.

The values of the four synthetic indexes (Table 2) shows that when the thermal resources are high, the water resources are low and the most fluctuating indicator is the bioclimatic one, whose spectrum is within the 9,9- 14.32.

Table 2. Evolution of climatic elements (2001-2013)

	Specification	Average	Year	Year
		2001-2011	2012	2013
Agroclimatic indices	The hydro-thermic coefficient CH)	0,75	0,97	0,50
	The real heliothermic index (IHr)	1,3	1,08	1,38
	The viticultural bioclimatic index (Ibcv)	9.9	11.2	14,32
	Index of the oenoclimatic aptitude (IAOe).	5231	5075	6493
	Huglin index	2392	2739,7	2358,2

Regarding the development of Huglin index values, it is noted that tends to increase, which exceeds the multiannual average in 2012, reaching a peak of 2739.76, conditions in which the vineyard, characterized by a warm temperate climate in general (IH4), acquires the appearance of a warm climate type (IH5) - (IS1, IH5, IF3).

The observations made show that the area in which the didactic-experimental field of U.S.A.M.V. Bucharest is found is favorable for growing varieties of grape-vines studied (registered in the south of Romania), and the elements of microclimate positively put their mark on the behavior of the studied varieties, although varieties are adapted to a cooler climate. Assigning the same number of buds per vine 32 buds/vine highlights their differentiated behavior in terms of quality and quantity of production, but its performance touch the limit required to obtain quality white wines (Table 3).

Table 3. Evolution of quality parameters on the experimental varieties

Varieties	Average weight of a grape (g)	Weight of 100 berries (g)	Yield (kg/vine)	Sugar (g/l)	Acidity (g/l tartaric)
Rhine Riesling	90,17	132	4,68	226,69	7,56
Müller-Thurgau	92,17	203	4,83	238,38	6,52
Silvaner	113,39	177	4,12	230,32	6,61
Bacchus	80,66	209	2,85	201,21	5,57
Phoenix	91,92	220	2,86	205,45	5,38
Ortega	119,68	174	2,99	212,89	5,67

The appreciation is based on their accumulated sugar levels, on the background of a pretty balanced acidity. The data show a highlight in quality of the varieties Rhine Riesling (226.69 g /l), Müller-Thurgau (238.38 g/l) and Silvaner (230,32g /l), but no other varieties are in imbalance, the minimum being registered at Bacchus variety (201.21 g /l). Appreciation of balance between the production of grapes and vine growth, made using Ravaz index, indicates that the most varieties are found in a balance with a slight imbalance registered at varieties Silvaner, Phoenix, Ortega, even the values are within the ideal highlights from this point of view (5-7).

Basically, wine year 2012 -2013 was a good wine year, favorable to the development and fruiting of varieties analyzed so they have been in steady growth and fruiting (Table 4).

Analysis of vegetation and productive balance index - VPBI (%) shows that varieties are in a veg-productive balance because they are close to the normal range of grape varieties for quality white wines (Table 4).

Highlighting the percentage share of the vegetative part of the vine to achieve the production of grapes, this index ranged from 11.58 to 24.38%.

We can say that the balance between vegetative growth and fruiting capacity was greater tilted in favor of fructification, except variety

Silvaner, where the index (24.38%) is in the average necessary to obtain high quality wines. Therefore, the higher the values recorded are, the more favorable and positive correlation is for the accumulation of large amounts of sugars in grapes.

Table 4. Overview of Ravaz index and vegetative and productive balance index – VPBI (%)

Experimental varieties	Index of Ravaz	Vegetative and productive balance index – VPBI (%)
Rhine Riesling	5,92	16,34
Müller-Thurgau	8,56	11,58
Silvaner	3,81	24,38
Bacchus	8,82	13,92
Phoenix	4,43	20,6
Ortega	4,04	14,85

Highlighting the percentage share of the vegetative part of the vine to achieve the production of grapes, this index ranged from 11.58 to 24.38%. We can say that the balance between vegetative growth and fruiting capacity was greater tilted in favor of fructification, except variety Silvaner, where the index (24.38%) is in the average necessary to obtain high quality wines.

Therefore, the higher the values recorded are, the more favorable and positive correlation is for the accumulation of large amounts of sugars in grapes.

Following the evolution of accumulated sugars according to the Huglin index in 2012-2013 vine year (Figure 1) there is a positive direct correlation in all varieties, and in addition, it was observed an increase from varieties potential, in general (Hillebrand et al., 1997). This is explained from the genetic origin of Varieties, the area of culture and not least less favorable climatic conditions.

Comparing the index values of Huglin from the areas of origin (Germany) with those recorded in the area of culture where the experience took place, we observe that Rhine Riesling variety accumulates 200 g/l sugars, index of Huglin being 1700^0C in cool area, making it possible to obtain white table wines; at $2358,2\ ^0C$ value of the same index in the south area of Romania, the amount is much higher - 226.69 g/l, which makes it possible to obtain quality white wines. This can be seen in other studied varieties, which leads us to affirm that the southern areas of Romania create the possibility of obtaining quality white wines from varieties of grapes analyzed.

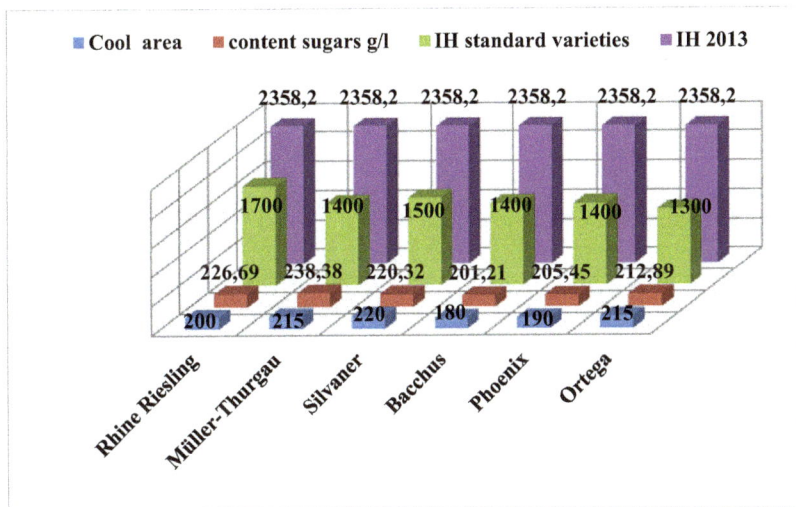

Figure 1. Correlation between Huglin index values and sugar contents g/l

CONCLUSIONS

The area in which took place the experience is favorable for cultivating the studied varieties (the entire south area of Romania) and specific factors (annual temperatures average, the ones from the vegetation period of grape-vines, rainfall) have a positive contribution on the behavior of the studied varieties.

The varieties are distinguished by a high degree of similarity between them in terms of quantitative and qualitative performance based on the genetic lineage, as follows: the first three are found as genitors of Bacchus variety and Müller-Thurgau variety is a result of interbreeding Rhine Riesling x Silvaner's.

Assigning the same number of buds per vine 32 buds/vine highlights their differentiated behavior in terms of quality and quantity of production, but their performance reach the limits necessary to obtain quality white wines.

Following the results, the varieties of german origin - Rhine Riesling, Müller-Thurgau, Silvaner, Bacchus, Phoenix and Ortega are in veg-productive balance in climate conditions of south area, and therefore can be successfully introduced into the wine culture in southern Romania, with great possibilities of obtaining outstanding production quantitatively and qualitatively.

REFERENCES

Belea, Gianina, Mihaela, 2008 - Research on optimization of the grape-vine vegetation to improve production quality, PhD Thesis, USAMV, Bucharest.

Celotti E., F., Battistuta, P., Comuzzo, B., Scotti, P., Poinsaut, R., Zironi, 2000 - Emploi des tanins oenologiques: expérience sur Cabernet Sauvignon, Revue des Enologues, France.

Celotti, E., G.C., De Prati, and S., Cantoni, 2001 - Rapid evaluation of the phenolic potential of red grapes at winery delivery: application to mechanical harvesting. Australian Grapegrower & Winemaker 449a,151-9.

Hillebrand W., H. Lott, F. Pfaff, 1997 – Taschenbuch der Rebsorten (pag. 58-68, 70-83, 192-193).

Huglin, P., 1978 - Nouveau mode d'évaluation des possibilités héliothermiques d'un milieu viticole. Comptes Rendus de l'Académie d'Agriculture, France 1117-1126.

Laget F., M.T. Kelly, Deloire A., 2008 - Indications of climate evolution in a mediterranean area considerations for the wine and viticulture sectors. Organisation Internationale de la Vigne et du Vin, Verona, Italia, le juin 2008.

Maccarone G., A., Scienza 1996 - Valutazione dell'equilibrio vegeto-produttivo della vite, l'Informatore Agrario, 46.

Tonietto J., Carbonneau A., 2004 - A multicriteria climatic classification system for grape-growing regions worldwide. Agricultural and Forest Meteorology 124, 81-97.

Ulrike Maab, Arnold Schwab, 2011 - Der Huglin - Index und der Wärmeanspruch von Rebsorten-Veröffentlichung in „Das deutsche Weinmagazin" 10/2011.

http://www.lwg.bayern.de/weinbau/rebenanbau_qualitaet smanagement/linkurl 18.pdf

20

THE IMPORTANCE OF TREES IN URBAN ALIGNMENTS. STUDY OF VEGETATION ON KISELEFF BOULEVARD, BUCHAREST

Elisabeta DOBRESCU, Claudia FABIAN

University of Agronomic Sciences and Veterinary Medicine Bucharest,
59 Marasti Blvd., District 1, Bucharest, Romania
Corresponding author email: veradobrescu@gmail.com

Abstract

Tree alignment plays an important role in shaping the busy urban roadways. These types of trees play a crucial role in rendering ecosystem services, while, at the same time, imparting a characteristic image of the place. The sustainable management of the species used in these alignments as well as the process of replacing them with other species are two of the major issues which fall under the scope of specialists concerned with ensuring a high quality urban environment. In light of the constantly changing climatic factors and an increasingly polluted urban environment, the studies on the alignments from Kiseleff Boulevard give rise to discussions on the choice of the most suitable tree species for an urban area with high traffic. Moreover, the use of species must be done after a rigorous selection and after studying the local environmental conditions. Apart from that, in order to preserve the efficient ecosystem services, it is essential to properly manage the trees that are currently found in the previously mentioned alignments. The tree's health, its visual impact, the shading capacity, the age are just a few basic elements that shape and restore the image of a boulevard that is a landmark of the urban area of Bucharest. Dendrometrical studies have brought forward the direct link between the age and the state of the tree (its health) on the one hand and the ability to adapt of the studied species to the specific conditions of the Bucharest urban environment on the other.

Keywords: Kiseleff Boulevard, urban environment, ecosystem services, dendrometrical studies.

INTRODUCTION

Over the centuries, trees have had a great importance in the formation and shaping of the public space, sketching and giving life to promenades, alleys, avenues, squares, gardens and parks (Forrest and Konijnendijk, 2005). Planting trees in alignment in public spaces from the European area was one of the key elements in the cities in terms of building a certain image and raising the quality of urban areas.

If at the beginning, the trees were elements that contributed to the design and organization of the urban space, they've gradually become key elements of the site's identity. Depending on the geographical typologies, local and European influences, personal affinities or preferences, boulevards have always had various species in the alignment which provided a special image. Ever since the sixteenth century, European cities have paid attention to the trees, using them to mark the way to cathedrals and churches. One example is Paris where, during King Henry II of France, it was mandatory to plant and take care of trees,

and also design and organize any other aligned planting of trees (Forrest and Konijnendijk, 2005).

Starting with the second half of the 19th century, promenades and boulevards from all around Europe transformed into favoured places where people gather to socialize. The indiscriminate popularity of the walks in these places revealed the vegetation's role in producing oxygen (Pellegrini, 2012), and the middle of the nineteenth century allowed the appearance of a true fact of society: healthy trees, sidewalk and pedestrians (Pellegrini, 2012). With the opening of walks for everybody, sidewalks are charged with the role of protecting pedestrians from carriages with horses and sidewalks are also the ones that will enable the introduction of tree alignment. One such example is the 19th century Paris during the transformation of the Haussmannien urbanism, when alignments of trees were made depending on three main characteristics: hygiene, comfort and aesthetics (Pellegrini, 2012). In the haussamannian Paris, the trees which most often encountered in road alignments are plane trees and chestnut trees, but also other species (Pellegrini, 2012) which could be found sporadically on avenues. Their planting was determined by the fashion of that time or certain preferences.

The situation in Bucharest proved to be similar to that of other European cities. Located in the northern part of Bucharest, the Kiseleff Boulevard is a historic roadway, which was built in the 1840s, at the same time with the urban changes Bucharest went through in the early nineteenth century (Fezi, 2010). Kiseleff represented the beginning of the road between Bucharest and Brasov, and at the same time the place outside the city where Mogoşoaia road ended (what we call Victoria Avenue after 1878). The documents which required the presence of this place where people were supposed to relax and socialize are the organic regulations (Toma, 2001). The place has been known since 1833 when the Great Steward (Romanian title assigned to the noble who was in charge of supervising the court) G. Filimon mentions it in a letter to General Kiselev. The alley was to be covered with sand and designed by "replanting the trees on the sides" (Toma, 2001). It was originally planted with *Tilia* sp. chosen for their shading capacity, and quickly enriched with *Robinia* sp. plantations. Thus, after 1846 Meyer proposes the extension of the *Tilia* sp. alley from Mogoşoaia road to the first round. From here until the second round (nowadays Arc de Triomphe) and further to Baneasa alley, it was overshadowed by an alignment of two rows of *Robinia pseudoacacia*. After 1851 the boulevard was extended to the third round (currently Press Square).

Due to the newly improvements that had been added to the street (mostly trees and gravel), it quickly became "the usual place of promenade" (Fezi, 2010) and of walking of the high society from Bucharest, offering opportunities to meet and socialize with other people, images which remained forever in the memory of the inhabitants (Toma, 2001). *Tilia* sp. were chosen as alignment trees and were planted in four rows. Their number was initially over 2000 (Vătămanu, 1973). Following the changes undertaken since the twentieth century, the boulevard increased and was enriched with new species of trees.

Based on these considerations, the study aims to establish the state of the inventoried trees, taking into account the fact that over time the trees used initially were replaced. If at the beginning the space chosen for this research was planted with *Robinia pseudoacacia* displayed in four rows, the boulevard contains nowadays alignments of *Fraxinus excelsior, Platanus hybrida, Aesculus hippocastanum, Tilia* sp., and other species used in urban areas now.

The role of the study is to present an inventory of plant material found in section A from the Press Square to the Arc de Triomphe and to establish after careful dendrometrical measurements the health of the inventoried species. This research started from the need to determine the steps to replace the species of trees after a certain period of time and also to find effective methods for using the species that will replace the damaged trees or those which are at the end of life.

The prevention of the physiological decline of the alignment leads to the maintenance of the urban image. Various studies have shown that trees which are in poor health affect the urban image and lead to a reduction of the space's social role and especially the ecological role.

MATERIALS AND METHODS

The inventory of species on Kiseleff Boulevard aimed trees throughout the entire chosen section, from Victory Square to the Press Square. This street is most certainly a historical landmark for urban transformation of the city. To improve the results of the research, the boulevard was divided into two sections. Section A from the Press Square to the Arc de Triomphe and section B from the Arc de Triomphe to Victory Square. But this article only regards section A, the remaining inventory thus contained in a larger study. The tree inventory was conducted using mixed teams made up of landscape architects and horticulture experts.

The working method in the inventory was to create sheets that contained both quantitative criteria such as: size, health, tree circumference at 1.3 m, but also qualitative criteria: aesthetic image, maintenance cutting or regeneration. The chosen qualitative and quantitative criteria

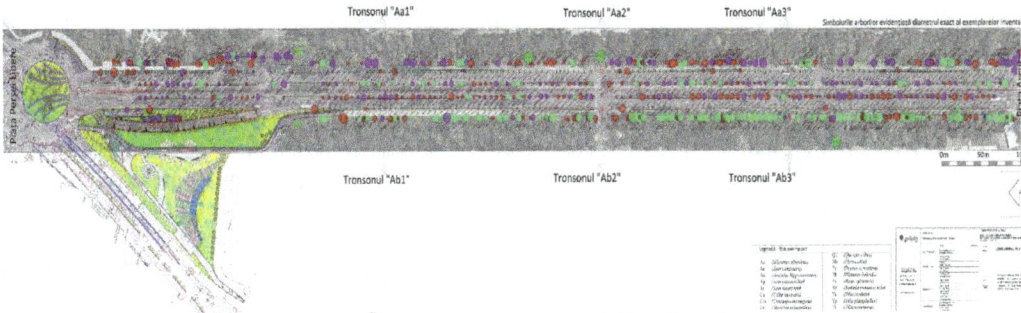

Figure 1. The section A of Kiseleff Boulevard

will complete the information about ecosystem services such as shading, carbon storage, storage of dust particles etc.

The cultural importance is also a criterion that is worth taking into account, because the species of trees from Kiseleff Boulevard have been replaced over time, and the newly planted species were not always adapted to Bucharest's urban environment. Thus, the selected criteria fall into four main groups:

▪ General features of the trees (the identification number of each tree, the circumference of the trunk, the trunk diameter, the insertion height of the treetop, the treetop diameter and the total height of the tree).

▪ The health of the tree (excellent, good, bad or very bad, dry) - which is determined only by the appearance of the tree and / or by determining diseases and pests that cause visible damage.

▪ The aesthetic value - relevant for determining the image of the boulevard as well as the possibility of decorating the chosen species. The argument for the introduction of this criterion is a qualitative one since cultural services are important for any recreation area. By determining aesthetic value, we can deduce the attractiveness or the monotony that can be induced by an alignment along a boulevard.

The maintenance status is justified by the need to intervene, types of interventions: cuts and treatments that can be performed on site depending on the chosen management plan.

The alignment referred to with the numeral 3 is actually the first line of trees which border the boulevard. In these generous green spaces, the number of trees and the number of trees species

are various and bring an extra ecological value to the boulevard (Figure 1).

The chosen criteria are closely linked to the health of the trees, but also their aesthetic value, due to the fact that proper caring for urban trees provides an extension of their lives and also add value for the recreation component. For the inventory of the alignment of the avenue trees as well as a more careful study, each section was subdivided (Aa1, Aa2, Aa3 etc.) depending on the position of the alignment in the plan. Section A has four well established alignment sections.

RESULTS AND DISCUSSIONS

In the crowded cities where there are densely designed alignments, studies have shown that it is necessary to maintain and continuously restore a healthy environment to ensure a better quality of life. All this can be possible with the help of urban green spaces that, aside from their environmental qualities, help build a unique identity to the respective areas (Heidt and Neef, 2008).

If it base our research on the idea that trees play a crucial role in preserving a healthy environment, their importance is clear from both the absorption and storage of carbon dioxide and other polluting compounds, but also from the way they help reduce density of the ground by preventing flow (Merse et al., 2009). Planted near busy highways, trees can cut down the sound by reducing noise and pollution. It is known that small spaces with trees and shrubs can absorb or neutralize 68 tonnes of dust per ha / year and streets, squares and small parks with trees can absorb between

20% and 25% of the amount of dust that is found in the areas free of trees (Heidt and Neef, 2008). Thus, on green areas 50-100 m wide, they affect air quality up to a distance of 300 m (Heidt and Neef, 2008). Small spaces with trees (Merse et al., 2009), especially street alignment and boulevards cause a slower growth and lower life expectancy, especially when planted in crowded areas (Merse et al., 2009). The effects of urban pollution on vegetation and trees are the decrease of the default period of life, a slower development and an unattractive and sad appearance of the green elements. In the case of the Kiseleff Boulevard, trees have always played an important role for the spatial configuration and for providing ecosystem services. Its embellishment proved to be quite costly. The boulevard was in a continuous change over time (Potra, 1990). Thus, after the inventory conducted in mixed teams, section A with its subdivisions Aa1-Ab1, Aa2 - Ab2, Aa3 - Ab3 has a total of 956 specimens of 18 species out of which only one species of coniferous wood: *Pinus sylvestris*, the remaining 17 species belonging to the group of deciduous trees.

The middle layout Aa1 - Ab1, which has an alignment of *Fraxinus excelsior*, we have a number of 283 trees and for Aa2 and Ab2 *Platanus hybrida* alignments, with a total of 324 trees. In the case of Aa3 and Ab3 alignments, where we took stock of the trees from the first row, their total number is 349 (Figure 2).

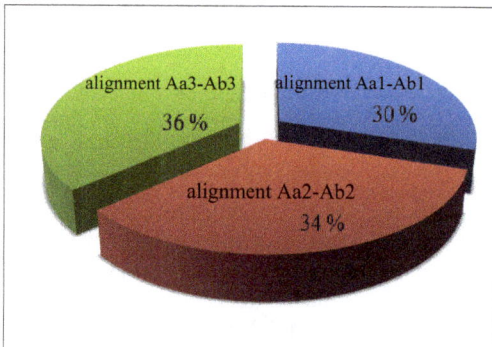

Figure 2. Total of the trees of the section A

As a result of the analysis of the three types of alignment, it was discovered that the diversity of species and volumes is mostly found in the

boulevard's sides, in the portions numbered Aa3 and Ab3.

The alignment Aa1 - Ab1 has 283 tress, out of which 278 are *Fraxinus excelsior* and only 5 specimens of *Platanus hybrida*. In percentage, for a clearer image of the make-up of the alignments, 98.23% are part of the *Fraxinus excelsior* and the remaining 1.77% trees are *Platanus hybrida*. Note that this middle alignment consists of trees that have almost reached maturity, a result of the measurements carried out. Therefore, depending on the height of the specimens they may be classified into three distinct categories 2 m – 6 m (23%), 7 m - 9.5 m (39%) and 10 - 12 m (38%).

Following evaluation, it has been determined that out of all trees from the species *Fraxinus excelsior*, a percentage of 17.67% are in a state of good health, while the majority of specimens, 45.58% are in a mediocre state of health. Trees with a poor health condition or even dried specimens are found at a rate of 36.74%. This means that for more than one third of the inventoried specimens, more frequent cutting actions need to be applied to allow them to regenerate and recover. The state of the trees is a reflection of the way we manage all maintenance processes as well as a picture of the way trees and other plants are chosen and whether they are adequately matched to the environmental conditions and requirements.

But no species is totally resistant to pollution, because pollution resistance can be determined by several indicators that are constrained by certain characteristics of pollutants, the plant's growth stage, health status, location (Sieghardt et al., 2005).

For the following two alignments Aa2 and Ab2, the dominant species is *Platanus hybrida*. It must be noted that the alignment Aa2 is made up mostly of *Platanus hybrida* trees (88%). Also, in this alignment, *Fraxinus excelsior* is located in a proportion of 10%, as well as other species 2%. There are 154 trees on the alignment Aa2 that are between 4 meters and 7.5 meters high and only 17 trees between 8 and 16 meters high. 96% of them are young trees (148 trees) in good condition and only 4% are not in a very good condition or they are dry. The majority (70%) of the 17 trees between 8 meters and 16 meters high are in poor conditions or very dry and only 6% are in good health.

It is easy to observe that most trees of *Platanus hybrida* are young specimen that were planted recently (the year of planting is usually 2013). It appears that trees are becoming increasingly fragile and their health status changes as they adapt increasingly difficult to urban pollution and its consequences.

In alignment Ab2 among the 153 trees we find species of *Platanus hybrida* - 145 trees with heights between 4 meters and 7 meters, 5 trees *Fraxinus excelsior* with heights of 8 meters and 12 meters and 3 specimens of *Aesculus hippocastanum* heights of 8 meters and 10 meters. Specimens of *Platanus hybrida* and those of *Aesculus hippocastanum* are young trees, while those of *Fraxinus excelsior* have almost reached maturity.

The health criteria reflects about the same situation as in alignment Aa2. The young species of *Platanus hybrida* (145 trees) are in good health, being planted throughout 2013, but specimens of *Fraxinus excelsior* and *Aesculus hippocastanum* are in bad health condition. Again, we can conclude that at the time of this study, namely in 2015, the majority of trees from the alignments Aa2 and Ab2 are newly planted ones and that they are trying to adapt to a boulevard with dense traffic conditions. But as they get older, their health deteriorates as well.

In section A, the last subdivisions Aa3 and Ab3 are actually the first lines of trees from the massive edges which complete this avenue's image. The trees inventoried in both Aa3 and Ab3 are mostly quite diverse, part of a large number of species and are found either in good health or in a bad condition.

Thus in Aa3, there are 175 trees of 16 of hardwood as *Acer* sp., *Tillia* sp., *Robinia pseudoacacia* etc.

For the Ab3 alignment, most of the 171 trees are *Tillia* sp., *Fraxinus excelsior* and *Aesculus hippocastanum*, but with a higher procentage regarding their condition (70% of them are in a good health state).

One of the results of the study show that the diversity encountered in certain areas has helped the adaptation process of the trees and generally speaking, the majority of them has reached maturity in good conditions.

An important role in urban areas is played by this diversity of species which helps to create and maintain a stable urban environment. Therefore species diversity in urban ecosystem management is one of the most important elements that should been encouraged as it is desirable to have a balance between biological diversity and production and maintenance costs of the plants (Heidt and Neef, 2008).

Depending on the adaptability of each tree, there could be a significant growth rate in the first years. However, there could also be trees which grow more slowly after they were planted and then, in a few years, they dry out and die (Bradshaw et al., 1995). Therefore, in the case of the species used for alignment, the criteria behind their management refers primarily to the possibility of survival and the selection of the best adapted trees, depending on the site's characteristics (Heidt and Neef, 2008). Managing the trees on Kiseleff Boulevard, section A, which was presented in this article also provides an updated and correct choice of tree species that are most suited to the conditions of urban pollution. The study observed that age and health are related and that most mature specimens identified at the site are in a poor state of health. This situation is caused not only by the characteristics of the urban environment, but also by the poor management from stakeholders.

CONCLUSIONS

It is therefore necessary to establish a vegetation management plan to anchor the current conditions of a polluted urban environment. Newly taken actions on trees alignment in European cities seek a closer study of the relationship between the area where the tree will be planted and the stresses to which specimens are exposed. It has been found that the life of trees is shorter in urban areas because of the high stress levels they face (Sæbø et al., 2005). Experts have identified many stress factors: from pollutants, physical damage, excessive cuts of treetops and the very small space for root development, to a soil that is poor in nutrients and is unable to absorb the water and oxygen needed for the roots (Sæbø et al., 2005).

For a better determination of the species which are well adapted to a particular local context, one can use the numerous studies that rely both on species selection and a more careful selection of genotypes. So the selection of urban trees is made after three criteria determining: the species' adaptation to environmental conditions of the site, the way its ecological functions adapt to urban conditions and a low cost of production and management (Sæbø et al., 2005).

The trees from street alignments must have several physical characteristics that recommend them for this type of use: they must have a strong growth of ramifications, well defined branches, with a steady and predictable growth and with a long life in order to lower or remove successive costs and a certain aesthetic value (Sæbø et al., 2005).The present study showed that a large proportion of all inventoried trees reach maturity in a mediocre state of health.

A tree that is in a poor or mediocre health state loses its environmental, social, cultural and economic benefits (Bradshaw et al., 1995). Therefore, encouraging a quality management which has among its concerns urban trees is desirable. Selecting and managing tree by tree translates into a better quality of life and reduction of urban pollution

REFERENCES

Bradshaw A.D., Hunt B., Walmsley T.J., 1995. Trees in the Urban Landscape: Principles and practice, Ed. University Press, Cambrige, 17-22.

Fezi B.A., 2010. Bucurestiul European, Ed. CurteVeche Publishing, Bucureşti, 49-53.

Forrest M., Konijnendijk C., 2005. A History of Urban Forests and Trees in Europe, in "Urban Forests and Trees: A Reference Book" by Konijnendijk, C.C., Nilsson, K., Randrup, Th.B., Schipperijn, J., Ed. Springer Science & Business Media, Verlag, Berlin, Heidelberg, 23-48.

Heidt V., Neef M., 2008. Benefits of Urban Green Space for improving urban climate, in Ecology, Planning, and Management of Urban Forests. International Perspectives for Improving Urban Climate, by Carreiro, M.,M., Song, Y.-C., Wu, J., Springer Science+Business Media, New York, 84-96.

Merse C., Buckley G.L., Boone C.G., 2009. Street trees and urban renewal: a Baltimore case study, in The Geographical Bulletin, vol.50(2):65-81.

Pellegrini P., 2012. Pieds d'arbre, trottoirs et piétons : vers une combinaison durable?, Développement durable et territoires, vol.3(2):1–16.

Potra G., 1990. Din Bucureştii de ieri, Ed. Ştiinţifică şi Enciclopedică, Bucureşti, 329-334.

Sæbø A., Borzan Z., Ducatillion C., Hatzistathis A., Lagerström T., Supuka J., García-Valdecantos J.L., Rego F., Slycken J. V., 2005. The Selection of Plant Materials for Street Trees, Park Trees and Urban Woodland, in "Urban Forests and Trees: A Reference Book" by Konijnendijk, C.C., Nilsson, K., Randrup, Th.B., Schipperijn, J., Ed. Springer Science & Business Media, Verlag, Berlin, Heidelberg, 257-280.

Sieghardt M., Mursch-Radlgruber E., Paoletti E., Couenberg E., Dimitrakopoulus A., Rego F., Hatzistathis A., Randrup T.B., 2005. The Abiotic Urban Environment: Impact of Urban Growing Conditions on Urban Vegetation, in "Urban Forests and Trees: A Reference Book" by Konijnendijk, C.C., Nilsson, K., Randrup, Th.B., Schipperijn, J., Ed. Springer Science & Business Media, Verlag, Berlin, Heidelberg, 281-324.

Toma D., 2001. Despre grădini şi modurile lor de folosire, Ed. Polirom, Iaşi, 144-164.

Vătămanu N., 1973. Istorie bucureşteană, Editura Enciclopedica Romana, Bucureşti, 104-109.

21

EFFECT OF SECONDARY METABOLITES PRODUCED BY DIFFERENT *TRICHODERMA* SPP. ISOLATES AGAINST *FUSARIUM OXYSPORUM* F.SP. *RADICIS-LYCOPERSICI* AND *FUSARIUM SOLANI*

Cristina PETRIȘOR, Alexandru PAICA, Florica CONSTANTINESCU

Research and Development Institute for Plant Protection,
Ion Ionescu de la Brad Blvd., No.8, District 1, Bucharest, Romania
Corresponding author email: crisstop@yahoo.com

Abstract

Secondary metabolites produced by filamentous fungi have different structure and function, and they are a source of novel compounds with pharmaceutical, agricultural and medicinal importance. Trichoderma spp. are considered to be an abundant source of secondary metabolites, some of them with applications in biological control, plant growth promotion, like aroma constituents or in plant immunity. The aim of this study was to assess the potential of volatile and non-volatile metabolites released from some antagonistic Trichoderma spp. isolates against pathogens Fusarium oxysporum f.sp. radicis-lycopersici and Fusarium solani which causes wilting for more cultures. In vitro studies have demonstrate that volatile compounds produced by T49 (67.69%), Tk14 (64.61%), T50 (61.53 %), T85 (60%) showed strong inhibitory effect on FORL growth compared with M14 isolate. The non-volatile tests revealed that three isolates of Trichoderma (T85, T50, Tal12) are the best which inhibited the growth of F.solani in vitro.

Key words: *volatile compounds, non-volatile compounds, biocontrol, antifungal effects.*

INTRODUCTION

Fusarium sp. is a soil borne fungal pathogen that attacks plants through roots at all stages of plant growth, is considered as one of the main soil-borne systemic diseases and the major limiting factor in the production of tomato both in greenhouse and field-grown (Srivastava et al., 2010; Borrero et al., 2004).
Various species of fungi described as antagonists of phytopathogenic fungi produce secondary metabolites with strong antifungal activity. *Trichoderma* species have been used widely as biocontrol agents because produce many antifungal metabolites, volatile and non-volatile that adversely affect growth of different fungi phytopathogens (Li et al., 2016; Barakat et al., 2014; Nagendra et al., 2011; Ajith and Lakshmidevi, 2010; Srivastava et al., 2010; Faheem et al., 2007; Vinale et al., 2006; Dennis and Webster 1971a; Dennis and Webster 1971b). Also, the potential of *Trichoderma* spp. to produce many volatile (e.g. pyrones, sesquiterpenes) and non-volatile secondary metabolites (e.g.) has been reviewed by Reino et al., 2008. *Trichoderma* spp. differ in their abilities to produce volatile and non-volatile secondary metabolites and their production

varies greatly between species and between isolates of the same species, depends on environmental conditions (Vinale et al., 2009). Many previous studies revealed that antimicrobial metabolites produced by *Trichoderma* spp. are effective against a wide range of phytopathogenic fungi, *Botrytis fabae*, *Fusarium* spp, *Rhizoctonia solani*, *Macrophomina phaseolina* (Chen et al., 2015; Barakat et al., 2014; Sreedevi et al., 2011).
Dubey et al., 2011 reported that secondary metabolites from culture filtrates with higher concentration of *Trichoderma viride*, *Trichoderma virens* and *Trichoderma harzianum* inhibited mycelial growth of *Fusarium oxysporum* f.sp. *ciceris*. Vinale et al., 2006 reported that volatile secondary metabolites play a key role not only in mycoparasitism by *Trichoderma harzianum* and *Trichoderma atroviride*, but also in their interactions with tomato and canola seedlings. Ajith and Lakshmidevi, 2010 reported the effect of volatile and non-volatile compounds produced by *Trichoderma* spp. against *Colletotrichum capsici*, a fungal pathogen responsible for anthracnose disease in bell peppers. The results of this authors showed that the volatile compounds produced by *Trichoderma* spp.

showed 30 to 67% inhibition of *Colletotrichum capsici,* whereas non-volatile compounds have the ability to control growth of *Colletotrichum capsici* by 21 to 68% at a concentration of 50% culture filtrate. In Romania there is little information available on the implication of second-dary metabolites produced by *Trichoderma* with inhibitory effect on different pathogens (Raut et al., 2014a; Raut et al., 2014b).

The aim of the present study was to determine, *in vitro,* effect of volatile and nonvolatile compounds produced by some *Trichoderma* isolates against fungal plant pathogens such as *Fusarium oxysporum* f.sp. *radicis-lycopersici* and *Fusarium solani.*

MATERIALS AND METHODS

The tested *Trichoderma* isolates as well as phytopahogens *Fusarium oxysporum* f.sp. *radicis-lycopersici (FORL)* and *Fusarium solani (F.solani)* used in this experiment belong to Microbial Collection of RDIPP.

Effect of volatile compounds produced by Trichoderma isolates on the mycelial growth of Fusarium

The effect of volatile metabolites produced by *Trichoderma* against *Fusarium* was tested using the inverted plate technique described by Dennis and Webster, 1971a. The mycelial disk (5 mm) of *Trichoderma* excised from the edge of 5 days old cultures was inoculated into the center of a Petri dish which containing PDA medium. The lid of each plate was replaced by the bottom of a plate containing PDA medium inoculated with a 5-mm-diameter mycelial disk of *FORL and F solani* so as test pathogens were directly exposed to antagonistic environment created by *Trichoderma*. Then, the two plates were sealed together with parafilm and incubated at 28°C for 6 days in the dark. The control sets did not contain the antagonist. Radial growth of pathogens was recorded after 6 days of incubation and percentage inhibition was calculated in relation to control.

Effect of non-volatile compounds produced by Trichoderma isolates on the mycelial growth of Fusarium

The production of non-volatile compounds by *Trichoderma* isolates against *Fusarium* was studied using the method described by Dennis and Webster (1971b). Initially, mycelia agar plugs (5mm diameter) removed from the edge of a 5 days old *Trichoderma isolates* mycelium were inoculated in 100 ml sterilized Potato Dextrose Broth in 250 ml conical flasks, and incubated at 28±2°C on a rotatory shaker set at 100 rpm for 10 days. The culture filtrate was filtered through Whatman paper for removing mycelial mats and then centrifugated at 3000 rpm 10 minutes.

The filtrate was added to molten PDA medium (at 40±3°C) to obtain a final concentration of 10% (v/v), 25% (v/v), 50% (v/v). Then PDA containing Petri dishes were inoculated with mycelial plugs (5 mm diameter) of *FORL* and *F. solani* at the centres. The dishes were incubated at 26±2°C until the colony reached the plate edge. Plates without filtrate served as control. There were three replicates for each treatment and the experiment was repeated two times. The percentage inhibition was calculated in relation to the control by the formula:

$$L=C-T/Cx100$$

where: L – inhibition of radial mycelial growth; C – the radial growth measurement of the pathogen in the control; T– the radial growth measurement of the pathogen in the presence of antagonists.

RESULTS AND DISCUSSIONS

A large variety of volatile secondary metabolites could be produced by *Trichoderma* spp. such as ethylene, hydrogen cyanide, alde-hydes and ketones, which play an important role in controlling various plant pathogens (Faheem et al., 2010; Siddique et al., 2012; Chen et al., 2015). The results for volatile metabolites activity against pathogens are presented in table 1 and figure1.

From our results, it is evident that volatile compounds produced by *Trichoderma* isolates studied decreased the mycelial growth of *FORL* and *F.solani* (table 1). The effects of volatiles produced by the *Trichoderma* isolates studied over the 6-day incubation period were different for *FORL* and *F. solani. In vitro* studies showed that the volatile compounds produced by T49, T50, Tk14, T85 and Tal12 signi-ficantly reduced mycelial growth and inhibited spore germination of *FORL* with inhibition percent between 58.46% and 67.69%.

Table1 Effect of volatile compounds produced by different *Trichoderma* strains on *Fusarium* mycelial growth

Pathogen	Antagonist *Trichoderma* strain	Growth inhibition (%)
Fusarium oxysporum f.sp. *radicis-lycopersici*	T50	61.53
	T49	67.69
	T85	60
	Tal12	58.46
	Tk14	64.61
	TK20	56.15
	M14	31.53
Fusarium solani	T50	55.38
	T49	32.30
	T85	33.84
	Tal12	66.15
	Tk14	48.76
	TK20	38.46
	M14	29.23

Volatile metabolites produced by Td49 were more efficient in reducing the mycelial growth of *FORL* by 67.69%, after 6 days of incubation, respectively than M14 being 31.53%. The data obtained show that the Tal12 (66.15%), T50 (55.38%) and Tk14 (48.76%) strains producing volatile compounds with significant effect in reducing growth of phytopatogenic fungi *F.solani*. Other strains studied (T49, T85, Tk20, M14) inhibit very weak the growth of *F.solani*, inhibition percentages ranging from 29.23% (M14) to 38.46% (Tk20).The high degree of growth inhibition of *FORL* by all of the strains suggests that the inhibitory effect observed in dual culture could mostly be attributed to volatile metabolites.

Many strains of *Trichoderma* have been reported to produce volatile compounds that inhibit the growth of pathogen fungi significantly. Studies of Zhang et al. (2014) showed that the volatile compounds produced by *Trichoderma harzianum* T-E5 have an significant inhibition of *FOC* mycelial growth but not by killing *FOC*. This investigation suggests that metabolites released by these *Trichoderma* species are toxic and fungistatic to *Fusarium*.

Some *Trichoderma* species (*T. viride* and *T.asperellum*) tested by Qualhato et al., (2013) produced volatile metabolites having significant effects on the mycelial growth and development of the *S. sclerotiorum, F.solani*.

Tapwal et al. (2011) reported that volatile compounds of *Trichoderma* spp. isolates significantly inhibited the mycelial growth and spore germination of *F.oxysporum*.

Research of Calistru et al. (1997) revealed that volatile metabolites produced by *Trichoderma harzianum* species can significantly suppress the growth of *Aspergillus flavus and Fusarium moniliforme* rather than mycoparasitism. Also, studies of Raut et al. (2014a) demonstrated that volatile metabolites produced by two *Trichoderma* strains displayed inhibitory effects on *R. solani* and *P.ultimum* pathogens growth.

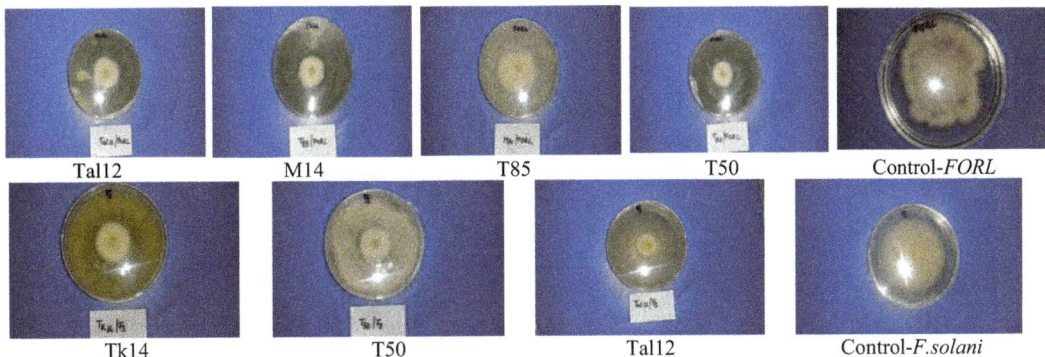

Figure 1. Plate assay for the influence of volatile metabolites from *Trichoderma* isolates on the mycelial growth of fungal pathogens

The effect of 10%, 25%, 50% filtrate concentration of *Trichoderma* isolates on *F. solani* radial growth has been show in figure 2. All concentrations of the metabolites reduced radial growth of *F.solani* in different percent.T50 and T85 and Tal12 isolates showed the highest inhibition of *F.solani* radial growth, however Tk20 and M14 had the lowest effect on growth of this pathogen. The efficiency of the non-volatile metabolites on the mycelial growth of the pathogenic strain of *F.solani* varied from 30% and 70%. This results are in accordance with of Raut et al., 2014b that supported that 25% and 50% filtrate concentration of *Trichoderma* isolate produced between 70-80% inhibition of *F.solani*.

The non-volatile secondary metabolites produced by *Trichoderma* isolates used in this study were found to be more effective in suppressing the mycelia growth of *Fusarium solani* when compared to FORL.

Our results is consistent with the results of Kavitha and Nelson, 2013 which supported that non-volatile compound of *Trichoderma* inhibited growth of *Fusarium javanicum* and *Fusarium oxysporum*. This results is supported by the previous reports of Hasan et al., 2012 which found that *Trichoderma harzianum* inhibited the radial growth of *Fusarium graminearum* by 43.33%.

The concentrated solutions (50%) of the *Trichoderma* culture filtrates suppressed the growth of *FORL* but weaker compared to *F.solani*. Neverthelss culture filtrates of *Trichoderma* diluted (10%, 25%) showed very low inhibition level (20%) of this pathogen.

Some investigations suggest that metabolites released by *Trichoderma* species are the most effective on *Fusarium culmorum* and can be used successfully to control *Fusarium* foot rot in wheat seedlings (El-Hasan et al., 2008). Also, Barakat et al. (2014) founded that the non-volatile secondary metabolites of *Trichoderma* species were more effective in suppressing the mycelial growth of *Botrytis fabae* when compared to volatile compound.

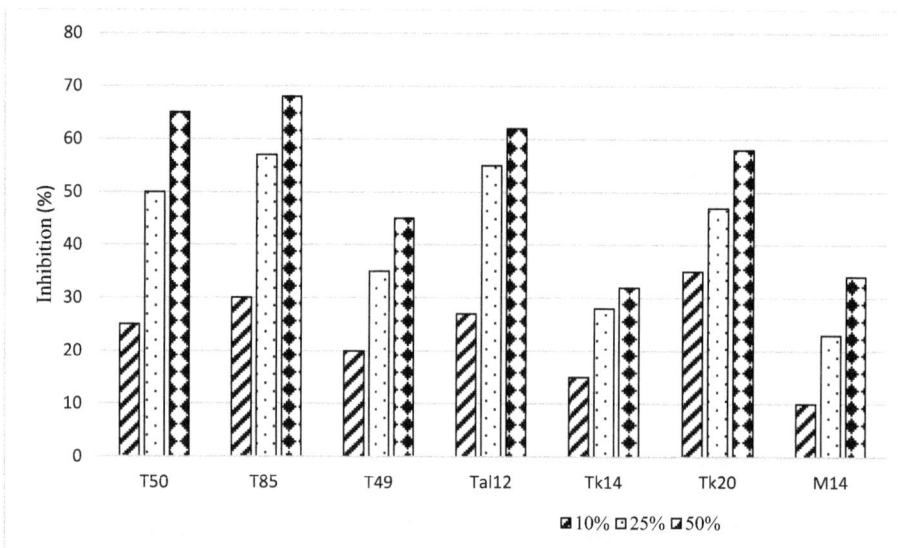

Figure 2. Effect of different concentrations of non-volatile metabolites produced by *Trichoderma* isolates on radial growth of *Fusarium solani* after 6 days of inoculation

CONCLUSIONS

Our results demonstrated the involvement of volatile and non-volatile compounds in the inhibition of *FORL* and *F.solani*

All *Trichoderma* isolate produced volatile compounds having significant effect in reducing the growth of *Fusarium*.

The volatile compounds produced by T49, T50, T85 and Tk14 isolates significantly reduced the radial growth of *FORL*.

The non-volatile compounds produced by filtrate concentrations (25% and 50%) of T85,

T50 and Tal12 inhibited between 50-70% radial growth of *F.solani*.

The present results showed that the ability of secondary metabolites production is different among isolates studied .

ACKNOWLEDGEMENTS

This paper is supported by Ministry of Research and Innovation through Core Programme and was financed from the project PN 16 29- 02-01/2016.

REFERENCES

Ajith P.S., Lakshmidevi N., 2010. Effect of volatile and non-volatile compounds from *Trichoderma* spp against *Colletotrichum capsici* incitant of anthracnose on bell peppers. Nature and Science, 8(9): 265-269.

Barakat F. M., Abada K. A., Abou-Zeid N. M., El-Gammal Y. H. E., 2014. Effect of volatile and non-volatile compounds of *Trichoderma* spp. on *Botrytis fabae* the causative agent of faba bean chocolate spot. American Journal of Life Science, 2(6-2):11-18.

Borrero C., Trillas M.I., Ordovas J., Tello J.C., Aviles M., 2004. Predictive factors for the suppression of *Fusarium* wilt of tomato in plant growth media. Phytopathology, 94:1094-1101.

Calistru C., McLean M., Berjak P., 1997. *In vitro* studies on the potential for biological control of *Aspergillus flavus* and *Fusarium moniliforme* by *Trichoderma* species: A study of the production of extracellular metabolites by *Trichoderma* species. Mycopathologia, 137: 115-124.

Chen L.J, Sun S.Z., Miao C.P, Wu K., Chen Y.W., Xu L.H., Guan H.L., Zhao L.X., 2015. Endophytic *Trichoderma gamsii* YIM PH 30019: a promising biocontrol agent with hyperosmolar, mycoparasitism and antagonistic activities of induced volatile organic compounds on root-rot pathogenic fungi of *Panax notoginseng*. Journal of Ginseng Research, 38:1-10.

Dennis C., Webster J., 1971a. Antagonistic properties of species groups of *Trichoderma* II. Production of volatile antibiotics.Trans. Br. Mycol. Soc., 57: 41-48.

Dennis C., Webster J., 1971b. Antagonistic properties of species groups of *Trichoderma* I. Production of non-volatile antibiotics. Trans. Brit. Mycol. Soc., 57: 25-39.

Dubey S.C., Tripathi A., Dureja P., Grover A., 2011. Characterization of secondary metabolites and enzymes produced by *Trichoderma* species and their efficacy against plant pathogenic fungi. Indian J Agr. Sci., 81:455–461.

El-Hasan A., Walker F., Buchenauer H., 2008. *Trichoderma harzianum* and its metabolite 6-pentyl-alpha-pyrone suppress fusaric acid produced by *Fusarium moniliforme*. J. Phytopathol,156: 79-87.

Faheem A., Razdan V.K., Mohiddin F.A., Bhat K.A., Sheikh P.A., 2010. Effect of volatile metabolites of *Trichoderma* spp. against seven fungal plant pathogens *in vitro*. J. Phytol, 2(10): 34-3.

Kavitha T., Nelson R., 2013. Exploiting the biocontrol activity of *Trichoderma* spp against root rot causing phytopathogens. ARPN Journal of Agricultural and Biological Science, 8(7):571-574.

Li Y., Sun R., Yu J., Saravanakumar K., Chen J., 2016. Antagonistic and Biocontrol Potential of *Trichoderma asperellum* ZJSX5003 against the maize stalk rot pathogen *Fusarium graminearum*. Indian J Microbiol, 56(3):318–327.

Nagendra B., Prasad A., Kumar M.R., 2011. Effect of non-volatile compounds produced by *Trichoderma* spp. on growth and sclerotial viability of *Rhizoctonia solani*, incitant of sheath blight of rice. Indian J. Fundamental Appl. Life Sci,1 (2) 37-42.

Qualhato F.T., Cardoso-Lopes F.A., Steindorff A.S., Brandao R.S., Amorim Jesuino S.R., Ulhoa C.J., 2013. Mycoparasitism studies of *Trichoderma* species against three phytopathogenic fungi: evaluation of antagonism and hydrolytic enzyme production Biotechnol. Letters, 35:1461–1468.

Reino J.L., Guerrero R.F., Hernandez-Galan R., Collado I.G., 2008. Secondary metabolites from species of the biocontrol agent *Trichoderma*. Pytochemistry Review, 7: 89–123.

Raut I., Doni-Badea M., Calin M., Oancea F., Vasilescu G., Sesan T., Jecu L., 2014a. Effect of volatile and non-volatile metabolites from *Trichoderma* spp. against important phytopathogens. Revista de Chimie, 65(11):1285-1288.

Raut I., Calin M., Vasilescu G., Doni-Badea M., Sesan T., Jecu L., 2014b. Effect of non volatile compounds of *Trichoderma* spp against *Fusarium graminearum, Rhizoctonia solani* and *Pythium ultimum*. Scientific Bulletin. Series F. Biotechnologies, Vol. XVIII 178-181.

Siddiquee S., Cheong B.E, Taslima K., Kausar H., Hasan M.M., 2012. Separation and identification of volatile compounds from liquid cultures of *Trichoderma harzianum* by GC-MS using three different capillary columns. Journal of Chromatographic Science 50:358-367.

Sreedevi B., Devi C., Saigopal D.V.R., 2011. Isolation and screening of effective *Trichoderma* spp. against the root rot pathogen *Macrophomina phaseolina* Journal of Agricultural Technology, 7(3): 623-635.

Srivastava R., Khalid A., Singh U.S., Sharma A.K., 2010. Evaluation of arbuscular mycorrhizal fungus, Fluorescent *Pseudomonas* and *Trichoderma harzianum* formulation against *Fusarium oxysporum* f. sp. *lycopersici* for the management of tomato wilt. Biol. Control, 53:24-31.

Tapwal A., Singh U., Teixeira de Silva J.A., Singh G., Grag S., Kumar R., 2011. *In vitro* antagonism of *Trichoderma viride* against five phytopathogens. Pest Technology, 5 (1): 59–62.

Vinale F., Ghisalberti E.L., Sivasithamparam K., Marra R., Ritieni A., Ferracane R., Woo S., Lorito M., 2009. Factors affecting the production of *Trichoderma harzianum* secondary metabolites during the interaction with different plant pathogens. Letters in Applied Microbiology, 48:705–11.

Vinale F., Marra R., Scala F., Ghisalberti E.L., Lorito M., Sivasithamparam K., 2006. Major secondary metabolites produced by two commercial *Trichoderma* strains active against different phytopathogens. Lett Appl. Microbiology, 43:143–148.

Zhang F., Yang X., Ran W., Shen Q., 2014. *Fusarium oxysporum* induces the production of proteins and volatile organic compounds by *Trichoderma harzianum* TE-5. FEMS Microbiol.Lett.359:116-123

22

ANTAGONISTIC ACTIVITY OF INDIGENOUS *PSEUDOMONAS* ISOLATES AGAINST *FUSARIUM* SPECIES ISOLATED FROM ANISE

Aleksandra STANOJKOVIĆ-SEBIĆ[1], Snežana PAVLOVIĆ[2],
Mira STAROVIĆ[3], Radmila PIVIĆ[1], Zoran DINIĆ[1],
Zorica LEPŠANOVIĆ[4], Dragana JOŠIĆ[1]

[1]Institute of Soil Science, Teodora Drajzera 7, 11000 Belgrade, Serbia
[2]Institute for Medicinal Plant Research "Dr Josif Pančić", 11000 Belgrade, Serbia
[3]Institute for Plant Protection and Environment, 11000 Belgrade, Serbia
[4]Military Medical Academy, 11000 Belgrade, Serbia
Corresponding author email: astanojkovic@yahoo.com

Abstract

Fusarium species are widely distributed and responsible for several plant diseases in different medicinal plants. Fungi of this genera cause very important economic losses in Serbian plantation. Antibiotic production by plant-associated microorganisms represents an environmentally acceptable method of disease control, especialy in cultivation of medicinal and aromatic plants. Among the plant growth promoting bacteria (PGPB), Pseudomonas have been recognized as the most frequent antagonists of plant fungal pathogens and antibiotic producers. This is probably due to the widely distribution of this diverse group of bacteria in temperate soils and their often predomination among bacteria from plant rhizosphere. In this study, we examined the antifungal activity of eleven indigenous Pseudomonas isolates (PB4, PB5, K38, Q34, PBA12, PD5, C7, C8, Q16P, K29 and K35) against eight phytopathogenic fungi belonging to genus Fusarium (Fusarium tricinctum, F. sambucinum, F. equiseti, F. heterosporum, F. sporotrichioides, F. semitectum, F. verticillioides and F. oxysporum), which had infected anise (Pimpinella anisum L., fam. Apiaceae), using in vitro growth inhibition tests. The obtained results demonstrated that all Pseudomanas isolates showed more or less pronounced antifungal activity, whereby the most pronounced activity was observed for K29 and K35 strains. F. oxysporum and F. verticillioides showed the highest sensitivity to antibiotic-producing Pseudomanas isolates. In general, it has been concluded that studied Pseudomonas isolates have potential in controlling plant diseases caused by Fusarium spp., whereby the bacterial isolates with the highest inhibitory potential will be selected for further experiments.

Key words: Pseudomonas, Fusarium spp., Pimpinella anisum, antifungal activity.

INTRODUCTION

The use of chemical fertilizers and pesticides has caused an incredible harm to the environment. These agents are both hazardous to animals and humans and may persist and accumulate in natural ecosystems. An answer to this problem is replacing chemicals with biological approaches, which are considered more environment friendly in the long term. One of the emerging research area for the control of different phytopathogenic agents is the use of biocontrol plant growth promoting rhizobacteria (PGPR), which are capable of suppressing or preventing the phytopathogen damage (Nihorembere et al., 2011).

Phytopathogenic fungi, as the most common plant pathogens, are capable of infecting different types of plant tissues. Among the main aims in agriculture is finding adequate strategies for their suppression. One of these strategies is biological control (biocontrol) of plant diseases that relies on the use of natural antagonists of phytopathogenic fungi (Heydari and Pessarakli, 2010).

A special place among the natural antagonists of phytopathogenic fungi belongs to rhizobacteria that show beneficial effects on plant growth (PGPR) (Zehnder et al., 2001). These bacteria use various mechanisms for their action: production of plant hormones, asymbiotic fixation of N_2, antagonism towards phytopathogenic microorganisms and the ability to solubilize mineral phosphates and other nutrients (Cattelan et al., 1999). Different isolates of fluorescent *Pseudomonas* species take prominent place in this respect.

Consequently, these isolates have been intensively studied.

Fluorescent *Pseudomonas* species are present in temperate and tropical soils, often dominant among rhizobacteria (Ayyadurai et al., 2007). They belong to PGPR because of the ability to colonize the roots of plants and stimulate growth by decreasing the frequency of diseases. Suppression of diseases includes the inhibition of pathogens by competition and/or by antagonism (Couillerot et al., 2009). The prominent feature of fluorescent *Pseudomonas* species is the production of antibiotics as inhibitory compounds that play a role in the suppression of diseases caused by phytopathogenic fungi (Haas and Défago, 2005). One of the best-studied antibiotics of fluorescent *Pseudomonas* species are phenazines, nitrogen-containing heterocyclic compounds (Fernando et al., 2005). The only known natural producers of phenazines are bacteria (Pierson III and Pierson, 2010).

Fluorescent *Pseudomonas* species are capable of inhibiting the phytopathogenic fungi that belong to genus *Fusarium* (Showkat et al., 2012). *Fusarium* spp. are a widespread cosmopolitan group of fungi and commonly colonize aerial and subterranean plant parts, either as primary or secondary invaders. Some species are common in soil and it is rare to find necrotic root of a plant in most agricultural soils that is not colonized by at least one *Fusarium* sp. (Nelson et al., 1983).

One of the hosts of *Fusarium* spp. is anise (*Pimpinella anisum* L., fam. Apiaceae). Anise is an aromatic plant which is used in traditional medicine (especially its fruits) as carminative, aromatic, disinfectant and galactagogue (Shojaii and Abdollahi Fard, 2012).

The aim of this study was to examine the antifungal activity of eleven indigenous *Pseudomonas* isolates against the eight phytopathogenic fungi belonging to genus *Fusarium*, which had infected anise (*Pimpinella anisum* L., fam. Apiaceae).

MATERIALS AND METHODS

The antifungal activity of the following indigenous *Pseudomonas* isolates: PB4, PB5, K38, Q34, PBA12, PD5, C7, C8, Q16P, K29 and K35, was examined against the phytopathogenic fungi belonging to genus *Fusarium* (*F. oxysporum, F. tricinctum, F. sambucinum, F. equiseti, F. heterosporum, F. sporotrichioides, F. semitectum, F. verticillioides*), which had infected anise (*Pimpinella anisum* L., fam. Apiaceae).

The examination was conducted on Waksman agar plates nutrient media, using *in vitro* inhibition tests. Overnight cultures of the tested *Pseudomonas* isolates, optimized to $1 \cdot 10^7$ cfu/ml were used to examine the influence of extracellular metabolites of cells (1 ml of cultures was centrifuged at 13000 rpm for 10 min and resuspended in the same volume of sterile saline solution).

Inoculation of Waksman nutrient media with the tested cultures of *Pseudomonas* isolates was done near the edges of Petri dishes and mycelia of the studied *Fusarium* species were placed in the center. Control variants contained only mycelia of *Fusarium* species on Waksman agar plates.

Observation and the measuring of zones of growth inhibition of mycelia around bacterial colonies were performed after seven days of incubation at 25°C (Nair and Anith, 2009). The percentage of growth inhibition of mycelia of *Fusarium* species was calculated by the formula: % Inhibition = [(Control - Treatment)/Control] x 100 (Ogbebor and Adekunle, 2005).

RESULTS AND DISCUSSIONS

Due to the soil-borne nature of the diseases caused by *Fusarium* species the use of chemical methods for the control of disease is rarely successful. Inconsistencies in biocontrol under varying environmental conditions have been a common limitation of soil-borne pathogens. The present research was conducted to evaluate the efficacy of indigenous *Pseudomonas* isolates against these pathogens.

Table 1 displays the data on *in vitro* antifungal activity of selected *Pseudomonas* sp. isolates toward *Fusarium* species, which had infected anise.

The obtained results imposed that all *Pseudomonas* isolates showed more or less pronounced antifungal activity, whereby the mycelial growth of *Fusarium* species was inhibited in the range of 3.33% (for

Pseudomonas isolates PB4 and PB5 toward *F. tricinctum*) to 77.78% (for *Pseudomonas* isolates K29 and K35 toward *F. oxysporum*).

The highest percentage of growth inhibition was caused by *Pseudomonas* isolates K29 (from 35.71% toward *F. equiseti* to 77.78% toward *F. oxysporum*) and K35 (from 37.50% toward *F. semitectum* to 77.78% toward *F. oxysporum*).

The lowest percentage of inhibition was caused by the following *Pseudomonas* isolates: PB4 (from 3.33% toward *F. tricinctum* to 48.89% toward *F. oxysporum*), PB5 (from 3.33% toward *F. tricinctum* to 51.11% toward *F. oxysporum*), PBA12 (from 13.33% toward *F. tricinctum* to 51.11% toward *F. verticillioides*),

PD5 (from 13.33% toward *F. tricinctum* to 53.33% toward *F. verticillioides*).

In general, *F. oxysporum* and *F. verticillioides* showed the highest sensitivity to antibiotic-producing *Pseudomanas* isolates.

Antifungal activity of indigenous *Pseudomonas* isolates was also confirmed in other investigation (Jošić et al., 2012). In addition, *in vitro* assays in previous studies (Velusamy et al., 2011; Shojaii and Abdollahi Fard, 2012) revealed high sensitivity of *F. oxysporum* to *Pseudomonas* sp. as in the present research. As pronounced by other authors (Karimi et al., 2012), PGPR can be used in the biocontrol of phytopathogens.

Table 1. Antifungal activity of selected *Pseudomonas* sp. isolates toward *Fusarium* species (F1 - *Fusarium tricinctum*; F2 - *F. sambucinum*; F3 - *F. equiseti*; F4 - *F. heterosporum*; F5 - *F. sporotrichioides*; F6 - *F. semitectum*; F7 - *F. verticillioides*; F8 - *F. oxysporum*)

Pseudomonas sp. isolates	*Fusarium* species							
	F1	F2	F3	F4	F5	F6	F7	F8
PB4	3.33[*]	25.71	21.43	20.00	35.56	18.75	42.22	48.89
PB5	3.33	25.71	28.57	8.00	40.00	37.50	42.22	51.11
K38	23.33	17.14	28.57	24.00	33.33	43.75	55.56	44.44
Q34	16.67	17.14	21.43	24.00	33.33	43.75	51.11	51.11
PBA12	13.33	14.29	28.57	8.00	22.22	37.50	51.11	48.89
PD5	13.33	5.71	21.43	12.00	22.22	43.75	53.33	46.67
C7	43.33	34.29	21.43	24.00	44.44	12.50	57.78	51.11
C8	36.67	28.57	21.43	20.00	44.44	25.00	53.33	51.11
Q16P	56.67	42.86	35.71	40.00	66.67	37.50	60.00	71.11
K29	66.67	54.29	35.71	44.00	64.44	43.75	66.67	77.78
K35	60.00	54.29	42.86	40.00	64.44	37.50	66.67	77.78

[*]Inhibition (in %)

CONCLUSIONS

Biological control of *Fusarium* species, one of the most aggressive isolates from medicinal plants in Serbia, isolated from anise, is an ecological method of plant protection.

Our investigation confirmed more or less pronounced antifungal activity of all tested *Pseudomonas* isolates, whereby the most pronounced activity was observed for K29 and K35 strains. Regarding the *Fusarium* species, the highest sensitivity to antibiotic-producing *Pseudomanas* isolates was observed for *F. oxysporum* and *F. verticillioides*.

Our findings impose that the studied *Pseudomonas* isolates have potential in controlling plant diseases caused by *Fusarium* spp., whereby the bacterial isolates with the highest inhibitory potential will be selected for further experiments.

ACKNOWLEDGEMENTS

This research was supported by the Ministry of Education, Science and Technological Development, Republic of Serbia, Project III46007.

REFERENCES

Ayyadurai N., Ravindra Naik P., Sakthivel N., 2007. Functional characterization of antagonistic fluorescent pseudomonads associated with rhizospheric soil of rice (*Oryza sativa* L.). Journal of Microbiology and Biotechnology, 17(6):919-927.

Cattelan A.J., Hartel P.G., Fuhrmann J.J., 1999. Screening for plant growth-promoting rhizobacteria to promote early soybean growth. Soil Science Society of America Journal, 63(6):1670-1680.

Couillerot O., Prigent-Combaret C., Caballero-Mellado J., Moënne-Loccoz Y., 2009. *Pseudomonas fluorescens* and closely related fluorescent pseudomonads as biocontrol agents of soil-borne phytopathogens. Letters in Applied Microbiology, 48(5):505-512.

Fernando W.G.D., Nakkeeran S., Zhang Y., 2005. Biosynthesis of antibiotics by PGPR and its relation in biocontrol of plant diseases. In: Siddiqui Z.A. (Ed.), PGPR: Biocontrol and Biofertilization, Springer Science, Dordrecht, The Netherlands, 67-109.

Haas D., Défago G., 2005. Biological control of soil-borne pathogens by fluorescent pseudomonads. Nature Reviews Microbiology, 3(4):307-319.

Heydari A., Pessarakli M., 2010. A review on biological control of fungal plant pathogens using microbial antagonists. Journal of Biological Sciences, 10(4):273-290.

Jošić D., Protolipac K., Starović M., Stojanović S., Pavlović S., Miladinović M., Radović S., 2012. Phenazines producing *Pseudomonas* isolates decrease *Alternaria tenuissima* growth, pathogenicity and disease incidence on cardoon. Archives of Biological Sciences, Belgrade, 64(4):1495-1503.

Karimi K., Amini J., Harighi B., Bahramnejad B., 2012. Evaluation of biocontrol potential of *Pseudomonas* and *Bacillus* spp. against *Fusarium* wilt of chickpea. Australian Journal of Crop Science, 6(4):695-703.

Nair C.B., Anith K.N., 2009. Efficacy of acibenzolar-S-methyl and rhizobacteria for the management of foliar blight disease of amaranth. Journal of Tropical Agriculture, 47(1):43-47.

Nelson P.E., Toussoun T.A., Marasas W.F.O., 1983. *Fusarium* species: an illustrated manual for identification. Pennsylvania State University Press, University Park, Pennsylvania, USA, 193.

Nihorembere V., Ongena M., Smargiass M., 2011. Beneficial effect of the rhizosphere microbial community for plant growth and health. Biotechnologie, Agronomie, Société et Environnement, 15(2):327-337.

Ogbebor N., Adekunle A.T., 2005. Inhibition of conidial germination and mycelial growth of *Corynespora cassiicola* (Berk and Curt) of rubber (*Hevea brasiliensis* Muell. Arg.) using extracts of some plants. African Jornal of Biotechnology, 4(9):996-1000.

Pierson III L.S., Pierson E.A., 2010. Metabolism and function of phenazines in bacteria: impacts on the behavior of bacteria in the environment and biotechnological processes. Applied Microbiology and Biotechnology, 86(6):1659-1670.

Shojaii A., Abdollahi Fard M., 2012. Review of pharmacological properties and chemical constituents of *Pimpinella anisum*. ISRN Pharmaceutics, 510795.

Showkat S., Murtaza I., Laila O., Ali A., 2012. Biological control of *Fusarium oxysporum* and *Aspergillus* sp. by *Pseudomonas fluorescens* isolated from wheat rhizosphere soil of Kashmir. IOSR Journal of Pharmacy and Biological Sciences, 1(4):24-32.

Velusamy P., Ko H.S., Kim K.Y., 2011. Determination of antifungal activity of *Pseudomonas* sp. A3 against *Fusarium oxysporum* by high performance liquid chromatography (HPLC). Agriculture, Food and Analytical Bacteriology, 1:15-23.

Zehnder G.W., Murphy J.F., Sikora E.J., Kloepper J.W., 2001. Application of rhizobacteria for induced resistance. European Journal of Plant Pathooogy, 107(1):39-50.

23

PRELIMINARY DATA ON PESTS OCCURRENCE ON SAFFLOWER CROP UNDER GREENHOUSE CONDITIONS

Aurora DOBRIN[1], Roxana CICEOI[1], Vlad Ioan POPA[1], Ionela DOBRIN[2]

[1]University of Agronomic Sciences and Veterinary Medicine of Bucharest, Laboratory of Diagnosis and Plant Protection of Research Center for Studies of Food Quality and Agricultural Products, 59 Marasti Blvd, District 1, Bucharest, Romania
[2]University of Agronomic Sciences and Veterinary Medicine of Bucharest,
59 Marasti Blvd, District 1, Bucharest, Romania

Corresponding author email: roxana.ciceoi@gmail.com

Abstract

Safflower is a very important oilseed crop with multiple uses in food, pharmaceutic, cosmetic, varnish and paint industry. The quality of safflower flowers and seed yield rely on successful and integrated pest management solution. The safflower crop was tested in Romania in the last decades and the results show a high adaptability of this species to our pedoclimatic conditions, which led to a yield higher than 2000kg/ha, for the studied varieties. Our observations were carried out in the Research Greenhouse of University of Agronomic Science and Veterinary Medicine from Bucharest, in 2016, on Carthamus tinctorius L., which represent the first attempt in growing safflower in greenhouse conditions in our country. The most damaging pests that were identified were Tetranychus urticae Koch. and Trialeurodes vaporariorum Westood, two threatening polyphagous pest all around the world, causing serious yield losses, especially in greenhouses. Their presence was associated with the high temperature in June and July. Besides the introductory review of the most important safflower pest in the world, this study gives new and important insights about the safflower crop response to associated greenhouse pests and allowed a closer analyze using the electronic microscopy of the white fly eggs and eggs hatching characteristics. Our observation on safflower might be a premise for new control strategies against the white fly.

Key words: Carthamus tinctorius, Tetranychus urticae, Trialeurodes vaporariorum, mature eggs, eggs hatching.

INTRODUCTION

Safflower (*Carthamus tinctorius* L.) is a very important multipurpose crop, that is used in producing herbal drugs, cosmetics, natural food coloring, high oleic and high linoleic oil, natural dye, oil for painting and animal feed.
Safflower is a drought and salt-resistant oilseed crop. Safflower grows on various types of soils, but the highest yield may obtain on clay, sandy lands with neutral pH, well drained in depth. Safflower crops on lands with excessive humidity have a high risk of contacting specific diseases and pests (Kizil et al., 2008; Amini, 2014; Hussain et al., 2016).
The most important pest insects feeding inside the flower heads of safflower mentioned until the present moment are *Acanthiophilus helianthi*, *Chaetorellia carthami*, *Trellia luteola*, *Larinus flavescens*, *Larinus liliputanus* and *Helicoverpa peltiger* (Saeidi and Nur Azura, 2011).

The safflower fly, *Acanthiophilus helianthi* Rossi (*Diptera*: *Tephritidae*) is one of the most important pests of safflower allover the world. Larval feeding can causes important losses, disrupt plant metabolism with a negative influence on number of flower buds and decrease quality and quantity of the crop yield (Riaz and Sarwar, 2013; Saeidi et al., 2013). In Iran both safflower fly and Silver Y moth cause major damage to the safflower crops. (Saeidi et al., 2011; Esfahani et al., 2012).
Safflower anthodium can be infested with the tephritid fruit flies as *Acanthiophilus helianthi* Rossi and *Chaetorellia carthami* Stackelberg, but there are five associated hymenopteran parasitoid species, namely *Bracon luteator* Spinola; *B. intercessor* Nees (*Braconidae*); *Eurytoma varicolor* Silvestri; *E. rtellii* Domenichini (*Eurytomidae*) and *Torymus rubi* (Schrank) (*Torymidae*) that are keeping the tephiritid populations under control (Basheer et al., 2014).

In Iraq, Israel, and Kirgizstan, the species from the genus *Chaetorellia* (*Diptera*: *Tephritidae*), especially the *C. carthami* Stackelberg was reported as a safflower pest. (Saeidi et al., 2015).

The most important pest insects feeding outside the safflower anthodium are *Oxycarenus pallens*, *Oxycarenus hyalipennis, Lygus* sp. (Saeidi and Nur Azura, 2011; Esfahani et al., 2012).

The most important pest insects feeding on the whole safflower plant are *Uroleucon compositae*, *Pleotrichophorus glandolosus*, *Brachycaudus helichrysi*, *Neoaliturus fenestratus*, *Euscelis alsius*, *Macrosteles laevis*, *Psammotettix striatus*, *Circulifer haematoceps*, *Thrips tabaci*, *Aeolothrips collaris*, *Haplothrips sp*, *Helicoverpa peltigera* (Saeidi and Nur Azura, 2011; Esfahani et al., 2012). The safflower aphid (*Uroleucon compositae* Theobald) is the major safflower pest in India because in high infestations can damage the crop completely. The yield losses of safflower due to aphids are reported to be 24,2 - 72%. (Esfahani et al., 2012; Singh and Nimbkar, 2016). Among the 36 species of pests damaging safflower in India, the safflower aphid, the capsule borer, *Helicoverpa armigera* (Hubner) and leaf eating caterpillar, *Perigea capensis* (Walker) are considered to be the most important pests of the crop (Esfahani et al., 2012; Saeidi et al., 2015).

In the Mediterranean region there were reports about *Acanthiophilus helianthi*, *Heliothis peltigera* SchiV. (*Noctuidae*), *Chaetorellia carthami* Stackelberg, *Ch. jaceae* R.D., *Terellia luteola* Wiedemann, *Urophora mauritanica* Macquart (*Tephritidae*), *Larinus grisescens* Gyll., *Larinus syriacus* Gyll., *Larinus orientalis* Cap., and *Larinus ovaliformis* Cap. (*Curculionidae*) on the Xower heads; and *Lixus speciosus* Mill. (*Curculionidae*), Agapanthia sp. (*Cerambycidae*), four *Chloridea* spp., *Plusia gamma* L. (*Noctuidae*), *Pyrameis cardui* L. (*Nymphalidae*), and *Cassida palaestina* Reiche (*Chrysomelidae*) damaging the safflower (Smith et al., 2006).

In central and northern Greece, *Botanophila turcica* (*Diptera*: *Anthomyiidae*) was reported recently for the first time on safflower. The larvae of this fly tunnel through the rosette meristem and root of the developing host plant, causing deformation of the developing leaves and occasionally plant losses. *B. turcica* has been reported to attack only rosettes of the invasive saffron thistle *Carthamus lanatus* L. and has, therefore, been suggested as a potential biological control agent of *C. lanatus* (Tsialtas et al., 2013).

Other reported pests with low impact are scarab beetle *Epicometis hirta*), Egyptian cotton leaf, *Spodoptera littoralis*), wireworms, *Limonius* spp., cotton boll worm, *Heliothi obsoleta Lasioderma serricorne, Stegobium penliceum and Trogodema* (Esfahani et al., 2012).

Another important pest category is represented by mites. They cause damage by sucking cell contents from leaves. At first, the damage shows up as a stippling of light dots on the leaves; sometimes the leaves take on a grey, yellow or bronze colour. Necrotic spots occur in the advanced stages of leaf damage. Spider mites are highly polyphagous pests (Godfrey, 2011; Fasulo and Denmark, 2016). *Tetranychus urticae* Koch (*Acari*, *Tetranychidae*) is an notorious mite species causing serious yield losses almost all over the world. It is considered to be a temperate zone species, but it is also found in the subtropical regions (Esfahani et al., 2012; Fasulo and Denmark, 2016; Jiao et al., 2016; Rector et al., 2016).

For an integrated pest management a very important key are the natural enemies that limit pests. In Egypt, the safflower capsule fly is attacked by three species of parasitoid wasps from the families of *Eulophidae* (*Pronatalia* sp.), *Torymidae* (*Antistrophophlex conthurnatus*) and *Pteromalidae* (*Pteromalus* sp.) (Saeidi et al., 2015).

The source of tolerance to aphids can be present in the locally available germplasm. Aphid resistance in safflower is reported to be under the control of both additive and nonadditive gene actions with a predominance demonstrated for nonadditive gene action (Singh and Nimbkar, 1993). Breeding for aphid resistance has been initiated recently in India since it is the most economical, time-tested, and eco-friendly method for controlling aphids. Aphid tolerant safflower keeps the environment safe by way of avoiding chemical usage (Esfahani et al., 2012; Singh and Nimbkar, 2016).

Important genera include the predatory mites, *Amblyseius*, *Metaseiulus*, and *Phytoseiulus*; the lady beetles, *Stethorus picipes*; the minute pirate bugs, *Orius*; *Scolothrips sexmaculatus, Leptothrips*; and the lacewing larvae, *Chrysopa*. *Galendromus occidentalis*. In greenhouses, the ghost ant, *Tapinoma melanocephalum* (Fabricius), a pest in itself, was also reported as a significant predator (Godfrey, 2011; Fasulo and Denmark, 2016).

Other pests control strategies could be: a very good sanitation, that is a key for controlling pests in greenhouses, weed control, clean up all debris from previous crops, temporary quarantine and inspection of all plants upon arrival from other greenhouse, and regular monitoring of stock plants used for propagation, seed selection and proper seed rate, respecting proper sowing time, varietal selection and crop rotation; insect grow regulators are a least-toxic pesticide control option for pests, bio rational pesticides and fertilizers levels (Greer and Diver, 1999; Gupta and Gupta, 2016).

The aim of this work is to present the associate arthropod pests identified on safflower crop grown in greenhouse in 2016.

MATERIALS AND METHODS

The observations were made in the Research Greenhouse of University of Agronomic Sciences and Veterinary Medicine from Bucharest on *Carthamus tinctorius* L. crop in greenhouse. For our country, it is a novelty to obtain safflower crop in the greenhouse. The sowing was done in 19 April 2016 and the harvest on 12 July 2016. During the growing period, we observed several pests affecting the leaves and anthodia. Both the pests and the infested leaves were collected in entomological jars once a week and after each inspection, the pests were immediately analysed at the stereomicroscope. After drying at room temperature, the samples with whitefly eggs exuvia were kept in laboratory, in plastic Petri dishes. The safflower infected leaves and anthodia were analysed with a Leica S8 APO stereomicroscope and with the Scanning Electron Microscope SEM FEI Inspect S50. For both observation methods, there is no sample preparation needed.

RESULTS AND DISCUSSIONS

Trialeurodes vaporariorum eggs analyse

The first greenhouse whitefly eggs were noticed immediately after the leaf emergence. Heavy infestations, between 27 and 48 eggs/cm^{-1} have been observed on 10 May, at 20 days after sowing (figure 1), in the 8 real leaves phenological growing stage.

Figure 1. Heavily infested safflower leaves with greenhouse whitefly eggs

The leaves were also very soon infested by *Tetranychus urticae* Koch (figure 2), the silk webbing on infested leaves being easily detectable and the spherical and translucent eggs being visible at the stereomicroscope. A density of 2 to 5 adult mites and 6 to 12 mite eggs on cm^{-1} has been estimated at the same phenological growing stage.

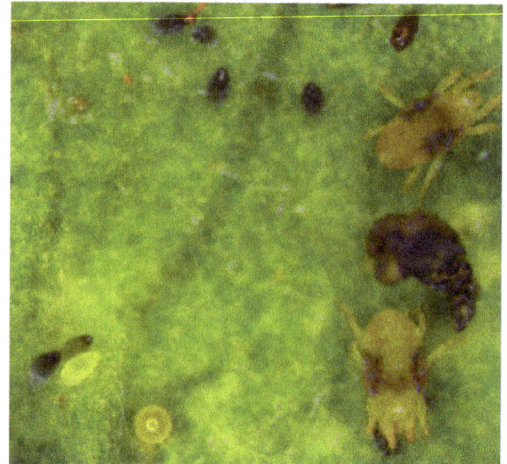

Figure 2. *Tetranychus urticae* eggs, adults and greenhouse whitefly eggs

In the literature it is often cited that the eggs of greenhouse whitefly are pale yellow when first laid and turn to darker colour, until black before hatching (figure 3).

We found no data about the whitefly eggshell neither about its description or its opening structures. This fact is usually undetected, as the leaves are usually covered by sooty mould.

The humidity and temperature conditions correlated with the morpho-anatomical safflower cuticle allowed *T. vaporariorum* eggs to remain on the lamina after the eggs hatching, so that we could observe and analysed the eggs shells opening with the Scanning Electron Microscope (figures 4 and 5).

Figure 3. Newly laid greenhouse whitefly eggs (translucent colour) and more mature ones (dark colour)

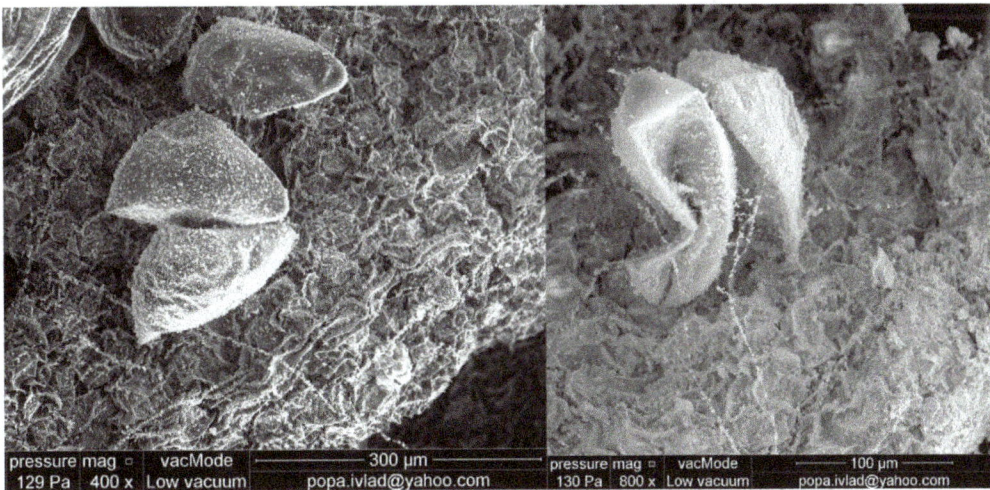

Figure 4. Greenhouse whitefly eggs and details of the safflower lower cuticle

Figure 5. The opening pattern of greenhouse whitefly eggs during hatching

The observed longitudinal opening of the egg shell could offer new insights for the integrated control measures against the greenhouse whitefly. On our knowledge, this aspect hasn't been discussed so far, so further studies are needed.

CONCLUSIONS

The humidity and temperature conditions proved to be an important factor, high temperatures facilitating the observation of new morphological aspects. In the same time, the morpho-anatomical safflower cuticle surface proved to be a perfect medium in preserving the whitefly egg shells.
This new information about the pests associated with *Carthamus tinctorius* L. crop are very useful in the context of food and nutritional safety and sustainable production of safflower in our country and new strategies are required to raise safflower productivity sustainably.

REFERENCES

Amini H., Arzani A., Karami M., 2014. Effect of water deficiency on seed quality and physiological traits of different safflower genotypes. Turkish Journal of Biology Vol.38, Issue: 2, 271-282, DOI: 10.3906/biy-1308-22.

Basheer A., Asslan L., Abdalrazaq F., 2014. Survey of the Parasitoids of the Tephritid Fruit Flies of the Safflower *Carthamus tinctorius* (*Asteracea*) in Damascus, Syria. Egyptian Journal of Biological Pest Control, Vol.24, 169-172.

Esfahani M.N., Alizadeh G., Zarei Z., Esfahani M.N., 2012. The Main Insect Pests of Safflower on Various Plant Parts in Iran. Journal of Agricultural Science and Technology A 2, 1281-1289, ISSN 1939-1250.

Fasulo T.R., Denmark H. A., 2016. Twospotted Spider Mite. *Tetranychus urticae* Koch (*Arachnida: Acari: Tetranychidae*). EENY150 , DPI Entomology Circular 89, Entomology and Nematology Department, UF/ IFAS Extension, 1-5.

Godfrey L. D., 2011. Spider Mites. PEST NOTES, Statewide Integrated Pest Management Program, Publication 7405, UC Statewide Integrated Pest Management Program, University of California, Davis, CA 95616, available online at: http://ipm.ucanr.edu/PMG/PESTNOTES/pn7405.html.

Greer L., Diver S., 1999. Integrated pest management for greenhouse crops.Pest management systems guide ATTRA 8 0 0 -3 4 6 -9 1 4 0, 1-34.

Gupta R.D., Gupta S.K., 2016. Chapter 1 – Strategies for Increasing the Production of Oilseed on a Sustainable Basis. Breeding Oilseed Crops for Sustainable Production. Opportunities and Constraints. 1–18, http://dx.doi.org/10.1016/B978-0-12-801309-0.0000 1-X.

Hussain M.I., Lyra D.A, Farooq M., Nikoloudakis N., Khalid N., 2016. Salt and drought stresses in Safflower: a review. Agronomy for Sustainable Development Vol. 36, Issue: 1, DOI: 10.1007/s13593-015-0344-8.

Jiao R., Xu C.X., Yu L.C., He X.Z., Qiao G.Y., He L.M., Li L.T., 2016. Prolonged coldness on eggs reduces immature survival and reproductive fitness in *Tetranychus urticae* (*Acari: Tetranychidae*). Systematic and Applied Acarology, 21(12), 1651-1661. http://doi.org/10.11158/saa.21.12.6; ISSN 1362-1971 (print), ISSN 2056-6069 (online).

Kizil S., Cakmak O., Kirici S., Inan M., 2008. A comprehensive study on safflower (*Carthamus tinctorius* L.) in semi-arid conditions. Biotechnol Biotec Eq. 22: 947- 953.

Rector B.G., Czarnoleski M., Skoracka A., Lembicz M., 2016. Change in abundance of three phytophagous mite species (*Acari: Eriophyidae, Tetranychidae*) on quackgrass in the presence of choke disease. Exp Appl Acarol.; 70: 35–43. doi: 10.1007/s10493-016-0060-3.

Riaz M., Sarwar M., 2013. A New Record of Safflower Fly *Acanthiophilus helianthi* (Rossi) of Genus *Acanthiophilus* Becker in Subfamily *Tephritinae* (*Diptera: Tephritidae*) from the Fauna of Pakistan. Research & Reviews: Journal of Agriculture and Allied Sciences, ISSN: E 2347-226X, P 2319-9857, 39-44.

Saeidi K.,Nur Azura A., 2011. A survey on pest insect fauna of safflower fields in the Iranian Province of Kohgiloyeh and Boyerahmad. African Journal of Agricultural Research Vol. 6(19), pp. 4441-4446, DOI: 10.5897/AJAR11.966, ISSN 1991-637X.

Saeidi K., Nur Azura A., Omar D., Abood F., 2011. Pests of safflower (*Carthamus tinctorious* L.) and their natural enemies in Gachsara, Iran. South Asian Journal of Experimental Biology, Vol 1, No 6 (2011) ISSN: 2230-9799, 286-291.

Saeidi K., Nur Azura A., Omar D., Abood F., 2013. Population Dynamic of the Safflower Fly, *Acanthiophilus Helianthi* Rossi (*Diptera: Tephritidae*) in Gachsaran Region, Iran. Entomol Ornithol Herpetol 2013, 2:1 Volume 1, Issue 1, ISSN: 2161-0983 http://dx.doi.org/10.4172/2161-0983.1000103.

Saeidi K., Mirfakhraei S., Mehrkhou F., Valizadegan O., 2015. Biodiversity of Insects Associated with Safflower (*Carthamus tinctorius*) Crop in Gachsaran. Iran. Journal of Entomological and Acarological Research, Vol 47, No 1, 26-30, DOI: 10.4081/jear.2015.1910.

Singh V., Nimbkar N., 2016. Chapter 7–Safflower. Breeding Oilseed Crops for Sustainable Opportunities and Constraints Production, 149-167, ISBN: 978-0-12-801309-0, http://dx.doi.org/10.1016/B978-0-12-801309-0.00007-0.

Smith L., Hayat R., Cristofaro M., Tronci C., Tozlu G., Lecce F., 2006. Assessment of risk of attack to saffower by *Ceratapion basicorne* (*Coleoptera*: *Apionidae*), a prospective biological control agent of *Centaurea solstitialis* (*Asteraceae*). Biological Control 36, 337–344.

Tsialtas I.T., Michelsen V., Koveos D.S., 2013. First report of *Botanophila turcica* (*Diptera*: *Anthomyiidae*) on safflower *Carthamus tinctorius* L. in Greece. Journal of Biological Research-Thessaloniki, Vol 19, 80-82 ISSN : 1790-045X.

DP 03: Morphological Identification of Spider Mites (*Tetranychidae*) Affecting Imported Fruits, 2014 available online at: https://www.nappo.org/files/3714/3782/0943/DP_03 _Tetranychidae-e.pdf.

24

EFFECTS OF DIFFERENT IRRIGATION TREATMENTS ON QUALITY PARAMETERS OF CUT CHRYSANTHEMUM

Arif TURAN[1], Yusuf UCAR[1], Soner KAZAZ[2]

[1]Süleyman Demirel University, Agricultural Faculty, Farm Structure and Irrigation Department, 32260, Isparta-Turkey, Email: arifturan43@gmail.com. yusufucar@sdu.edu.tr

[2]Ankara University, Agricultural Faculty, Horticulture Department, 06100, Dışkapı-Ankara-Turkey. Email: skazaz@ankara.edu.tr

Corresponding author email: yusufucar@sdu.edu.tr

Abstract

This study was carried out to determine the effects of different irrigation intervals and water amounts on yield and quality parameters of cut chrysanthemum. Spray cut chrysanthemum (cv. 'Bacardi') plant was used as a plant material. Class A pan was placed in the greenhouse to determine the amount of irrigation water values. Irrigation treatments consisted of three irrigation intervals (I_1: 2-, I_2: 4-, and I_3: 6-day) and four crop-pan coefficients (k_{cp1}: 1.20=T_1, k_{cp2}: 0.90=T_2, k_{cp3}: 0.60=T_3, and k_{cp4}: 0.30=T_4). The irrigation water amounts applied to the experimental treatments ranged from 249.7 to 517.9 mm, and seasonal evapotranspiration ranged from 340.9 to 560.5 mm. Different irrigation water amounts and irrigation intervals had statistically significant effects on flower stem length, stem diameter, stem weight, the number of flowers, the vase life and root length of chrysanthemum. Stem length varied between 52.36-79.81 cm, stem diameter varied between 4.62-7.69 mm, stem weight varied between 32.48-123.61 g and root length varied between 18.88-24.22 cm. The optimum irrigation scheduling was T_1I_1, in which the longest flower stem and the highest stem weight were obtained.

Key words: Chrysanthemum, Class A Pan, Evapotranspiration, Irrigation interval, Water deficit.

INTRODUCTION

The total production area of ornamental plants worldwide is 1.573.167 ha according to the data of 2013. Some 651.800 ha of it is composed of cut flowers and pot plants. The important production regions according to land areas are Asia, North America, Europe, South America, Africa, and the Middle East. The continent with the largest production area for cut flowers and pot plants worldwide is Asia-Pacific (468.000 ha) (Anonymous, 2013). Chrysanthemum is one of the major cut flowers in the world. The demand for the flower reached 35% of the overall market request, second only to roses (Steen, 2010). As in all plants, irrigation is an essential practice for chrysanthemum growing, but its adequate handling has been neglected by growers, resulting in growing loss and consequent productivity and quality decreases in the final product (Farias et al., 2009). In order to irrigate more extensive areas with the available water resources, such factors as soil,

plant, and water resource must be taken into consideration. In addition, the values of plant water consumption under either sufficient or deficient water conditions should be known throughout the growing season of plants and water-yield relationships should be formed accordingly. These data can be obtained by making a large number of investigations for each plant (Doorenbos and Kassam, 1979). To generate the data concerned, Conover (1969), Harbaugh et al. (1985), Parnell (1989), Kiehl et al. (1992), Schuch et al. (1998), Rego et al. (2004), Conte e Castro et al. (2005), Fernandes et al. (2006), Budiarto et al. (2007), Farisa et al. (2009), Waterland et al. (2010) and Villalabos (2014) made investigations on irrigation and flower quality in the chrysanthemum plant. The majority of the investigations concerned are in the form of pot studies, and they are studies in which the plant quality was determined in different soil moisture tensions. Unlike the above-mentioned studies, this study aimed to determine the effects of different irrigation intervals and water amounts on yield and quality parameters in the chrysanthemum plant under greenhouse conditions in the Mediterranean climatic zone.

MATERIALS AND METHODS

The research was conducted in a polyethylene-covered greenhouse of 255 (6 m x 42.5 m) m^2 on the Research and Application Farm of the Faculty of Agriculture at Süleyman Demirel University (lat. 37.83° N, long. 30.53° E, altitude 1,020 m) in 2011 (in Isparta, Turkey). Some characteristics of the greenhouse soil (in 0- to 50-cm depths) were as follows: texture: clay loam; bulk density: 1.32-1.41 g cm^{-3}; field capacity: 24.80-27.01%; permanent wilting point: 7.08-8.51%, and total available water holding capacity in 0- to 50-cm soil depths: 123.6 mm (Table 1).

Table 1. Some Properties of the Soil in the Greenhouse Soil

Soil Depth	FC		WP		BD	AWHC	
cm	%	mm	%	mm	g cm^{-3}	%	mm
0-25	24.80	81.8	7.08	23.4	1.32	17.7	58.4
25-50	27.01	95.2	8.51	30.0	1.41	18.5	65.2
Total		177.0		53.4			123.6

FC: Field capacity, WP: Wilting pointh, BD: Bulk Density, AWHC: Available water holding capacity.

The mean daily temperature ranged from 20 to 30°C in the greenhouse but from 15 to 25°C outside the greenhouse in 2011. The relative humidity was 70-80% in the greenhouse but 50-70% outside the greenhouse (Figure 1) (DMI, 2011). Spray cut chrysanthemum (*Chrysanthemum morifolium* cv. 'Bacardi') was used as the plant material in the research. Uniform rooted cuttings were planted on 20 June 2011 into plots (1-m length, 1-m width) with five rows (20×12.5 cm spacing, 40 plants/m^2), and each plot contained 40 plants. Plants were grown under long day (LD) conditions until the plant height reached 0.3 m, followed by short day (SD) period up to harvesting. SD (08:00-17:00) period was enforced by using a blackout screen (Kofranek, 1980; Kazaz et al., 2010; Lin et al., 2011). Fertilization was applied to each treatment at equal amounts as follows: (ppm): N: 200, P: 20, K: 150, Ca: 80, Mg: 25, Fe: 3.0, Mn: 0.5, Cu: 0.02, Zn: 0.05, B: 0.5, Mo: 0.01 (Yoon et al 2000). Standard cultivation practices for flower bud removal, supporting system, disease

and pest control as used for commercial standart spray cut chrysanthemum production in Turkey were employed for growing the crops during the experiment. The practice of pinching was not applied to the plants in the study.

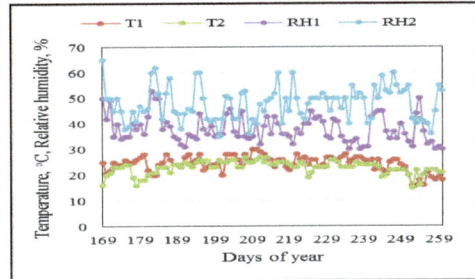

Figure 1. Temperature and relative humidity values at inside and outside of greenhouse. (T_1 and RH_1: Temperature and relative humidity at inside of greenhouse, T_2 and RH_2: Temperature and relative humidity at outside of greenhouse)

All the water which evaporated from Class A Pan (CAP) for 25 days after planting (DAP) was applied equally to all the treatments as irrigation water to ensure the root development and full survival of seedlings. The application of different irrigation intervals and irrigation water amounts was initiated 25 days after planting (DAP). The irrigation treatments were arranged as three different irrigation intervals (I_1:2-, I_2:4-, and I_3:6-day) and 4 different crop-pan coefficients (T_1:$k_{cp}1$=1.20, T_2:$k_{cp}2$=0.90, T_3:$k_{cp}3$=0.60, and T_4:$k_{cp}4$=0.30). The experiment was conducted according to the randomized plots experimental design with 3 replications.

The CAP placed in the greenhouse was utilized to determine the irrigation water amounts (Allen et al., 1998). Irrigation treatments were based on the evaporation data (Ep, mm) obtained from a CAP located inside the greenhouse. Irrigation water amount was calculated using Equation 1. Irrigation water was applied to each irrigation treatment by measuring it with a water meter.

$$IW= A \times E_{pan} \times k_{cp} \qquad [1]$$

In the equation, IW denotes the irrigation water (mm), A the plot area (m^2), E_{pan} the amount of cumulative evaporation at the irrigation interval (mm), and k_{cp} the crop-pan coefficient.

The irrigation applications were carried out with the drip irrigation method. The dripper and lateral space was 20 cm, whereas the dripper discharge was 2 l/h (Uçar et al., 2011). The soil water content in the root zone of the plant was measured by means of watermarks (Irrometer, Model; Watermark200SS, USA). The watermarks were placed in the depths of 15 and 40 cm from the soil surface, with each experimental plot containing 2 watermarks. The watermarks were calibrated, and the calibration equation was found as $Pw=48.626 \times kPa^{-0.302}$ ($R^2=0.97$) (Pw: Soil moisture as the percentage of dry weight; kPa: Watermark readings).

Plant water consumption was computed by using Equation 2 according to the fundamental principle of water budget by considering the soil moisture values measured before each irrigation application (Allen et al., 1998):

$$ET = I + P - RO - DP + CR \pm \Delta SF \pm \Delta SW \quad [2]$$

In the equation, ET denotes plant water consumption (mm), I the irrigation water applied (mm), P precipitation (mm), RO surface runoff (mm), DP deep percolation (mm), CR capillary rise (mm), ΔSF subsurface runoff (mm), and ΔSW the change in the moisture content of the root zone (mm). Precipitation (P), surface runoff (RO), capillary rise (CR) and subsurface runoff (ΔSF) were neglected in the calculations. The chrysanthemum plant is shallow-rooted, and its effective root depth is about 30 cm. Thus, the values of the watermark placed at the 15th cm were taken into consideration in the computations of plant water consumption, while the deep percolations were examined from the watermark at the 40th cm in depth. The moisture values above the field capacity in the root zone of the plant were considered deep percolation. When the watermark reading limit was exceeded (199 kPa), soil samples were collected from the experimental treatments and the soil moisture content was determined with the gravimetric method.

The flowers were harvested on September 15, 2011, when the flower in the middle opened completely and the surrounding flowers displayed full development. Stem length, stem diameter, stem weight, the number of flowers, vase life and root length were determined.

The obtained data were subjected to an analysis of variance by means of MINITAB 16 computer software, and the LSD Multiple Comparison test was applied by means of MSTAT-C computer software in order to compare the averages.

RESULTS AND DISCUSSIONS

Irrigation Water and Evapotranspiration: The values of irrigation water, percolated water, and plant water consumption applied according to the experimental treatments are provided in Table 2. All the water which evaporated from CAP for 25 DAP (160.3 mm) was applied to all treatments as irrigation water to ensure the root development and full survival of seedlings. During the growing period, 517.9, 428.5, 339.1 and 249.7 mm of water was applied to treatments T_1, T_2, T_3, and T_4, respectively. The total amount of evaporation was 458.3 mm (Table 2).

Table 2. Evaporation and irrigation water values in the treatments

Treatments	Evaporation (from CAP)	IW_1	IW_2	IW
T_1		160.3	357.6	517.9
T_2	458.3*	160.3	268.2	428.5
T_3		160.3	178.8	339.1
T_4		160.3	89.4	249.7

*: 160.3 mm of evaporation had been measured before making a transition to scheduled irrigation. IW_1: The irrigation water amount applied to the experimental treatments before making a transition to scheduled irrigation (mm), IW_2: The irrigation water amount applied according to the k_{cp} coefficients after making a transition to scheduled irrigation (mm), IW: Total irrigation water (mm).

The values of evapotranspiration measured according to the experimental treatments are presented in Figure 2. The highest evapotranspiration took place in T_1 treatments, where 1.2 times the water which evaporated from the evaporation pan was applied as the irrigation water (I_1T_1: 560.5 mm, I_3T_1: 553.4 mm, and I_2T_1: 552.7 mm), followed by T_2 (I_1T_2: 504.6 mm, I_2T_2: 491.3 mm, and I_3T_2: 486.7 mm), T_3 (I_1T_3: 427.4 mm, I_2T_3: 423.1 mm, and I_3T_3: 415.2 mm), and T_4 (I_1T_4: 345.7 mm, I_2T_4: 342.5 mm, and I_3T_4: 340.9 mm). In the study, it is seen that the evapotranspiration varied at different irrigation intervals even if the same amount of irrigation water was applied. Since the soil surfaces of the treatments with short irrigation intervals were continuously wet, the values of plant water consumption measured in these treatments were higher. The deep percolation values ranged from 27.72 to 18.90 mm according to the experimental treatments. Since the irrigation water amount applied at the beginning of the experiment was higher than the values of plant water consumption, the majority of deep percolation (18.90 mm) had taken place before making a transition to scheduled irrigation. After making a transition to scheduled irrigation, no deep percolation occurred in treatments T_3 and T_4 (Figure 2).

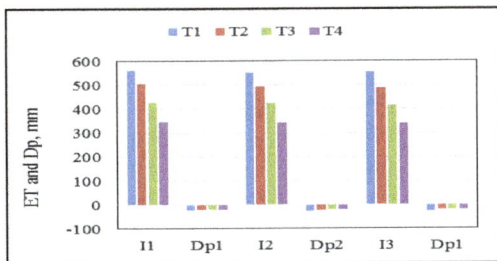

Figure 2. Evapotranspiration and deep percolation values according to the experimental treatments (T_1, T_2, T_3 and T_4: The level of irrigation water amount; I_1, I_2 and I_3: Irrigation interval; Dp_1, Dp_2 and Dp_3: Deep percolation; ET: Evapotranspiration)

Quality Parameters: Different irrigation intervals and irrigation water amounts significantly affected stem length, stem diameter, number of flower, stem weight, and root length at %1 level, and affected vase life at %5 level (Table 3).

Stem length: Growing conditions (temperature, light, photoperiod, relative humidity, CO_2, and planting density) have significant effects on plant height, the number of flowers per plant, and flower size that are among the important quality criteria in chrysanthemum (Carvalho & Heuvelink, 2001). The main climatic factor used to control plant height is temperature (Carvalho et al., 2002), and the optimum temperature requirement of chrysanthemum is 18-20°C (van der Ploeg & Heuvelink, 2006). In this study, however, the temperature of the interior of the greenhouse ranged from 20 to 30°C.

The increased irrigation water amount caused significant increases in stem length. The longest stem (75.03 cm on average) was recorded in T_1 treatments, to which the largest amount of irrigation water was applied, followed by T_2 (70.99 cm), T_3 (65.21 cm), and T_4 (57.22 cm) with the smallest amount of irrigation water application. The highest stem length in T_1 with the largest amount of irrigation water application was obtained from I_1 (79.81 cm). Stem lengths were 73.75 and 71.52 cm in I_3 and I_2, respectively (Table 4). The differences between I_1 and I_2 and between I_1 and I_3 were statistically significant, while the difference between I_2 and I_3 was insignificant. Likewise, the longest stem in T_3 treatments was determined in I_2 (67.93 cm), followed by I_1 (67.18 cm) and I_3 (60.52 cm). Although there was a difference between I_1 and I_2, it was not statistically significant. In T_2 and T_3, the highest stem length was obtained from I_1. When the same amount of water was applied at different irrigation intervals, its effect on stem length was not the same. This led to an interaction between the irrigation intervals and irrigation amount. The longest stem (70.96 cm) was recorded in I_1 with a 2-day irrigation interval, followed by I_2 (67.44 cm) and I_3 (62.94 cm). It was also stressed by Harbaugh et al. (1985) that stem length generally increased with an increase in the irrigation water amount applied. In the study concerned, they stated that the plant height was 62 cm in the treatment of 0.16 cm/day, 76 cm in the treatment of 0.24 cm/day, 86 cm in the treatment of 0.31 cm/day, 92 cm in the treatment of 0.40 cm/day, and 97 cm in the treatment of 0.47 cm/day. These data are in agreement with our results. Stem length is one of the most important indicators for the market value in chrysanthemum, as in the other cut flower species. Although varying by country, the branches which are 70-80 cm long are generally preferred in chrysanthemum (Kazaz, 2010). Chrysanthemums are classified when their flower stem lengths are 60-75 cm according to the classification in the USA but when their flower stem lengths are 50-70 cm according to the classification in England (Mengüç, 1996). All experimental treatments according to the English classification and all treatments other than I_2T_4 and I_3T_4 according to the American classification are included in the good class.

Table 3. The results of variance analysis of mean values of spray cut chrysanthemum quality parameters

Variation Sources	df	Mean square error					
		Stem length	Stem diameter	Stem weight	Number of flower per plant	Vase life	Root length
Replication (R)	2	19.45	0.16	74.70	6.02	5.86	0.10
Irrigation Interval (II)	2	194.22**	4.31	552.40**	54.28**	12.19*	2.96**
Irrigation water amount (IW)	3	537.44**	5.37**	8211.40**	440.99**	10.74*	38.09**
II*IW	6	22.60**	0.13**	191.80*	13.43**	4.05**	0.26
Error	22	6.03	0.12	60.0	4.00	1.89	0.41
Total	35	2307.86	28.49	28358.90	1612.22	106.22	130.97

df: degrees of freedom,*P<0.05 and **P<0.01

Stem weight: Stem weight ranks first among the most important quality criteria which are taken as the basis in the marketing of chrysanthemums worldwide. At the flower auction of the Netherlands (FloraHolland), the stem weights range from 45 to 105 g depending on the stem length (65, 70, and 72 cm) in spray chrysanthemums. In addition, the optimal stem weight is 70 g (Anonymous, 2010).

In terms of the irrigation water amounts, the highest stem weight was found in T_1 (108.72 g on average), followed by T_2 (83.79 g on average), T_3 (60.09 g on average), and T_4 (38.67 g on average). In terms of the irrigation intervals, the highest stem weight was obtained from I_1 (78.89 g on average), followed by I_2 (74.07 g on average) and I_3 (65.49 g on average). I_1 and I_2 were not statistically different in either T_1 or T_2, ranking first and second in terms of stem weight, while I_3 was different (P<0.05). In a study reported concerning stem weight, Harbaugh et al. (1985) determined the plant stem weight as 93 g in the treatment of 0.16 cm/day, 127 g in the treatment of 0.24 cm/day, 138 g in the treatment of 0.31 cm/day, 149 g in the treatment of 0.40 cm/day, and 168 g in the treatment of 0.47 cm/day in different daily irrigation water applications. Higher stem weights were obtained in the treatments treated with a large amount of water in our study, which is similar to these results. It was reported that the stem weight of a chrysanthemum of a high quality ranged from 25 to 105 g according to the classification criterion of the Dutch flower auction (Anonymous, 2010), while it was reported in Japan that the chrysanthemums

which were 80-90 cm long should weigh 55-100 g (Yoon et al., 2000). Even though the stem weights in all experimental treatments are included in the good class according to the Dutch classification criterion, the stem weights of the flowers obtained from the experimental treatments other than T_4 are included in the good class according to the Japanese classification system.

Number of flower per plant: There were differences in the number of flowers in terms of both irrigation intervals and the irrigation water amounts applied. The difference between T_1 (28.60 flowers/plant) and T_2 (26.99 flowers/plant) was statistically insignificant. On the other hand, T_3 (20.19 flowers/plant) and T_4 (13.31 flowers/plant) were statistically different from each other. The differences in the number of flowers per branch between I_1 (24.64 flowers/plant) and I_2 and between I_1 and I_3 were significant, whereas the difference between I_2 (21.65 flowers/plant) and I_3 (20.53 flowers/plant) was insignificant (P<0.05).

Vase life: Irrigation interval and irrigation water amount had significant effects on vase life (P<0.01). The longest vase life among the irrigation intervals was determined in I_3 (17.82 days on average). However, the difference between I_1 (16.00 days) and I_2 (16.42 days) was insignificant. Although the difference among the irrigation water treatments was significant, there was no linear correlation either between the decrease and increase in irrigation water or between the increase and decrease in vase life. The longest vase life among all experimental treatments was recorded in I_3T_1 (19.67 days).

Table 4. Mean values and significance groups of quality parameters of spray cut chrysanthemum

Treatments	Stem length, cm			Average	Stem diameter, cm			Average
	I_1	I_2	I_3		I_1	I_2	I_3	
T_1	79.81 a	71.52 bc	73.75 b	75.03 A	7.69	6.77	6.68	7.05 A*
T_2	74.29 b	73.57 b	65.12 de	70.99 B	7.10	6.73	5.98	6.61 B
T_3	67.18 d	67.93 cd	60.52 fg	65.21 C	6.74	5.78	5.72	6.08 C
T_4	62.58 ef	56.73 g	52.36 h	57.22 D	6.11	5.03	4.62	5.25 D
Average	70.96 A	67.44 B	62.94 C		6.91 A	6.08 B	5.75 C	
$LSD_{0.01}$	LSD_I:2.079, LSD_T:2.401, LSD_{IxT}:4.158				LSD_I:0.295, LSD_T:0.340			
	Stem weight, gr				Vase life, day			
T_1	123.61 a	101.00 b	101.54 b	108.72 A	16.00 de	17.67 a-d	19.67 a	17.78 A
T_2	85.82 c	95.24 bc	70.31 d	83.79 B	14.00 e	16.33 cd	17.00 b-d	15.78 B
T_3	60.27 d	62.36 d	57.64 de	60.09 C	18.33 a-c	16.00 de	18.67 ab	17.67 A
T_4	45.86 ef	37.69 fg	32.48 g	38.67 D	15.67 de	15.67 de	16.33 cd	15.89 B
Average	78.89 A	74.07 A	65.49 B		16.00 B	16.42 B	17.82 A	-
$LSD_{0.01}$	LSD_I:6.558, LSD_T:7.573, LSD_{IxT}:13.120				LSD_I:1.158, LSD_T:1.337, LSD_{IxT}:2.317			
	Number of flower per plant, numbers				Root length, cm			
T_1	29.41 ab	26.30 bc	30.09 a	28.60 A	18.93	18.57	19.54	19.01 D
T_2	29.24 ab	26.29 bc	25.44 c	26.99 A	20.67	21.14	21.82	21.21 C
T_3	23.38 cd	21.20 d	15.98 ef	20.19 B	21.71	22.68	23.07	22.49 B
T_4	16.53 e	12.80 fg	10.60 g	13.31 C	23.40	23.93	24.22	23.85 A
Average	24.64 A	21.65 B	20.53 B	-	21.18 B	21.58 B	22.16 A	
$LSD_{0.01}$	LSD_T:1.693, LSD_T:1.955, LSD_{IxT}:3.387				LSD_I:0.5429, LSD_T:0.6268			

*The difference among the averages is significant at 5% level.

Stem diameter: Stem diameter is an important criterion for determining the resistance of a branch. In terms of the irrigation water amounts, the thickest stem occurred in T_1 (7.05 mm on average), followed by T_2, T_3, and T_4. The stem diameters in these treatments were 6.61 mm, 6.08 mm, and 5.25 mm on average, respectively. When the irrigation intervals were examined, the thickest stem as found in I_1 (6.91 mm), followed by I_2 (6.08 mm on average) and I_3 (5.75 mm on average). The thickest stem was recorded in I_1T_1 (7.69 mm), to which the largest amount of water was applied at a 1-day interval, whereas the thinnest stem was obtained from I_3T_4 (4.62 mm) with the smallest amount of water application at a 6-day interval.

Root length: The longest root length was determined at T_4 (23.85 cm) which was applied the least irrigation water and also the lowest root length was determined at T_1 (19.01 cm) applied highest irrigation water. When the consideration irrigation interval, the highest root length was measured I_3 (22.16 cm). On the contrary, other quality parameters, when irrigation water amount and irrigation interval increased, the root length was reduced. In other words, the lowest root length was obtained from T_1 which had the most irrigation water amount and highest irrigation interval. It is thought, due to the plants had not water stress and could easily get water from soil, the root growth in T_1 was better than applied less irrigation water such as T_4 or T_3.

CONCLUSIONS

The irrigation water amounts applied under experimental conditions ranged from 249.7 to 517.9 mm, while the plant water consumption varied between 340.9 and 560.5 mm. The large amount of irrigation water applied increased the plant water consumption, and its effect was reflected positively on the quality parameters;

hence, a longer stem and a higher stem weight were obtained from the treatments with a larger amount of water application and high plant water consumption accordingly. In the study, the longest stem (79.81 cm), the thickest stem diameter (7.69 mm) and the highest stem weight (123.61 g) were obtained from combination I_1T_1, while the largest number of flowers per plant (30.09 flowers/plant) and the longest vase life were determined in combination I_3T_1 (19.67 days). When stem length and stem weight are particularly considered in terms of marketable products, the optimum irrigation scheduling is I_1T_1. When it is intended to save water, treatment I_1T_2 or I_2T_2 might be selected as the irrigation scheduling. In this case, the reduction in flower quality will be minute.

ACKNOWLEDGEMENTS

This research was supported as a Master Thesis by Suleyman Demirel University Unit of Scientific Research Project (Project No: 2934-YL-11).

REFERENCES

Allen R.G., Pereria L.S., Raes D., Smith M., 1998. Crop Evapotranspiration. Guidelines for Computing Crop Water Requirements. Food and Agriculture Organization Irrigation and Drainage Paper No. 56. Rome, Italy.

Anonymous, 2010. Product specification chrysanthemum indicum group. Ducth flower auction association (VBN) 7, Hollanda. http://www.vbn.nl/ (Erişim tarihi: 06.01.2013)

Anonymous, 2013. International Statistics Flowers and Plants 2013. AIPH-Union Fleurs 2013, 61, 165, The Netherlands.

Budiarto K., Sulyo Y., Dwi E.S.N., Maaswinkel R.H.M., 2007. Effects of irrigation frequency and leaf detachment on chrysanthemum grown in two types of plastic house. Indonesian Journal of Agricultural Science 8(1): 39-42.

Carvalho S.M.P. Heuvelink E., 2001. Influence of greenhouse climate and plant density on external quality of chrysanthemum (Dendranthema grandiflorum (Ramat.) Kitamura): first steps towards a quality model. Journal of Horticultural Science & Biotechnology 76: 249-258.

Carvalho S.M.P., Heuvelink E., Cascais R., Van Kooten O., 2002. Effect of day and night 408 temperature on internode and stem length in chrysanthemum: is everything explained 409 by DIF? Annals of Botany 90:111-118.

Conover C.A., 1969. Responses of Pot-Grown Chrysanthemum morifolium 'Yellow Delaware' to Media, Watering and Fertilizer Levels. Proceeding of the Florida State Horticultural Society, 82, 425-429.

Conte e Castro A.M., Macedo Junior E.K., Zigiotto D.C., Braga C.L., Sornberger A., Baldo M., Grisa S., Bianchini M.I.F., Sausen C., 2005. Effect of Irrigation Layers on Varities of Chrysanthemum for Cuttuing and on Soil Characteristics. Scientia Agraria Paranaensis, 4(2), 75-80.

DMİ. 2011. Devlet Meteoroloji İşleri Genel Müdürlüğü. Isparta MeteorolojiBölge Müdürlüğü Kayıtları. Isparta.

Doorenbos J., Kassam A.H., 1979. Yield Response to Water. Food and Agriculture Organization, Irrigation and Drainage Paper No. 33, 193. Rome.

Farias M.F., De Saad J.C.C, Denise M.C., 2009. Effect of soil-water tension on cut chrysanthemum floral quality and longevity. Applied Research & Agrotechnology 2(1): 141-145.

Fernandes A.L.T., Folegatti M., Pereira A.R., 2006. Valuation of different evapotranspiration estimate for (Chrysanthemum spp.) cultivated in plastic greenhouse. Irriga 11(2): 139-149.

Harbaugh B.K., Stanley C.D., Price J.F., 1985. Tricle Irrigation Rates for Chrysanthemum Cut Flower Production. Proceeding of the Florida State Horticultural Society, 98, 110-114.

Kazaz S., Aşkın M.A., Kılıç S., Ersoy N., 2010. Effects of day length and daminozide on the flowering, some quality parameters and chlorophyll content of Chrysanthemum morifolium Ramat. Scientific Research and Essays 5(21): 3281-3288.

Kiehl P.A., Lieth J.H., Burger D.W., 1992. Growth Response of Chrysanthemum to Various Container Medium Moisture Levels. Journal of American Society for Horticultural Science 114(2): 224-229.

Kofranek A.M., 1980. Introduction to Floriculture, Second Edition, Edited by R.A. Larson, Academic Press, 3-45, New York.

Lin L., Li W., Shoa J., Luo W., Dai J., Yin X., Zhou Y., Zhao C., 2011. Modelling the effects of soil water potential on growth and quality of cut chrysanthemum (Chrysanthemum morifolium) Scientia Horticulturae 130: 275-288.

Mengüç A (1996). Kesme Çiçek Yetiştiriciliği 3 (Kasımpatı). Anadolu Üniversitesi Yayınları No. 904, Açıköğretim Fakültesi Yayınları No. 486, 112-126, Eskişehir.

Parnell J.R., 1989. Ornamental Plant Growth Responses To Different Application Rates of Reclaimed Water. Proceedings of the Florida State Horticultural Society, 102, 89-92.

Rego J.L., Viana T.V.A., Azevedo B.M., Bastos F.G.C., Gondim R.S., 2004. Effects of irrigation levels on the chrysanthemum. Agronomic Science Magazine 35(2): 302-310.

Schuch U.K., Redak R.A., Bethke J.A., 1998. Cltivar, fertilizer and irrigation affect vegetative growth and susceptiibility of Chrysanthemum to western flower thrips. J. Amer. Soc. Hort. Sci 123(4): 727-733.

Steen M., 2010. A world of flowers: Dutch flower market and the market of cut flowers. J Appl Hortic 12: 113-121.

Uçar Y., Kazaz S., Aşkın M.A., Aydınşakir K., Kadayıfçı A. Şenyiğit U., 2011. Determination of irrigation water amound and interval for carnation (Dianthus caryophyllus L.) with pan evaporation method. Hortscience 46(1): 102-107.

Van der Ploeg A., Heuvelink E., 2006. The influence of temperature on growth and development of chrysanthemum cultivars: a review. Journal of Horticultural Science & Biotechnology 81(2): 174–182.

Villalobos R., 2014. Reduction of irrigation water consumption in the Colombian Floricculture with the use of tensiometer. http://irrigationtoolbox.com/ReferenceDocuments/TechnicalPapers/IA/2007/P1642.pdf (Erişim tarihi: 31.10.2014).

Waterland N.L., Finer J.J., Jones M.L., 2010. Abscisic acid applications decrease stomatal conductane and delay wilting in drought-stresses chrysanthemums. HortTechnology 20(5): 896-901.

Yoon H.S., Goto T., Kageyama Y., 2000. Developing a nitrogen application curve fır spray chrysanthemum grown ın hydroponic system and ıts practical use ın NFT system. Journal of the Japanese Society for Horticultural Science 69(4): 416-422.

25

CHEMICAL CONSTITUENTS OF THE ESSENTIAL OIL OF *ARTEMISIA SANTONICA* L. (*ASTERACEAE*) ECOTYPES FROM ROMANIA

Monica Luminiţa BADEA, Aurelia DOBRESCU, Elena DELIAN,
Ioana Marcela PĂDURE, Liliana BĂDULESCU

University of Agronomic Sciences and Veterinary Medicine of Bucharest,
59 Marasti Blvd., District 1, Bucharest, Romania
Corresponding author email: badea.artemisia@gmail.com

Abstract

Artemisia L. (Asteraceae) genus includes many species being the largest genus from Anthemideae tribe, as well as one of the largest genus from Asteraceae family. Of these species, A. santonica is a perrenial herb or small shrubs and have numerous uses in various field such as medicine, nutrition and industry (phytochemicals). Due to its characteristic features in terms of chemical composition, and their usefulness, the purpose of this paper is to present recent results regarding the chemical composition of essential oil extacted by hidrodistillation from A. santonica. The plant material were collected from different areas of Romania. The qualitative and quantitative essential oil analysis was performed by gas chromatography/mass spectrometry (GC/MS). The results expressed as percentage show the presence of chemical compounds represented in majority by α -thujona, β -thujona, borneol, eucalyptol and camphor.

Key words: essential oils, Artemisia, Asteraceae, chromatography, ecotype.

INTRODUCTION

Artemisia L. (*Asteraceae*) genus comprised almost 500 species being the largest genus from *Anthemideae* tribe and one of the largest genus from *Asteraceae* family (Bohm and Stuessy, 2001; Watson et al., 2002). Based on ethnobotanic studies, most species of *Artemisia* genus have numerous uses in traditional medicine, nutrition and industry (phytochemicals).

Artemisia santonica is a perennial herb or small shrubs, frequently aromatic. Stems 20-60 cm, cylindrical, woody in the lower part, glabrous or grey-tomentose. The leaves are usually grey-tomentose or glabrescent, rarely white, inferior leaves short petiolate, leaves slightly auriculate at the basis, with ovate lamina- 2-3 pennately-divided, with cu norow lacinia, rounded at the top; capitula small, numerous, usually pendent, in racemose, paniculate or capitate inflorescences, rarely solitary. Involucral bracts are disposed in few rows. Receptacle is flat to hemispherical without scales, sometimes hirsute. Florets are yellow, all tubular. Achenes is obovoid, absent or sometimes a small scarious ring (Figure 1).

The ethnobotanical and bio-chemical studies revealed the fact that *A. santonica* extracts proved to be benefic antioxidant effects, also in the anthelmintic treatment, in disorders of the digestive and urinary tract, and as poultices and infusions calms the cough and the cephalalgia. In the French kitchen is used as a spice. The extracts of *Artemisia* used in large doses can become toxic (Badea and Delian, 2014; Githiori et al., 2006; Tandon et al., 2011).

The objective of this study was to bring new information regarding the composition of essential oil extracted from the species *A. santonica*, collected from different areas (ecotypes) of Romania.

MATERIALS AND METHODS

Artemisia santonica L. plants were harvested during full blossoming, from natural populations in different areas of Romania: Slănic-Prahova (Prahova County), Ocna-Sibiului (Sibiu County), Slobozia (Ialomiţa County) and Plopu (Tulcea County).

The essential oil extraction was realized from *Herba*. Fresh herbal parts of the collected plants were subjected to hydrodistillation for 3h using a Singer-Nickerson equipment to produce oil. The separation and identification of components has been carried out using an Agilent gas chromatograph, equipped with quadruple mass spectrometer detector. A capillary column DB-5

(25 m length x 0.25 mm i.d. and 0.25 µm film thickness) and helium as carrier gas were used. The initial oven temperature was 60°C, then rising to 280°C at a rate of 4°C /min. The NIST spectra bank was used for to identify the essential compounds, which were verified with the Kovats indices.

Figure. 1. Morphological aspect
of *Artemisia santonica* L.

RESULTS AND DISCUSSIONS

GC-MS analyses of the oils were carried out according to a procedure that has been described above. The yield and composition of the essential oils of *Artemisia santonica* ecotypes are presented in Table 1.

The essential oil extracted from the *A. santonica* ecotypes, coming from the 4 ecotypes (Slănic-Prahova, Ocna-Sibiului, Slobozia and Plopu) contained a number of different chemical constituents: 15 (in Plopu) and 33 (in Ocna-Sibiului). Only five substances are common to all 4 ecotypes: α-thujona, β - thujona, eucalyptol, terpinen-4-ol and germacrene D, also three of them are considered as majority.

After Burzo (2008), the major compounds at the species *A. santonica* are represented by eucalyptol, camphor, cis-verbenol, borneol.

Comparing the substances founded in the majority of the essential oils extracted from the 4 ecotypes, was established that α-thujona had a limit of variation between 25.73% (Ocna-Sibiului) and 70.09% (Slobozia) compared to β-thujona whose variation limit was between 5.22% (Plopu) and 17.28% (Slobozia).

The eucalyptol was determined in proportion of 1.64% at the ecotype Plopu and 9.08% at the one from Ocna-Sibiului.

Those three majority substances (α-thujona, β-thujona and eucalyptol) varied widely, depending on the ecotype. Thereby, their total amount had the lowest value at the ecotype Ocna-Sibiului (45.12%), had intermediate values at the ecotypes Slănic-Prahova (58.32%) and Plopu (71.24%), and the highest value was measured at the ecotype Slobozia (91.31% from the total of the substances identified).

A small variation was measured in the case of terpinen-4-ol (0.50% at the Plopu ecotype and 2.73% at the one from Ocna-Sibiului) and at germacren D (0.14% at the Plopu ecotype and 1.14% at the one from Slobozia).

The camphor and the borneol are not present in composition of the essential oil from Slobozia ecotype, but these substances were quite well represented in the composition of the essential oil that came from the other three ecotypes. Thus, the camphor level varied between 4.71% (Plopu) and 24.65% (Slănic-Prahova) and the borneol values ranged between 1,57% at the Slănic-Prahova and 24.13% at the ecotype Ocna-Sibiului (Table 1).

Major compounds from this species (α-thujona, β-thujona, borneol and camphor) represent 66.99% from the total of the substances identified in the essential oil from the Ocna-Sibiului, 77.87% from the Slănic ecotype, 87.37% -Slobozia and 94.43% from the Plopu ecotype.

Table 1. Composition of essential oil extracted from 4 ecotypes of *Artemisia santonica*
(% from total identified substances)

Compounds	Slănic-Prahova	Ocna-Sibiului	Slobozia	Plopu
ethyl methyl butyrate	0.39	-	0.15	-
dimethyl metilen ciclohexane	-	-	0.48	-
santona -trien	0.09	-	-	-
tricyclen	0.30	-	-	-
α-pinene	0.15	0.54	-	-
camphene	4.70	1.93	-	0.35
sabinene	0.31	1.63	0.44	-
β-pinene	0.28	0.33	-	-
dehydro -cineole	-	0.27	-	-
β-felandren	-	-	-	0.14
1- octen-3-ol	-	-	-	0.18
α-terpinene	0.18	0.71	-	-
cimene	-	1.23	0.70	0.12
β -cimen	0.76	-	-	-
eucalyptol	6.67	9.08	3.94	1.64
artemisia ketone	2,87	-	-	-
y-terpinene	-	0,26	0,28	-
terpinolene	-	0,30	-	-
α-terpinolene	0.09	-	-	-
α-thujona	44.80	25.73	70.09	64.56
β-thujona	6.85	10.31	17.28	5.22
izothujol	-	0.33	-	0.22
crisantenona	0.43	-	-	-
trans-pinocarveol	-	1.28	0.42	0.33
camphor	24.65	6.82	-	4.71
sabina ketone	-	0.30	-	-
borneol	1.57	24.13	-	19.94
pinocarvone	0.55	-	-	-
terpinen-4-ol	0.66	2.73	0.66	0.50
α-thujenal	-	0.23	-	-
α-terpineol	-	0.53	-	0.18
myrtenol	0.61	0.67	-	0.28
isobornil formate	-	-	0.25	-
cumin aldehide	0.06	0.11	0.10	-
carvone	0.13	-	-	-
trans-chrysantenyl acetate	0.07	0.51	-	-
bornyl acetate	0.24	0.29	-	-
myrtenil acetate	3.53	0.96	0.16	-
α-copaen	-	0.17	-	-
trans -pinocarvil acetate	-	-	0.14	-
germacrene D	0.39	0.91	1.14	0.14
β-selinene	-	0.19	-	-
elixen	0.15	-	0.92	-
spatulenol	-	0.11	-	-
cariofilen oxid	-	0.15	-	-
tau-muurolol	-	0.53	-	-
α-cadinol	-	0.21	-	-
selinene-4-ol	-	0.33	-	-

Concerning the minority of chemical compounds, can be affirmed that were registered significant differences between the ecotypes. Thereby, the α-pinene, the camphene, the β-pinene, the α-terpinene, the trans-chry-santhenyl acetate, the bornyl acetate and the myrtenyl acetate were determined in plants found in the ecotypes Slănic-Prahova and Ocna-Sibiului, while the sabinene and the cumin aldehyde were founded in the essential oil extracted from the plants located in the Slanic-Prahova, Ocna-Sibiului and Slobozia ecotypes, missing at the ecotype Plopu.

The γ-terpinene has been identified at the plants from Ocna-Sibiului and Slobozia ecotype missing at the plants from Slanic-Prahova and Plopu.

The compounds ethyl methyl butyrate and elixen, have been determined in the essential oil resulted from the plants of the Slănic-Prahova and Slobozia, but were missing from the Ocna-Sibiului and Plopu ecotypes (Table 1).

Also, there were a minority of chemical compounds specific for each ecotype, as much:
- the essential oils of Slănic Prahova ecotypes contained santonia-trien, tricyclen, β-cimen, crisantenona, artemisia ketone, pinocarvone;
-the plants from Ocna-Sibiului ecotype contained dehydro-cineole, terpinolene, sabina ketone, α-thujenal, and β-selinene.
- in Slobozia ecotype contained dimethyl metilen ciclohexane, cimene, isobornyl formate, trans-pinocarvyl acetate, myrtenil acetate;
- the plants from Plopu ecotype contained α-felandren and 1-octen-3-ol;

The plants of *A. santonica* L. are considered halophiles, the essential oil extracted from the plants grown on a saline soil, with a high concentration of NaCl (Ocna-Sibiului and Slănic-Prahova) recorded a greater number of chemical components (27-33), comparative with the others two ecotypes analyzed (15-16).

CONCLUSIONS

The main substances identified in *Artemisia santonica* are represented by α-thujona, β-thujona, borneol, eucalyptol and camphor.

The total of the substances that were a majority varied between: 66.99% at the ecotype Ocna Sibiului, 77.87% at Slănic-Prahova, 87.37% at Slobozia and 94.43% at Plopu ecotype, from the total of all substances identified.

The chemical composition of essential oil extracted from the species *Artemisia santonica* varied depending on the ecotype.

ACKNOWLEDGEMENTS

We warmly thank to Prof. dr. Burzo Ioan from the Faculty of Horticulture, Bucharest

REFERENCES

Badea M.L., Delian E., 2014. *In vitro* antifungal activity of the essential oils from *Artemisia* spp. L. on *Sclerotinia sclerotiorum*, Romanian Biotechnological Letters 19(3): 9345-9352.

Badea M.L., Zamfirache M.M., 2011. Anatomical research on *Artemisia santonica* and *Artemisia scoparia* (*Asteraceae*). Analele ştiinţifice ale Universităţii „Al. I. Cuza" Iaşi, s. II a. Biologie vegetală , LVII (2): 21-24.

Bohm B. A. and Stuessy T.F., 2001. Flavonoids of the Sunflower Family (*Asteraceae*) Springer-Verlag Publishing House, Wein, 831pp.

Bremer K., 1994. *Asteraceae*, cladistics and classification, Timber Press: Portland. Carnegie Institution of Washington: Washington, DC.

Burzo I., Ciocârlan V., Delian E., Dobrescu A., Bădulescu L.,2008. Researches regarding the essential oil composition of some *Artemisia* L. species. Analele Stiinţifice ale Universităţii „Al. Ioan Cuza" Iaşi - Secţiunea II a. Biologie Vegetală, Editura Univ.Alex.Ioan Cuza Iaşi, LIV(2): 86-91.

Ciocârlan V., 2009. Flora ilustrată a României, Editura Ceres, Bucureşti.

Dr.Schar-Common Name: Santonica | Scientific Name: Artemisia Santonica available on: *http://doctorschar.com/archives/artemisia-santonica/*

Githiori J.B., Athanasiadou S., Thamsborg S.M., 2006. Use of plants in novel approaches for control of gastrointestinal helminths in livestock with emphasis on small ruminants, 139(4): 308–320.

Săvulescu T., Ghişa E., Grinţescu I., Guşuleag M., Morariu I., Nyarady E.I., Prodan I., 1964. Flora R.P.R., Editura Academiei Române, vol.IX, Bucureşti.

Tandon V., Yadav A.K., Roy B., Das B., 2011. Phytochemicals as cure of worm infections in traditional medicine system, Emerging Trends in Zoology: 351–378.

Tutin T. G., 1976. *Artemisia* L.: In: Tutin T. G., Burges N. A., Chater A. O., Edmondson J. R., Heywood V. H., Moore D. M., Valentine D. H., Walters S. M., Webb D. A. (Eds.), "Flora Europaea" (Plantaginaceae to Compositae and (Rubiaceae), Cambridge: Cambridge University Press, 4: 178-186.

Valles J. and McArthur E.D., 2001. *Artemisia* Systematics and Phylogeny: Citogenetic and Molecular Insights. In: McArthur, E. D.; Fairbanks, Daniel J., comps. 2001. Shrubland ecosystem genetics and biodiversity: proceedings; 2000 June 13–15; Provo, UT. Proc.RMRS-P-21. Ogden, UT: U.S. Department of Agriculture, Forest Service, Rocky Mountain Research Station.

Watson L. E., Bates P. L., Unwin M. M., Evans T. M., and Estes J.R., 2002. Molecular phylogeny of Subtribe *Artemisiinae (Anthemideae: Asteraceae)*, including *Artemisia* and its allied and segregate genera. BMC Evolutionary Biology. 2: 17(12 PP) Springer-Verlag: Wien.

26

HPTLC FINGERPRINT USE, AN IMPORTANT STEP IN PLANT-DERIVED PRODUCTS QUALITY CONTROL

Corina BUBUEANU, Alice GRIGORE, Lucia PÎRVU

National Institute for Chemical-Pharmaceutical R&D (ICCF-Bucharest),
Vitan Road 112 Sector 3, Bucharest, ROMANIA
Corresponding author email: corina.bubueanu@yahoo.com

Abstract

Romanian flora comprises a significant number of vegetal species, some of these already established in terms of chemical composition and pharmacological potential. Also, currently on the national profile market there are many plant-derived products for internal or external use that contain standardized extracts or herbal powders. So, it is very important for the plant-derived products production process to have quality control methods for raw materials and for finished products. In the context of complex chemical composition of vegetal raw material, depending on the cultivating region, the climate (temperature, humidity, light and wind), the harvest time and plants part used, the chromatographic fingerprint can certify the species, chromatographic fingerprint of vegetal products being one of the most simple and feasible method of quality control. Accordingly, High-performance thin-layer chromatography (HPTLC) has become one of the most important tool for quality control of plant-derived products on basis of its simplicity and accurately, as well. It can serve as a tool for instance, depending on the chromatographic conditions, HPTLC method can identify numerous classes (16) of vegetal compounds. Given these, this paper was aims at presenting HPTLC chromatographic phenolic fingerprint of some valuable Romanian vegetal species as follows: Juglans regia L. - walnut, Morus nigra L. - mulberry tree, Althaea officinalis L.- marshmallow, Carum carvi L.- caraway, Crataegus monogyna Jacq. - hawthorn, Tilia cordata Mill. - linden, Achillea millefolium L. - yarrow determined by HPTLC, species present in many plant-derived products.

Key words: chromatographic phenolic fingerprint, HPTLC, Romania, vegetal species.

INTRODUCTION

Romania has different bio-geographic regions and ecosystems, being considered a link between Europe and Central Asia. The flora and fauna of Romania represent a renewable resource of significant value if protected and exploited on a sustainable basis. Romania possesses about 50 % of Europe's flora and fauna with more than 3,500 plant species (Pârvu, 1997).

With this abundance of resources, the local producers that uses native flora as raw material, are in a continuous expansion. In the same time, the interest of the consumers for the plant-derived products is increasing, given the fact that these products have a lower price and no side effect comparable with synthetic drugs (Stoia and Oancea, 2013).

On the national market there is an important number of plant derived products for both internal and external administration. For the producers is very important to have reproducible raw material in the terms of chemical composition.

Therefore, is very important for the production process to have simple and feasible quality control methods for raw material and for finished products. In the context of complex chemical composition of vegetal raw material, depending on the cultivating region, the climate (temperature, humidity, light and wind), the harvest time and plants part used, the chromatographic fingerprint can certify the species, being an important step in plant-derived products quality control.

The High performance thin layer chromatography (HPTLC) has become one of the most important tools for quality control of plant-derived products on basis of its simplicity and accurately, fingerprint chromatography being accepted by the World Health Organization as an identification and quality evaluation technique for vegetal raw material (Alaerts, et al 2007).

Through, HPTLC technique, depending on the chromatographic conditions, can be identify numerous classes (16) of bioactive compounds. These bioactive compounds are often secondary metabolites, commercially important and find use in numerous plants - derived products. Among these, phenolic compounds occupy one of the first places due to their health promoting properties (Ghasemzadeh and Ghasemzadeh, 2011). The consumption of products that contain phenolic compounds has been often associated with decreased risk of developing several diseases. Phenolic compounds are known for their action in an important number of biological activities with antibacterial, antioxidant and anti-inflammatory properties. Most of the literature data regarding the biological activity of phenolic compounds refers to antioxidant properties witch may be due mainly redox properties, that allow them to act as a reducing agent and as a hydrogen donor (Kamatou et al 2010; Samec, et al 2010; Rice-Evans al 1996).

Given the importance of the these compounds, HPTLC chromatographic fingerprint of raw material can be an important step in quality control of production process.

The species selected for this study are some of the most valuable Romanian species, which are present in the plant derived products (Table 1).

Table1. Plant material

Latin name	Common name	Family	Part used
Juglans regia L	walnut	*Juglandaceae*	*Juglandis folium et pericarpium*
Morus nigra L.	mulberry tree	*Moraceae*	*Mori folium*
Althaea officinalis L	marshmallow	*Malvaceae*	*Althaeae folium*
Carum carvi L.	caraway	*Apiaceae*	*Carvi fructus*
Crataegus monogyna Jacq.	hawthorn	*Rosaceae*	*Crataegi folium cum flores*
Tilia cordata Mill.	linden	*Tiliaceae*	*Tiliae flores*
Achillea millefolium L.	yarrow	*Asteraceae*	*Millefolii flores*

MATERIALS AND METHODS

Raw material - was purchased from the local store in the form of tea products.

Sample preparation: the samples were prepared by extraction with ethanol 50 % (v/v) - vegetal material/ solvent rate -1/15 m/v for 5 minutes at boiling temperature of the solvent. The solution was filtered and frozen until analysis.

HPTLC Analysis for phenols: According to TLC Atlas - Plant Drug Analysis (Wagner, H. and Bladt S. 1996) was performed a densitometric HPTLC analysis for the development of characteristic fingerprint profile for phenolic compounds. 3-3.5µl of the samples and 1-3µl of references substances (10^{-3} M quercetin, rutin, hyperoside, chlorogenic acid, caffeic acid, rosmarinic acid, ferulic acid, apigenin - 7-glucoside - Sigma-Aldrich) were loaded as 10 mm band length in the 20 x 10 Silica gel 60F254 TLC plate using Hamilton-Bonaduz, Schweiz syringe and CAMAG LINOMAT 5 instrument. The mobile phase was constituted of ethyl acetate-acetic acid-formic acid-water 100:11:11:27 (v/v/v/v). After development, plates were dried and derivatized in Natural products–polyethylenglycol reagent (NP/PEG) (Sigma-Aldrich) reagent. The fingerprints were evaluated at 366 nm in fluorescence mode with a WinCats and VideoScan software.

RESULTS AND DISCUSSIONS

Figure 1 shows chromatographic phenolic fingerprint of S1-*Juglandis folium et pericarpium*, S2-*Mori folium,* S3-*Althaeae folium*, S4-*Carvi fructus,* S5-*Crataegi folium cum flores*, S6-*Tiliae flores,* S7-*Millefolii flores* and references substances, S8-ferulic acid, S9-hyperoside, S10- apigenin-7- glycoside, S11-caffeic acid, S12-chlorogenic acid, S13-rutin, S14-rosmarinic acid, S15-quercetin.

Juglans regia, walnut, is a large, deciduous tree that grows to 25-35 m in high. *Juglandis folium et pericarpium* consists in the leaves of the tree and the green husk of the nuts of *Juglans regia.* The tree was brought from Persia by the Romans. Now, grows in the South-East Europe,

East Asia, Himalaya and China. The chemical composition of the leaves and green husk includes phenolic compounds and juglone. The active principles have bactericidal, astringent, slightly hypotensive, hypoglycemic, soothing, healing, emollient, antidiarrheal properties (Istudor 1998; Pârvu, 1997; Cosmulescu and Trandafir, 2011). All parts of *Juglans regia* specie (as green walnuts or husk, shells, kernels, seeds, bark and leaves) are used in plant-derived products, for healthcare improvement (Stampar, et al., 2006).

In this study, the HPTLC fingerprint (S1) of *Jugladis folium et pericarpium* revealed the presents of flavonoid glycosides as orange spots, with hyperoside (rate of flow values Rf~0.67) and avicularin (Rf~0.89) as major compounds. Neochlorogenic acid, as blue spot (Rf~0.58) and kaempferol-3-arbinoside (Rf~0.92), as green spot were also present. (Figure 1, Figure 2). The obtained results were compared with literature data (Wagner and Bladt, 1996).

Ten phenolic compounds contained by leaves, determined by High-performance liquid chromatography with photodiode, were reported: 3- and 5-caffoylquinic acids, 3- and 4-p-coumaroylquinic acids, p-coumaric acid, quercetin 3-galactoside, quercetin 3-pentoside derivative, quercetin 3-arabinoside, quercetin 3-xyloside and quercetin 3-rhamnoside, (Pereira et al., 2007).

Morus nigra, mulberry tree, is a large, deciduous tree that grows to 12-15 m in height. *Mori folium* consists in the leaves of the tree. It is found in East, West and South East Asia, South Europe, South of North America and in some areas of Africa. In medicinal, economical, industrial and domestic fields, mulberries have enormous importance. *Mori folium* is commonly used for sudorific, antidiarrheal, adjuvant in the treatment of diabetes, myocardial dystrophy activities (Watson and Dallwitz, 2007; Pârvu, 1997).

This qualitative study determined that *Mori folium* chromatographic profile (S2) shows the presence of rutin as orange spot (Rf~0.41), chlorogenic acid as blue spot (Rf~0.51), hyperoside as orange spot (Rf~0.68), apigenin-7-glycoside as orange spot (Rf~0.75) (Figure 1, Figure 3).

Mori folium phenolic compounds identify by HPLC are p- hydroxybenzoic acid, vanillic acid, chlorogenic acid, syringic acid, p-coumaric acid, m-coumaric acid (Memon et al., 2010).

Althaea officinalis, marshmallow, is a perennial plant with erect, woody stems, 60-120cm high. The plant is indigenous to western Asia and Europe, and is naturalized in the USA. *Althaeae folium* are the leaves of marshmallow that contain as major chemical compounds, phenols. The leaves are used for emollient, antidiarrheal and soothing actions (Pârvu, 1997; WHO Monograph, 2003).

In our study, in marshmallow leaves (S3) were identified apigenin-7-glucoside (Rf~0.75), ferulic acid (Rf~0.94) and another two blue major spots that according to Wagner and Bladt, (1996) and based on the relationship spot color - Rf are caffeic acid derivates (Figure 1).

Carum carvi L. is one of the oldest spices cultivated in Europe. Caraway is growing on 20-30 cm stems. *Carvi fructus* are the fruits of the plants that have as active principles volatile oils and phenolic compounds. Carvi fructus acts as a carminative, against spasmodic gastro-intestinal complains, irritable stomach, indigestion, lack of appetite and dyspepsia and relieving flatulent colic (Pârvu 1997; Thippeswamy, et al., 2013).

In our results phenolic compounds revealed in *Carvi fructus* (S4) are chlorogenic acid (Rf~0.51), rosmarinic acid (Rf~0.89) as blue spots and hyperoside (Rf~0.67) as orange spots (Figure 1, Figure 4).

Determined by HPLC, caraway contained a mixture of phenolic acids including gallic acid, catechuic acid, caffeic acid, cinnamic acid, ferulic acid and flavonols such as quercetin and kaempferol (Thippeswamy and Rajeshwara 2014).

Crataegus monogyna Jacq., hawthorn, is a shrub or a small tree of 5-14 m height. It is found in Europe, North Africa and western Asia. *Crataegi folium cum flores* are the leaves and flowers of the hawthorn that have as active principles phenolic compounds. The plant is used for cardiac insufficiency (Pârvu, 1997).

On our hydroalcoholic extract (S5) only vitexin -2-O-rhamnoside (Rf~0.43) as yellow - green spot, hyperoside (Rf~0.68) as orange spot, caffeic acid (Rf~0.93) as blue spot were identified.

Methanolic extracts of hawthorn were reported to have as phenolic compounds rutin, vitexin - 2-O-rhamnoside, caffeoyl quinic acids, hyperoside, luteolin-5-O-glucoside, vitexin and cafeic acid, determined by TLC (Wagnera and Baldt, 1996). The different results between the extracts may be due to the different extraction solvent and also because of the extraction time (Figure 1).

Tilia cordata, linden, is a deciduous tree that grows to 20-40 m in height. Linden is found in Europe from England to Scandinavia, Russia, Spain, Italy, Greece, Romania, Bulgaria, Turkey and western Asia. *Tiliae flores* are the flowers of the tree, having as major active principles volatile oil and phenolic compounds. Flowers have diaphoretic, antipyretic, emollient, expectorant, sedative, anxiolytic, decongestant, anti-inflammatory and diuretic activities (Rodriguez-Fragoso et al., 2008; Pârvu 1997).

In Figure 1 (S6) - fingerprint of *Tilia cordata* flowers are present flavonoids as rutin (Rf ~0.41) and hyperoside (Rf~0.68) as well asflavonoid glycoside derived from quercetin, myricetin and kaempferol (Wagner and Bladt 1996) and ferulic acid (Rf~0.94) (Figure 5). The major flavonoids found in the flowers extract by HPTLC were: quercetin-3,7-di-O-rhamnoside, kaempferol-3,7-di-O-rhamnoside and kaempferol 3-O-(6"-p-coumaroyl glucoside) or tiliroside (Negri et al., 2013).

Achillea millefolium, yarrow, is a perennial herb, 30–90 cm in height, with aromatic odor and grayish- green colour. *Millefolii flores* consists in the dried flowering tops of *Achillea millefolium L.* The plant is Native to Asia, Europe and North America, now being widely distributed and cultivated in the temperate regions of the world. Major chemical constituents of the flowers are essential oil (0.2-1.0 %) and phenolic compounds.

It is used for internal administration for loss of appetite, dyspeptic ailments, common cold, such as mild spastic discomfort of the gastrointestinal tract, as a choleretic and for the treatment of fevers and for external administration for skin inflammation and wounds (WHO Monograph, 2003; Pârvu, 1997).

In (S7) we have found the following compounds: rutin (Rf~0.41), hyperoside (Rf~0.68), as well as flavonoid glycoside (yellow - orange spots Rf~0.19-0.25) and chlorogenic acid (Rf~0.51) (Figure 1, 6).

Eight phenolic compounds were identified by HPLC in extracts from yarrow flowers: chlorogenic acid and flavonoids, namely vicenin-2, luteolin-3',7-di-O-glucoside, luteolin-7-O-glucoside, rutin, apigenin-7-O-glucoside, luteolin, and apigenin (Benetis et al., 2008).

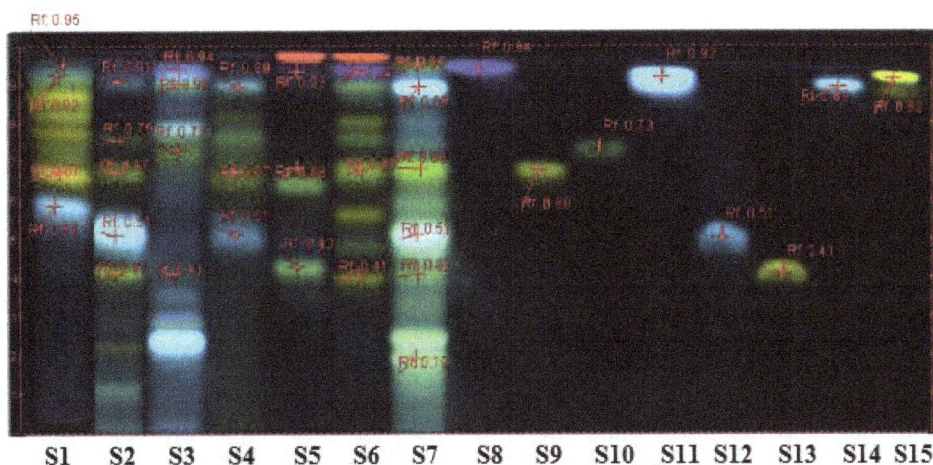

Figure 1. Chromatographic fingerprint of the species comparative with references substances

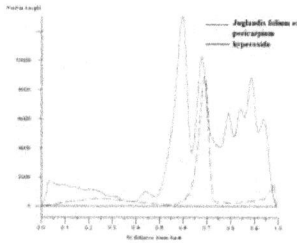

Figure 2. Profile comparison *Juglandis folium et pericarpium*/references substances

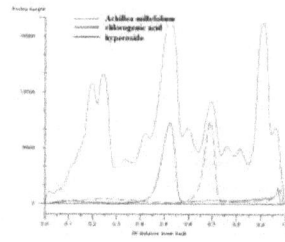

Figure 3. Profile comparison *Mori folium*/ references substances

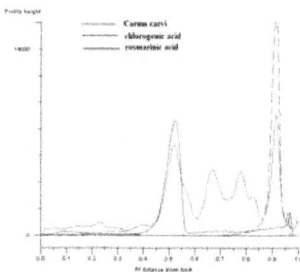

Figure 4. Profile comparison *Carum carvi*/ references substances

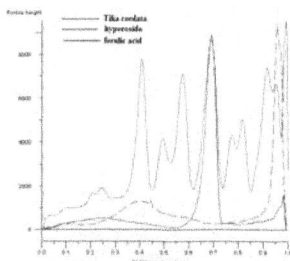

Figure 5. Profile comparison *Tilia cordata*/ references substances

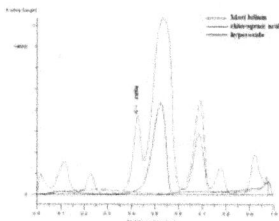

Figure 6. Profile comparison *Achillea millefolium*/ references substances

CONCLUSIONS

To ensure consumer health protection, the quality and safety of vegetal raw material, particularly those used for plant-derived products, must be determined.

With a market in continuous expansion, the competition between the producers is getting stronger every day. Therefore, in quality control management are needed safe, easy and not very expensive methods as HPTLC, especially as this method is accepted at international level.

REFERENCES

Alaerts G., Matthijs, N. Smeyers-Verbeke J., Vander Heyden Y., 2007. Chromatographic fingerprint development for herbal extracts: A screening and optimization methodology on monolithic columns Journal of Chromatography A, 1172 (1–8)

Benetis R, Radusiene J, Janulis V. 2008. Variability of phenolic compounds in flowers of *Achillea millefolium* wild populations in Lithuania. Medicina (Kaunas); 44(10):775-81

Cosmulescu S, Trandafir I. 2011. Variation of phenols content in walnut (*Juglans regia L.*) Southwest. J. Horticult., Biol. Environ., 2 pp. 25–33.

Ghasemzadeh Ali and Ghasemzadeh Neda, 2011. Flavonoids and phenolic acids: Role and biochemical activity in plants and human. Journal of Medicinal Plant Research Vol. 5(31), pp.6697-6703.

Istudor Viorica 1998. Farmacognozie Fitochimie Fitoterapie, Vol I, Editura Medicala, Bucuresti;

Kamatou GPP, Viljoen AM, Steenkamp P 2010. Antioxidant, Antiinflammatory activities and HPLC analysis of South African Salvia species. Food. Chem., 119: 684-688;

Memon Ayaz Ali, Najma Memon, Devanand L. Luthria, Muhammad Iqbal Bhanger, Amanat Ali Pitafi 2010. Phenolic acids profiling and antioxidant potential of mulberry (*Morus laevigata W., Morus nigra L., Morus alba L.*) leaves and fruits grown in Pakistan - Pol. J. Food Nutr. Sci. Vol. 60, No. 1, pp. 25-32

Negri, G, Santi, D. Tabach, R. 2013. Flavonol glycosides found in hydroethanolic extracts from *Tilia cordata*, a species utilized as anxiolytics. Rev. Bras. Pl. Med., Campinas, v.15, n.2, p.217-224,

Pârvu Constantin, 1997. Universul plantelor Mica enciclopedie – Ed. Enciclopedica , Bucuresti

Pereira Jose Alberto, Ivo Oliveira, Sousa Anabela, Valenta Patricia, Andrade Paula B., Ferreira Isabel C.F.R., Ferreres Federico, Bento Albino, Seabra Rosa, Estevinho Letıcia 2007. Walnut (*Juglans regia* L.) leaves: Phenolic compounds, antibacterial activity and antioxidant potential of different cultivars Food and Chemical Toxicology 45, 2287–2295.

Rice-Evans C, Miller NJ, Paganga G 1996. Structure-antioxidant activity relationship of flavonoids and phenolic acids. Free Rad. Bio. Med., 20: 933-956.

Rodriguez-Fragoso Lourdes, Reyes-Esparza Jorge, W. Burchiel Scott, Herrera-Ruiz Dea, Torres Eliseo. 2008. Risks and benefits of commonly used herbal medicines in Mexico Toxicology and Applied Pharmacology 227 125–135;

Samec D, Gruz J, Strnad M, Kremer D, Kosalec I, Grubesic RJ, Karlovic K, Lucic A, 2010. Antioxidant and antimicrobial properties of *Teucrium arduini L. (Lamiaceae)* flower and leaf infusions (*Teucrium*

arduini L. antioxidant capacity). Food Chem. Toxic., 48: 113-119.

Stampar, F., Solar, A., Hudina, M., Veberic, R., Colaric, M., 2006. Traditional walnut liqueur – cocktail of phenolics. Food Chemistry 95, 627–631.

Stoia Mihaela, Oancea Simona, 2013. Herbal dietary supplements consumption in Romania from the perspective of public health and education, Acta Medica Transilvanica,;2(2):216-219

Thippeswamy N.B., K. Akhilender Naidu, Rajeshwara N. Achur, 2013. Antioxidant and antibacterial properties of phenolic extract from *Carum carvi* L. Journal of Pharmacy Research 7 352- 357

Thippeswamy N. B. and Rajeshwara N. Achur, 2014. Inhibitory effect of phenolic extract of *Carum carvi* on inflammatory enzymes, hyaluronidase and trypsin World Journal of Pharmaceutical Sciences ISSN (Print): 2321-3310; ISSN (Online): 2321-3086)

Wagner, H. and Bladt S., 1996. Plant Drug Analysis, A thin layer chromatography atlas. Springer, New York, 359

Watson L., Dallwitz M.J., 2007. Moraceae In: The families of flowering plants: Descriptions, illustrations, identification, and information retrieval, 1992 onward http://delta-intkey.com

WHO Monograph, 2003, vol. 2, 4

27

NEW INVASIVE INSECT PESTS RECENTLY REPORTED IN SOUTHERN ROMANIA

Constantina CHIRECEANU, Andrei TEODORU, Andrei CHIRILOAIE

Research and Development Institute for Plant Protection Bucharest,
8 Ion Ionescu de la Brad, District 1, 013813 Bucharest, Romania
Corresponding author email: cchireceanu@yahoo.com

Abstract

This work presents the results of the field survey in 2016 referring to invasive insect pests newly recorded from the Southern Romania. Five non-European insect species, belonging to Hemiptera and Lepidoptera Orders damaging diverse plants were identified, as follows: three polyphagous species in Auchenorrhyncha group, Orientus ishidae, Phlogotettix cyclops and Acanalonia conica; one true bug species Leptoglossus occidentalis harmful to conifer seeds, and one leaf miner species Phyllocnistis vitegenella pest to grapevine. In this work we included summarized data related to some aspects of species origins, distribution in Europe and Romania and preference to the host plants.

Key words: invasive species, insect pests, Southern Romania.

INTRODUCTION

Phytophagous insects form a very important category of invasive species with a great rate of penetration in European territory. They are well represented by a wide spectrum of species belonging to numerous different taxonomic groups. According to Roques (2016), since 2000s the number of new phytophages that enter Europe is about 11.5 species per year. The same author reported that most of the exotic phytophagous species that established in Europe by 2014 are associated with woody plants, out of which insects represent 83.5%. Unintentionally introduced species have a greater velocity of spreading comparative to those deliberately introduced (Roques et al., 2016). Global costs associated with invasive insects reached a minimum of US $ 70.0 billion per year, and costs related to the health sector exceeded US $ 6.9 billion per year (Bradshaw et al., 2016).

Favored by natural factors (i.e. global warming, food chain) as well as intense human activities (i.e. commercial exchange, travel, tourism), the alien invasive species penetrate at a rapid pace from year to year in Romania as in other countries around the globe. In an attempt to detect the new exotic invasive species that entered Romania as soon as possible, and put together new accumulated data for a better understanding of the species distribution and richness, many research communities in institutions from different parts of the country were involved in issues related to invasive species.

The purpose of our research was to collect, identify and disseminate useful information on the new allochthonous invasive insect pests recorded in the Southern Romania in 2016, in order to bring an essential contribution to improve the available knowledge on this field to the country. Within our work, five invasive insect species, new for South part of Romania were detected, two of them being considered as the first report for Romanian pest fauna.

MATERIALS AND METHODS

The collecting of insects was performed in 2016 within the framework of the fields' survey program conducted by the Research-Development Institute for Plant Protection Bucharest in order to discover untimely pest insects having the status of invasive species. The sampling in 2016 was carried out in urban areas in Bucharest and rural areas in Ilfov and Giurgiu counties from the South part of Romania. Insects were captured with yellow sticky traps and by direct collection. To identify the insects in our samples we used the morphological features and illustrations describing adult specimens in relevant references in the literature. Using an Olympus camera connected to a TSZ 61 stereomicroscope we were able to take pictures of collected insects.

Geographical coordinate data for the sampling points were taken using the application Convertor online coordinate GPS: (http://www.calculatoare.ha-ha.ro/convertor_coordonate_gps_adresa.php; WGS84).

RESULTS AND DISCUSSIONS

The phytophagous insect species classified as invasive species, alien to Romania, that we encountered in the Southern Romania, during the vegetative season in 2016, are presented in Table 1.

Orientus ishidae (Matsumura, 1902) (Hemiptera: Cicadomorpha, Cicadellidae, Deltocephalinae) - the mosaic leafhopper species (Figure 1);

This species has East Asian origins. It has been recorded in the Northern Italy since 1998 (Guglielmino, 2005), and after that it rapidly spread to other European countries (EPPO, 2015).

In Romania, the first adults of *O. ishidae* have been described from the South part of the country (Bucharest) in 2016, by Chireceanu et al. (2017).

Table 1. Taxonomic position of invasive insect species detected in the Southern Romania in 2016; plant species on which the insects were found or traps were placed

Order	Family	Species	Common name	Host plants
Hemiptera	Cicadellidae	*Orientus ishidae* (Matsumura, 1902)	The mosaic leafhopper	Woody plants (*Crataegus monogyna, Malus domestica, Prunus avium, Ziziphus jujube*)
		Phlogotettix cyclops (Mulsant & Rey, 1855)		Woody plants (*Crataegus monogyna, Ziziphus jujube, Juglans regia*)
	Acanaloniidae	*Acanalonia conica* (Say, 1830)	The green cone-headed planthopper	Woody plants (*Crataegus monogyna, Malus domestica*)
	Coreidae	*Leptoglossus occidentalis* (Heidemann, 1910)	The western conifer seed bug	Building balcony
Lepidoptera	Gracillariidae	*Phyllocnistis vitegenella* (Clemens, 1859)	Grape leaf miner	Wild grapevine Hybrid *Vitis* sp.

Figure 1. Adult of *Orientus ishidae*

A number of 63 adult specimens were trapped in the course of the year 2016, on yellow sticky traps settled on apple (*Malus sylvestris*), sweet cherry *(Prunus avium),* Chinese date (*Ziziphus jujube*) and common hawthorn (*Crataegus monogyna*) trees present in urban areas of Bucharest.

The species of *M. sylvestris* (N44°30'05"/E26°4'35") and *P. avium* (N44°30'02"/E26°4'37") composed two research orchards of the Research-Development Station for Fruit Growing (RDSFG) in the Northern part of Bucharest; *Z. jujube* trees were enclosed in the experimental field of the USAMV Bucharest in the Northern part of Bucharest (N44°28'11"/E26°04'12") and in the 'D. Brândză' botanical garden of Bucharest in the central part of the capital (N44°26'17"/E26°03'52"); the shrubs of *C. monogyna* were present in a non-managed area (N44°30'15"/E26°04'02") near to the Research-Development Institute for Plant Protection and in the botanical garden of the USAMV Bucharest (N44°26'17"/ E26°03'52"), both institutions being situated in the North part of Bucharest.

O. ishidae is a polyphagous pest that lives on various plant species from woody plants and deciduous trees groups and feeds on the plant phloem. By reason of its strategy for feeding, the *O. ishidae* leafhopper is considered of great economic importance, because this was found to be associated to the spreading of some phytopathogenic microorganisms such as phytoplasmas (pathogens that live as obligate parasites in the phloem sieve tubes of plants and in insect vectors) to grapevine. The insect is a common presence in the vineyards in countries from West and Central Europe. Adults of O. *ishidae* collected in grapevine affected by yellows - type diseases, have been found infected with pathogens of the disease *Grapevine flavescence dorée* (GFD) (Mehle et al., 2011; Trivellone et al., 2015). Recent studies conducted in Italy (Lessio et al., 2016) have shown that *O. ishidae* was competent to transmit the GFD phytoplasma to grapevine. As regards the risk of FD to be spread by *O. ishidae* to grapevine, is believed to be real. This is because, in many references in the literature, *O. ishidae* is compared with *Scaphoideus titanus*, the main natural vector known to spread the GFD disease. Both species are Deltocephalinae leafhoppers belonging to the same family of Cicadellidae, and it is expected that *O. ishidae* to have the model of life cycle and behavior similar to *S. titanus*.

Phlogotettix cyclops (Mulsant & Rey, 1855) (Hemiptera: Cicadomorpha, Cicadellidae, Deltocephalinae); Syn. *Jassus cyclops* Mulsant & Rey, 1855 (Figure 2).

Figure 2. Adult of *Phlogotettix cyclops*

This leafhopper species is originally from Asian and Russian regions, spread in countries from Central and South East Europe on various plant species such as raspberry, fruit trees and grapevine (Chuche et al., 2010), this being known as a polyphagous species. The leafhopper has one generation per year and overwinters as eggs laid into plant tissues. First knowledge with regard to description of the presence of *P. cyclops* on grapevine was provided by Chuche et al. (2010) on Bordeaux vineyards. Like majority of the species in the Cicadellidae family, *P. cyclops* leafhopper is specialized in feeding on the phloem tissue of plants. From this reason, many scientists suspected this species as a possible new vector that may spread the phytoplasma pathogens to cultivated plants. The fruit trees and grapevine, crops of high economic importance, are mainly considered vulnerable to the risk of this insect. Wuu-Yang Chen et al. (2011) in East Asia has diagnosed the *P. cyclops* species as a serious vector with potential to transmit a strain of phytoplasma belonging to the 16SrI group. Reisenzein (2015) in Austria has found this cicadellid to be infected with phytoplasma of the *Flavescence dorée* disease, a grave systemic disease of grapevine in Europe. *P. cyclops* is cited in many reports in the European literature as present in Romania, based on the reports of Dlabola from 1977 and 1981 (www.faunaeur.org). After these reports, no article has been published on this species in Romania so far. In our field survey, 170 specimens of *P. cyclops* have been caught on yellow sticky traps settled on shrubs of common hawthorn (*Crataegus monogyna*) present in the area situated near the Plant Protection Institute (Northern Bucharest) (N44°30'15"/E26°04'02"), on trees of Chinese date (*Ziziphus jujube*) (N44°26'17"/E26°03'52") planted in the national Botanical Garden 'D. Branza' (Central part of Bucharest) and in the walnut (*Juglans regia*) orchard (N44°30'12"/ E26°15'49") in the Didactic Farm of USAMV Bucharest in Ilfov County at 15 km away from Bucharest.

Acanalonia conica (Say 1830) (Hemiptera: Fulgoromorpha, Acanaloniidae) - the green cone-headed planthopper (Figure 3).
This species that originates from North America was found for the first time in Europe in 2002, in Switzerland (Günther and Mühlethaler, 2002). In our field survey conducted in 2016, the adults of *A. conica*

captured on yellow sticky traps accounted 21 specimens. The yellow sticky traps were placed on trees of common hawthorn occurring in the non-administered area (N44°30'15"/E26°04'02") around the court of the Plant Protection Institute and on an apple orchard (N44°30'05"/E26°04'35") that belongs to the experimental field of the Research Station in Pomiculture. Both sampling locations are situated in the North part of Bucharest.Within the captures of *A. conica* during this survey activity, we revealed for the first time the presence of this species in the Bucharest zone and in Romania as well (Chireceanu et al., 2017). In the examination of the traps, we noticed that the adults of *A. conica* captured were associated with those of another invasive planthopper species, the flatid *Metcalfa pruinosa. A. conica* is a univoltine species and overwinters as egg stage. Adults are bright green in color and measure 10 mm long. A typical feature of this species is the conical vertex of the head; they can be found during June-September (Aldini et al., 2008).

Figure 3. Adult of *Acanalonia conica*

In many reports in the European literature, *A. conica* is not considered a severe pest of cultivated plants because this generally develops small populations that are not able to produce essential damages. However, the grapevine and ornamental plants are indicated among the host plants with economic importance that are affected by this insect (D'Urso and Uliana, 2006).

Phyllocnistis vitegenella (Clemens, 1859) (Lepidoptera: Gracillariidae) - grape leaf miner (Figure 4b).

This is a North American species, reported in Europe for the first time in 1995, in the northeast of Italy (Cara and Jermini, 2011). Then, it has also been confirmed in Slovenia (2004), Switzerland (2009) and Hungary (2014) (Cara and Jermini, 2011; Szabóky and Takács, 2014). Adult of 3 mm, is distinguished by a brilliant white color with characteristic brown stripes very finely and two black dots on the apex of the wing. Larva produces visible injuries on leaves consisting in distinctive galleries (mines), very long and sinuous on the upper side of the leaves of plants belonging to Vitaceae family (Figure 4a). The average number of mines per leaf is maximum 4 (Lips and Jermini, 2013). The micromoth miner has 3-5 generations per year and spends the winter as adult.

For Romania, the first detection of the pest was on grapevine in Moldova region in 2013 (Ureche, 2016).

Research conducted in Switzerland focused on the grape leaf miner behavior, showed its preference for leaves of the shoots laterally disposed on vine plants. At the *Merlot* cultivar, the leaves of lateral shoots were mined over 3 times more than those of main shots (Lips and Jermini, 2013). The same authors regarded *P. vitegenella* as a minor pest of grapevine because this did not induce considerable negative effects and accordingly the control measures against its population are not necessary.

In our study conducted in the Southern Romania in 2016, we obtained *P. vitegenella* adults from mined leaves sampled from the grapevine rootstocks in a field previously planted with vine plants for research proposes within framework of the RIDPP Bucharest, and also from mined leaves sampled from two plots of hybrid grapevines in house gardens (Naipu village, Giurgiu County). In addition, we gathered *P. vitegenella* adults from leaves with mines collected from wild vines found in spontaneous flora around some vineyards in Vrancea County, on which we investigated in 2016 with regard to the presence of the American grapevine leafhopper *Scaphoideus titanus*. We have not discovered any sign of the attack of *P. vitegenella* on vine plants in the noble vineyards that we have monitored.

From the samples collected in the south part of the country and maintained in laboratory conditions, some parasitoids (Figure 4c) belonging to the Eulophidae family have resulted.

European research focused on parasitoids of the grape leaf miner *P. vitegenella* have indicated a parasitic rate of the pest up to 33%.The parasitoids associated with this species are framed to Eulophidae and Ichneumonidae families.

Figure 4. (a) Mined grapevine leaf; (b) Adult of *Phyllocnistis vitegenella*;
(c) Parasitoid wasp from the Eulophidae family

Leptoglossus occidentalis (Heidemann, 1910) (Hemiptera: Heteroptera: Coreidae) - the western conifer seed bug (Figure 5).

L. occidentalis is a plant-feeding bug commonly called "leaf-footed bugs" or "squash bugs". It is a pest from North America, observed for the first time in Europe (Northern Italy) in 1999 (Taylor et al., 2001), from where this rapidly spread across the European continent, reaching up to Sweden and Ireland and even to Ukraine and Russia (Putchkov, 2013).

Figure 5. Adult of *Leptoglossus occidentalis*

It is a large true bug of approx 20 mm long (adults), easily recognizable by the hind legs strongly developed with a characteristic enlargement zone on the hind tibiae and long

femurs serrated on the inside; thin visible white lines in the shape of inverted 'W' in the middle of the wings (Ruicănescu, 2009); the first segment of antennas is thicker, slightly curved, orange-brown with a black longitudinal line inwards. *L. occidentalis* has one or two generations annually depending on the altitude of area where it is living (Tamburini et al., 2012), and survives the winter conditions as adult, commonly in large groups, under coniferous trees bark, but also in people's homes and other buildings, from where they emerge in the spring season giving birth to the summer generation. It is a polyphagous insect on conifer species in the Pinaceae family (Tamburini et al., 2012). To date, the western seed bug has not been reported to produce important damage to conifers in Europe, but some overwintering adults of this species were detected to bear spores of the fungus *Diplodia pinea*, and from this point of view they are suspected to play a role in spreading of this pathogen (Tamburini et al., 2012). Instead, it is considered a nuisance pest to people, as the overwintering form enters into people's homes and walk unhindered on the walls (Rabitsch, 2008). In the Romanian territory, the presence of *L. occidentalis* (two females) was for the first time reported in the central part of the country, in 2008 and 2009 by Ruicănescu (2009). In the following years, the seed bug was reported in several other Romanian regions, in the Southeastern (Şerban, 2011), the Northwestern (Rădac and Petrovich, 2016), the Northeastern, Central and Southern parts (Olenici and Duduman, 2016) of the country. The pest had a low density, and the collecting points have always been associated with the conifer trees present around.

In our research, *L. occidentalis* (one male) was collected on October 28th, 2016, by the first author of this paper, on geranium plants in the balcony of her bloc apartment in the Southern area of Bucharest (44°23′34″/ 26°6′42″).

The presence of *L. occidentalis* inside of buildings in urban areas does not appear to be unusual because all Romanian papers on this subject previously published indicated the cover spaces as common collecting places for this species.Our results in this study confirm the presence of *L. occidentalis* in the Southern Romania and it may be considered the second record of this pest for this area and the first record for Bucharest so far. Capturing of the adults of the *L. occidentalis* bug during the October and December months is explained by the fact that the insects are seeking shelter for wintering (Olenici and Duduman, 2016).

CONCLUSIONS

The field survey and sample evaluation performed by us in areas from the Southern Romania in 2016, led to obtaining relevant knowledge on the presence of five non-European important insect species, belonging to the Hemiptera and Lepidoptera Orders, such as: three polyphagous species in Auchenorrhyncha group, *Orientus ishidae, Phlogotettix cyclops* and *Acanalonia conica;* one true bug species *Leptoglossus occidentalis* harmful to conifer seeds, and one leaf miner species *Phyllocnistis vitegenella* pest to grapevine.

The results obtained within this work suggest that the activities of the field survey are critical for the early detection of alien invasive species that could unexpectedly penetrate into new territories, unaffected by them until then. Further data on the spreading in other areas as well as the incidence on plants for the new detected invasive species will be essential for the Romanian territory in the near future, so that the surveys in the field would be extended to many other crops, mainly those of economic importance, such as fruit trees, grapevine, ornamentals and conifers.

ACKNOWLEDGEMENTS

The study was performed in the framework of the project PN 16-29-01-01, financial support by the Ministry of Research and Innovation.

REFERENCES

Aldini R. N., Mazzoni E., Mori N., Ciampitti M., 2008. On the distribution in Italy of the Nearctic hopper Acanalonia conica, with ecological notes. Bulletin of Insectology, 61(1):153-154.

Bradshaw C. J. A., Leroy B., Bellard C., Roiz D., Albert C., Fournier A., Barbet-Massin M., Salles J-M., Simard F., Courchamp F., 2016. Massive yet grossly underestimated global costs of invasive insects. Nature Communications 7, 12986, DOI:10.1038/ncomms12986.

Cara C., Jermini M., 2011. La mineuse américaine Phyllocnistis vitegenella, un nouveau ravageur de la vigne au Tessin. Revue suisse Viticulture, Arboriculture, Horticulture, 43:224-230.

Chireceanu C., Teodoru A., Gutue M., Dumitru M., Anastasiu P., 2017. Two new invasive hemipteran species first recorded in Romania: Orientus ishidae (Matsumura 1902) (Cicadellidae) and Acanalonia conica (Say 1830) (Acanaloniidae). Journal of Entomology and Zoology Studies, 5(2): 824-830

Chuche J., Danet J.-Luc., Thiery D., 2010. First description of the occurrence of the leafhopper Phlogotettix cyclops in a Bordeaux vineyard. Journal International des Sciences de la Vigne et du Vin, 44:161-165.

Dlabola J., 1977. Chorologische Ergünzungen zur Zikadenfauna des Mittelmeergebdetes (Horn.: Auchenorrhyncha). Acta Entomologica Musei Nationalis Pragae, 33(1- 21):21- 31.

Dlabola J., 1981. Eregbnisse der Tschehoslowaikisch Iranisohen Entomologischen Expeditionen nach dem Iran (1970 und 1973) (Mit Augaben über einige Salmelresultate in Anatolienl Homoptera: Auchenorrhyncha m. Ten). Acta Entomologica Musei Nationalis Pragae, 40:127- 311.

D'Urso V., Uliana M., 2006. Acanalonia conica (Hemiptera, Fulgoromorpha, Acanaloniidae), a Nearctic species recently introduced in Europe. Deutsche entomologische Zeitschrift, 53(1):103-107.

Günthart H., Mühlethaler R., 2002. Provisorische Checklist der Zikaden der Schweiz (Insecta: Hemiptera, Auchenorrhyncha). Denisia, 4(176):329-338.

Lessio F., Picciau L., Gonella E., Andrioli M., Tota F., Alma A., 2016. The mosaic leafhopper Orientus ishidae: host plants, spatial distribution, infectivity, and transmission of 16SrV phytoplasmas to vines. Bulletin of Insectology, 69(2):277-289.

Lips A., Jermini M., 2013. Harmfulness of the American grape leaf miner Phyllocnistis vitegenella on the grapevine 'Merlot' (Vitis vinifera). Federal Department of Economic Affairs, Education and Research EAER, Agroscope.

Mehle N., Rupar M., Seljak G., Ravnikar M., Dermastia M., 2011. Molecular diversity of 'flavescence dorée' phytoplasma strains in Slovenia. Bulletin of Insectology, 64(suppl.): S29-S30.

Olenici N., Duduman M. L., 2016. Noi semnalări ale unor specii de insecte forestiere invazive în România. Bucovina Forestieră, 16(2), 161-174.

Putchkov P. V., 2013. Invasive true bugs (Heteroptera) established in Europe. Український ентомологічний журнал, 2(7):11-28.

Rabitsch W., 2008. Alien True Bugs of Europe (Insecta: Hemiptera: Heteroptera). Zootaxa 1827:1-44,

Rădac I. A., Petrovici M., 2016. Studies regarding the true bugs fauna (Insecta, Heteroptera) in Cefa Nature Park. Acta Oecologica Carpatica, 9:121-132.

Reisenzein H., 2015. Epidemiological studies on flavescence dorée (FD) IOBC meeting, Vienna, 1-27. http://www.faunaeur.org/distribution_table.php (last update 29 August 2013; 23.01.2017 visited).

Roques A., Auger-Rozenberg M. A., Blackburn T. M., Garnas J., Pyšek P., Rabitsch W., Richardson D. M., Wingfield M. J., Liebhold A. M., Duncan R. P., 2016. Temporal and interspecific variation in rates of spread for insect species invading Europe during the last 200 years. Biol Invasions, 18:907-920.

Ruicănescu A., 2009. Leptoglossus occidentalis Heidemann, 1910 (Heteroptera: Coreidae) in Romania. In Rakosy L., Momeu L. (Eds.), Neobiota in România. Presa Universitară Clujeană, Cluj-Napoca, 153-154.

Şerban C., 2011. Leptoglossus occidentalis Heidemann, 1910 (Heteroptera: Coreoidea): a new record for the invasive true bugs fauna of Romania. In Murariu D.C., Adam G., Chişamera E., Iorgu L. O., Popa O. P. (Eds.), Annual Zoological Congress of "Grigore Antipa" Museum. Book of abstracts, "Grigore Antipa" National Museum of Natural History, Bucharest, Romania, 102.

Szabóky C., Takács A., 2014. The first occurrence of American grape leaf miner (Phyllocnistis vitegenella Clemens, 1859-Gracillariidae) on grapevine (Vitis vinifera) in Hungary. Növényvédelem, 50(10):467-469.

Tamburini M., Maresi G., Salvadori C., Battisti A., Zottele F., Pedrazzoli F., 2012. Adaptation of the invasive western conifer seed bug Leptoglossus occidentalis Trentino, an alpine region (Italy). Bulletin of Insectology, 65(2):161-170.

Taylor S. J., Tescari G., Villa M., 2001. A Nearctic pest of Pinaceae accidentally introduced into Europe: Leptoglossus occidentalis (Heteroptera: Coreidae) in northern Italy. Entomological News, 112(2):101-103.

Trivellone V., Filippin L., Jermini M., Angelini E., 2015. Molecular characterization of phytoplasma strains in leafhoppers inhabiting the vineyard agroecosystem in Southern Switzerland. Phytopathogenic Mollicutes, 5 (suppl.):S45-S46.

Ureche C., 2016. First record of a new alien invasive species in Romania: Phyllocnistis vitegenella Clemens (Lepidoptera: Gracillariidae). Acta Oecologica Carpatica, IX: 133-138.

Chen WY., Huang YC., Tsai ML., Lin CP., 2011. Detection and identification of a new phytoplasma associated with periwinkle leaf yellowing disease in Taiwan. Australasian Plant Pathology 40: 476. DOI 10.1007/s13313-011-0062-x.

EPPO Reporting Service no. 05-2015 Num. article: 2015/098. Orientus ishidae: a potential phytoplasma vector spreading in the EPPO region.

28

INITIAL PLANTING DESIGN OF THE CAROL I PARK IN BUCHAREST

Ileana Maria PANȚU

University of Agronomic Sciences and Veterinary Medicine of Bucharest,
59 Marasti Blvd., District 1, Bucharest, Romania
Corresponding author email: ileana.pantu@gmail.com

Abstract

This paper aims to analyze the planting composition of the Carol I park in Bucharest at the beginning of the 20th century. The Carol I park is important for the history of Romanian landscape architecture and stands also as an example for the design of parks nowadays. Conceived in 1906 as a national park to host an international event, the "General Romanian Exhibition", The Carol I park was transformed throughout the 20th century to represent symbolically different political powers. A royal showcase at first, it then became a tool of communist propaganda in the '60s. The park's compositional style also changed with its ideological mutations and its planting design followed. It is essential to analyze and understand the evolution of parks and especially the evolution of their planting design in order to valorize and preserve and/or restore national landscape heritage. The first stage of the history of the Carol I park demonstrated an elegant Belle Époque style with both Romantic and geometric areas designed by the French landscape architect Édouard Redont, specially invited for the abilities he had already demonstrated in Romania. The composition of the park was centred on a generous circulation space with lots of pavilions, interspersed with water elements. Elegantly planted in a geometric style with a lot of attention to detail, this axis has continued visually towards a hill over a sinuous lake. On this hill, the eye is drawn to the Palace of the Arts. Below it, an elegant coniferous composition engulfed a graceful Romantic grotto with a high cascade guarded by a sculptural ensemble. The alleys and the vegetation near the lake and on the hill were designed in the French landscape style, in contrast with the geometry of the axis zone and its side alleys, with alignments of trees, pruned shrubs and flower platbands.

Key words: Belle Époque style, geometric style, planting design, public park, Romantic style.

INTRODUCTION

This paper will analyse the planting design of the first stage of the evolution of the oldest and one of the most important public parks in Bucharest, the park Carol I. Conceived in 1906 as a national park to host an international event, the "General Romanian Exhibition", a tribute to Carol I, who had then reigned for 40 years (Parusi, 2007; Potra, 1990; Panțu, 2011), the park was transformed in the '30s (Panțu, 2011) and more radically in the '60s in order to eradicate its royal symbols and become a tool of communist propaganda (Panțu, 2015). Thus, the park changed in compositional style, including its planting design, over its ideological mutations. The history and morphological transformations of the Carol I park demonstrate all of the Romanian landscape styles of the first half of 20th century in just one park (Panțu, 2012).

The French landscape architect Édouard Redont was invited to design the park due to the great abilities he had already demonstrated

in Romania (Teodorescu, 2007). He created an elegant *Belle Époque* park in a mixed style with predominant Romantic style - French landscape style, and a geometric part from the entrance zone to the heart of the park (Panțu, 2011) (Figure 1).

RESULTS AND DISCUSSIONS

Redont conceived a mixed style planting design, with a large part in the Romantic style, where he grouped the plants so as to resemble natural landscapes (Figure 1).

The composition of the park was centred on a generous circulation space with lots of pavilions – a large parterre with two lateral alleys shaded on both sides by alignments of linden trees (Figure 1). Redont treated in an architectural, geometric manner, in a classical style, the plantation along this main axis: the symmetry of the parterre with bowling green rhythmed with spherical box and yew, borders of ornamental leaf plants, flower platbands, tree alignments etc. He designed the rest of the park in a Picturesque, Romantic style: trees and shrubs in free shaped masses, in groups or isolated, composed in order to create varied sequences, profusions, clever effects of light and shade etc. The main axis continued visually over a sinuous lake towards a hill with the imposing building Palace of the Arts that dominated the entire park. Below it, a grotto with cascade and a sculptural ensemble was the heart of the Romantic area (Figure 1).

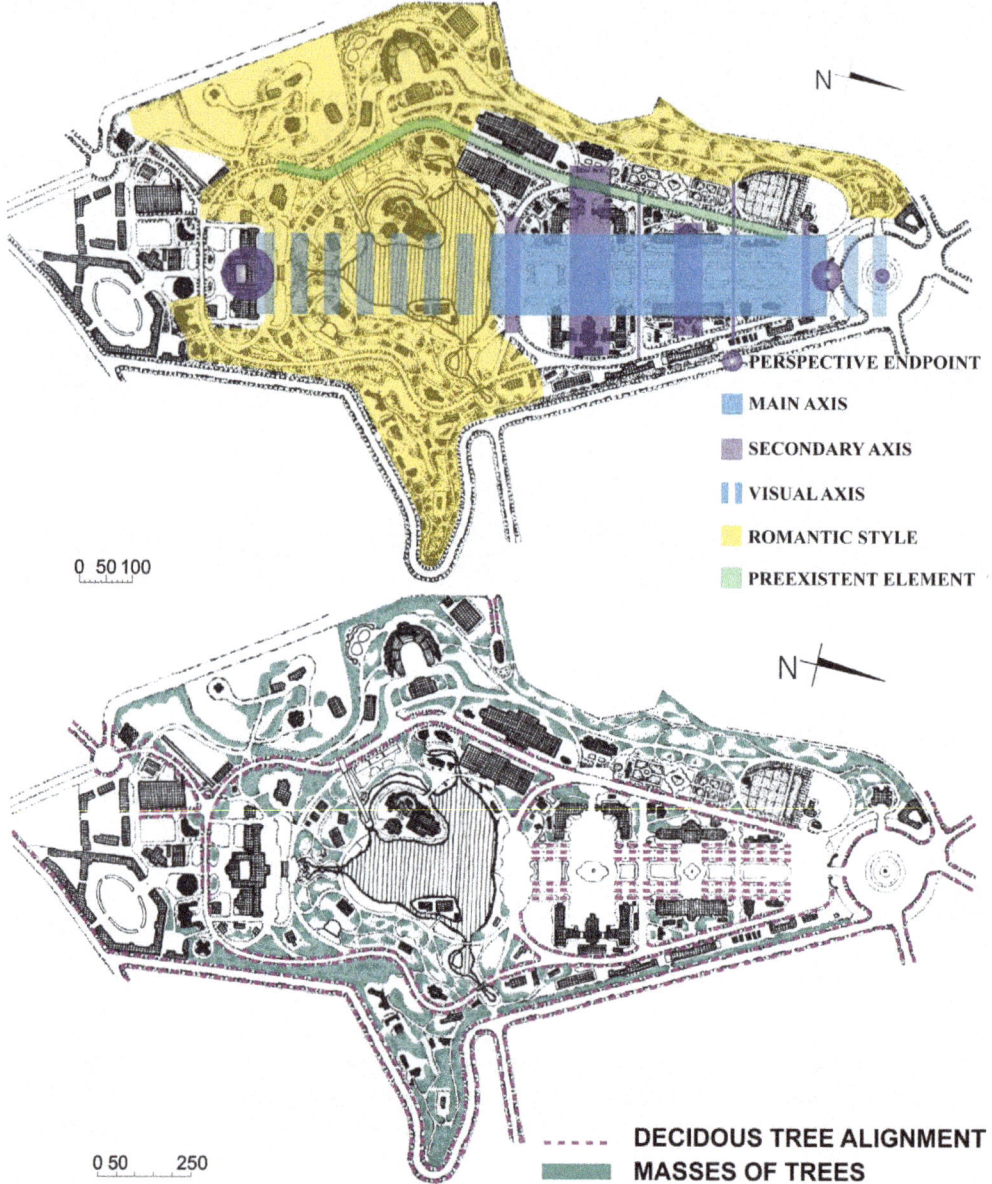

Figure 1. Composition and vegetation structure of the park Carol I in 1906 (Panțu, 2011; Marcus, 1958)

In order to create the planting design, Redont diligently searched Bucharest and discovered valuable mature tree species. Their vitality, their habit, the colour of their foliage, their blossom or fruit offered total success to one who knew how to find the right place to set them off. Redont's inventory stands as evidence for that (Teodorescu, 2007). In addition to the plants brought from nurseries, these species numbered about 6,000 resinous trees, 4,200 high, mature deciduous trees, 90,000 young trees (Teodorescu, 2007; Bulei, 1990), 48,200 shrubs, 49,200 forest plants, 8,400 plants with varied flowers, 98,000 flourished plants and 3,500 kg of grass seed (Potra, 1990; Panțu, 2011). All of these species conferred to park visitors all the benefits of a vegetation at least partially mature, such as shade so welcome in the hot Bucharest summers.

In the general framework, coniferous outnumbered by far deciduous trees (1.43 times), unusual for a park situated in the plain (Figure 1). Nevertheless, the large difference in level in the park site (Filaret Hill) amplified by Redont, by modelling the landscape and the usage of rockeries and abrupt vegetation created a mountain-like image and so allowed the resinous trees to dominate the view (Panțu, 2011, 2012).

Redont retained in the park's composition the former road that traversed the site as an important side alley with its double alignment of *Populus alba* - the oldest trees in the park, giving a lot of shade, necessary in the Bucharest summers (Figure 2). These white poplar lines are easily distinguished in all the early aerial views of the park. (Figures 3-5).

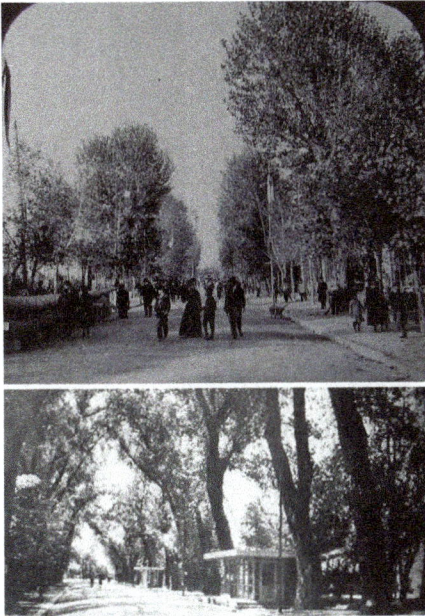

Figure 2. Former Filaret Avenue in 1906 (Zaharia, 1906) and late '50s (Marcus, 1958)

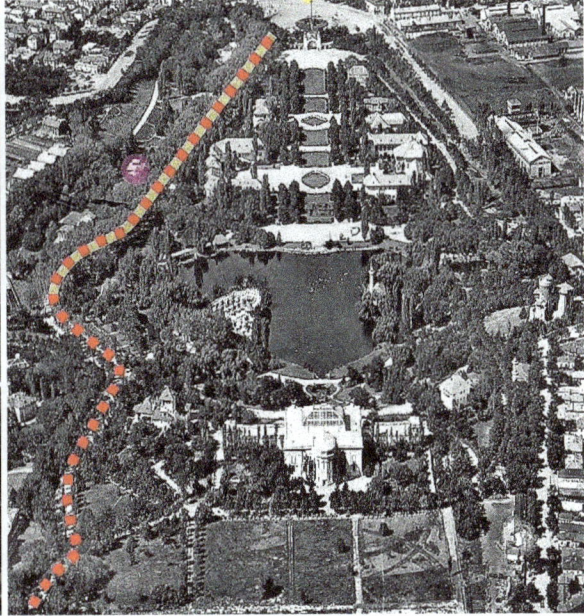

Figure 3. Visual analysis of the park Carol I in 1927: red - alley on the former Filaret Avenue with its white poplars alignment - yellow (Panțu, 2012; Drăgan, 2006)

The roads edging the parterre in the main axis were flanked on both sides by alignments of linden trees – young at plantation time (Figures 1, 8). On the outside margins of the axis, Redont put a line of more widely spaced columnar poplars alongside the linden trees, creating interesting vertical accents (Figures 4, 5, 7, 8). He used these species also in other areas of the park, especially on the lake shore, conceiving great compositional effects (Figures 14, 19).

On the eastern side of the park Redont designed an alley symmetrical to the former Filaret Avenue. He shaded it with a quadruple alignment of chestnuts a few of which remain today (Figure 4).

The axis was underlined by the horizontality of the linden alignments foliage, of the columnar poplars curtain and of the old white poplars in the background, as well as by the linearity of the parterre and of the box borders and box spheres that rhythmed its long edges (Figures 4, 5, 7, 14).

Redont placed spherical box not only along the axial parterre sides, where it highlighted the classical style, but also along secondary alleys to smooth the transition to the Romantic areas of the park (Figure 6). Some of these spheres still remain nowadays thanks to the well-known longevity of box (Figure 6).

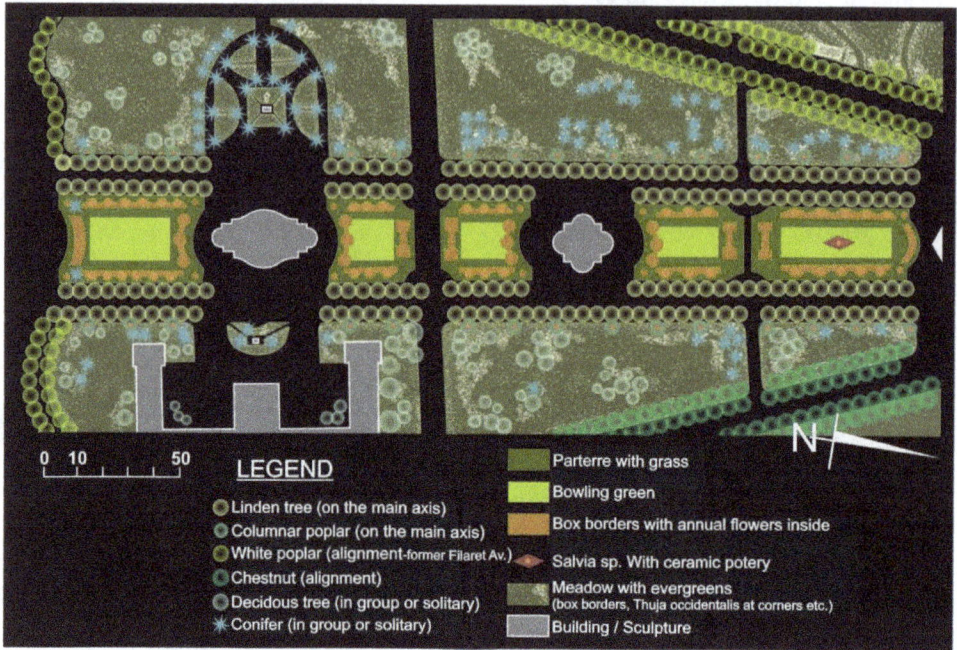

Figure 4. The main axis plan (Panţu, 2012; Marcus, 1958)

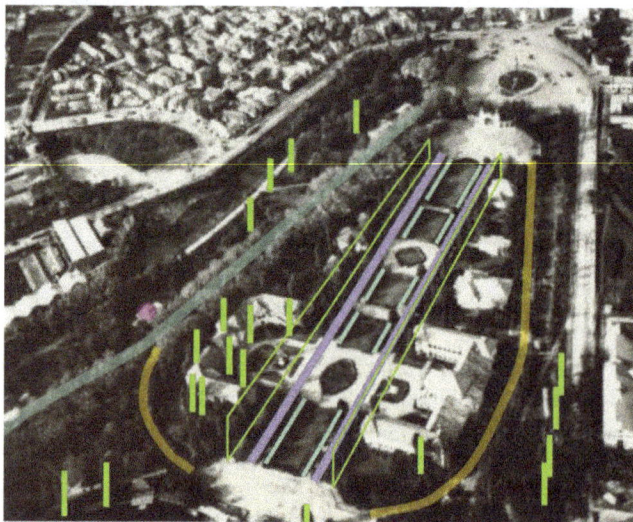

white poplar alignment - 4 lines (former Filaret Av.)
columnar poplar alignment
columnar poplar - Populus nigra 'Italica
linden tree alignment - 2 lines each side
dwarf box hedge underlining axis horizontality
chestnut alignment - 4 lines

Figure 5. Analysis of 1930 vegetation (Panţu, 2012)

Figure 6. Box spheres along curved alleys in 2011

We believe that the quadruple linden alignment on the main axis was geometrically pruned in 1935, on the occasion of another important exhibition that brought also other transformations to the park analysed in the paper *Carol I Park in Bucharest in the '30s – Celebrate Bucharest Month* (Panțu, 2011). The crowns formed long prisms, typical of the classical French style (Figure 7). Our arguments are that the intention of geometrization of the linden trees does not appear in the images until 1935 (Figures 3, 5), and becomes evident in the photographs from 1935 onwards (Figure 7). It is possible that the pruning of the alignments was part of Redont's vision for the future, as it was not possible in 1906 when the lindens where far from mature. The intention of pruning was still visible in the late '50s, but subsequently lost.

Figure 7. The main axis with the pruned linden trees, 1938

The vegetation details in Figure 8 from 1906 are in the same range of classic elements. The linden trees are young and the parterre with geometrical bowling green is contoured with small hedges in light coloured foliage plants and flowers. Those borders are punctuated with sphere shaped yew or box. The parterre corners were accentuated with *Canna indica*. Later, in the'50s, the ornamental details of the central parterre retained the same French classical spirit (Figure 9): strips marking its contour in *Begonia* and double borders in dwarf box, with accents of *Canna indica* and *Yucca filamentosa* etc.

Figure 8. The axial parterre and the Industries Pavilion at left to the entrance (Zaharia, 1906)

Figure 9. Planting design in a corner of the axial parterre in the '50s (Marcus, 1958)

Figure 10. Decorative basin on the main axis in 1928

In 1928 the neoclassical outline of the decorative basins was valorised by floral bands of *Canna indica* and *Tagetes* (Figure 10). The border of the basin in front of the Royal Pavilion was also accompanied by *Canna indica* (Figure 11).

Figure 11. Decorative basin on the main axis in front of the Royal Pavilion

We believe that there was a predilection for *Canna indica,* that can also be identified in other designs of the time, especially around basins, not only in Romania, but also in France, where, in our view, this fashion emanated at the end of 19th century. For example, The Buffet of the Avenue in Bucharest (a heritage building from 1892 by arh. Ion Mincu) (Figure 12) and a fountain in an important plaza in Bordeaux (Figure 13).

Figure 12. The Buffet of the Avenue in Bucharest

Figure 13. Fountain on Allées de Tourny, Bordeaux

Figure 14. The main axis from the Palace of the Arts (Romanian National Library Archive)

Linden alignment created prisms accentuating axiality and horizontality
Vertical accent of columnar poplar - *Populus Nigra 'Italica*
Coniferous vertical accent
Built vertical accent
Water jet vertical accent
Horizontal ligne in box
Rhythm elements in pruned box and yew accentuating axiality, horizontality
Background - white poplar alignment accentuating the horizontality

Figure 15. Visual analysis in 1935 (Panţu, 2012)

The most visible vertical accent in the Carol I park in the '30s was the minaret clearly distinguished not only on the sky by the contrast with the dark colour of its top, but also against the background vegetation by the difference in texture and colour and on the lake. Its verticality was underlined by the lake's horizontality and by dark foliage and a slender conifer that Redont placed in front (Figure 15). So, the minaret constituted an interesting landmark visually, but also culturally, as a Romantic park folly.

Vertical accents were placed also on the other side of the lake – unaligned, as a Romantic figure. Two spruces asymmetrically disposed framed the perspective of the axis towards the main entrance. The composition was balanced to the visual axis and so were all the vegetal elements (Figure 15).

The main axis continued visually over the sinuous lake towards a hill – the Romantic area. Here a balanced composition took the place of symmetry. Each element had something to balance it on the other side of the visual axis: the two slender spruces on the lake shore, the weeping trees on their side, the voluminous conifers with compact foliage – pruned yews or compact horticultural forms – the bold cypresses, the small shrubs at the ends of the Romantic bridge imitating tree branches (Eiffel pattern) etc. (Figures 14, 16). The group of three weeping trees - *Ulmus glabra 'Pendula* - formed an interesting presence on the lake shore (Figures 14, 16). Redont placed this variety of elm elsewhere in the park also, for its spectacular shape.

Figure 16. Views from the grotto towards the main entrance around 1910, in 1928 and in the '50s (Panţu, 2012; Octavian and Georgescu, 1999; A.F. Iliescu archive)

Beneath the Palace of Arts, around the grotto, Redont conceived a vegetal composition in conifers only, appropriate to the rockery articulating different height levels and so creating a mountain-like image on the steep slope. Dense habit, voluminous conifers were balanced disposed to the main axis (Figure 17).

Figure 17. Coniferous composition on the hill beneath the Palace of the Arts around 1910 and later, in 1937

Redont designed the plantation in Romantic style also in the surrounding park areas, on the lake shore and on the slopes. He composed it in elegant tree and/or shrub masses and groups and solitary specimens with which he created savant landscape sequences, in a clever scenography (Figures 17-19).

The lake shore was decorated with black and columnar poplars, willows, bold cypress spectacularly colouring the park landscape in autumn (Figures 16, 18, 19).

Figure 18. The lake, the mosque and the palace

Figure 19. The eastern lake shore in the '50s

Under the communist regime, in 1960, the Carol I park underwent radical transformations and lost a great part of its Romantic style in favour of a monumental geometric style. As a consequence, a lot of the plantation from 1906 disappeared. The percentage of original plantation from 1906 that has lasted until nowadays is small, about only 10%. I believe that this was caused also by the passing of time over 100 years, lack of vegetal protection strategy, bad heritage park management and faulty maintenance.

CONCLUSIONS

The percentage of the 1906 plantation that had lasted until nowadays is small, about only 10%. I believe that this was caused by the radical transformations in 1960, passing of time over 100 years, lack of vegetal protection strategy, bad historical park management and faulty maintenance.

Planting design is most important for overall composition and shows the characteristics of respective styles. It is primordial to understand the initial design of the Carol I park in order to comprehend, then preserve and/or restore the national landscape heritage.

REFERENCES

Bulei I., 1990. Atunci cand veacul se nastea... Eminescu Publishing House, Bucharest.

Drăgan V.E., 2006. Studiu Istoric pentru Parcul Carol I. Analiza valorilor culturale ale imobilului şi ale zonei învecinate. Reguli de intervenţie pentru conservarea valorilor culturale şi integrarea în zonă.

Marcus R., 1958. Parcuri şi grădini din România. Tehnica Publishing House, Bucharest.

Octavian T., Georgescu M.P., 1999. Inter-Bellum Bucharest, CD, NOI Media-Print, Bucharest & D.O.R. Kunsthandel, Olsberg.

Panţu I.M., 2011. Carol I Park in Bucharest at the Beginning of the 20th Century. in USAMV Bulletin Horticulture, 68 (I)/2011, Cluj-Napoca, 400-407.

Panţu I.M., 2011. Carol I Park in Bucharest in the '30s – Celebrate Bucharest Month. Scientific Papers. Series B. Horticulture – vol. LV 2011, Bucharest, 310-315.

Panţu I.M., 2012. Parcul public bucureştean din secolul XX. Influenţa modelului francez. Doctoral thesis

Panţu I.M., 2015. Carol I Park in Bucharest in the Second Half of the 20th Century. Scientific Papers. Series B. Horticulture – vol. LIX 2015, Bucharest, 385-392.

Parusi, Gh., 2007. Cronologia Bucureştilor. Compania Publishing House, Bucharest.

Potra, G., 1990. Din Bucureştii de ieri. Ştiinţifica şi Enciclopedica Publishing House, Bucharest.

Teodorescu V. Z., 2007. Un parc centenar. Parcul Carol I. Muzeul Municipiului Bucuresti Publishung House, Bucharest

Zaharia A., 1906. Expoziţiunea Generală Română din 1906 fotografiată de Al. Zaharia. Stereoscopic photos set.

29

EVALUATION *OF HELIOTROPIUM GREUTERI* FOR MORPHOLOGICAL CHARACTERISTICS AND POTENTIAL USE AS AN ORNAMENTAL PLANT

Akife DALDA ŞEKERCİ[1], Tuğçe TECİRLİ[1], Osman GÜLŞEN[1]

[1] Erciyes University, Department of Horticulture, Kayseri, Turkey

Corresponding author email: akifedalda@erciyes.edu.tr

Abstract

Boraginaceae is an important family distributed worldwide that includes herbs, subshrubs, shrubs, or trees. The paper aims to introduce a potential ornamental plant in the Kayseri city of Turkey. Belonging to the family Boraginaceae, the plants of Heliotropium greuteri naturally occur. In this study, 50 genotypes were examined for two morphological features (leaf size and flower diameter). The data obtained in the study were analysed with SPSS statistical software. Heliotropium gruteri genotypes indicated quite different characteristics studied in this region. Leaf size is highly variable (11- 42 cm) and leaf sizes of genotype 2, 9 and 17 are the largest; genotypes 37, 42 and 44 mm have smaller leaves. Looking at the flower top diameter, genotypes had values between 8.09 and 12.37 mm. Genotypes 25, 39 and 9 have the highest flower top diameters and genotypes 31, 10, 48 have the smallest flowers. No significant difference between genotypes in terms of flowers top diameter was detected. There was no correlation between leaf length and diameter of flower diameter. Thus, with fragrant flowers, it is suitable for use as a ground cover, which provides a decorative appearance to their environment, which can grow even in rocky areas, affected by harsh winds. They are able to tolerate moving and attracts honey bees as well. Because of the many features of H. greuteri such as drought tolerance and fragrance, it can have potential as ornamental plant. They can be suggested in parks, road sites, cemeteries and all low input areas as well as high input areas.

Key words: Heliotropium greuteri, Boraginaceae, ornamental plants

INTRODUCTION

Boraginaceae family of *Lamiales* order has 100 genera and 2.000 species of tropical, subtropical and temperate regions. *Boraginaceae* is an important family distributed worldwide that includes herbs, subshrubs, shrubs and trees. Within the family *Boraginaceae*, *Heliotropium* species exhibit great variation in many features of biological interest including habitat preferences, physiognomy and morphological traits (Al-Turki, 2001). The genus *Heliotropium* is part of the *Heliotropieae* tribe and *Heliotropioideae* subfamily; it is a euryspecific genus with 250 species (Saad-Limama, 2005). South-West and Central Asia are major centers of diversity in the genus *Heliotropium*. Most species of *Heliotropium* grow in areas with an arid and semi-arid climate, mostly on dry soils, gypsum hills, sandy and gravelly deserts, disturbed soils, eroded slopes, as weeds in cultivated lands and wastelands, along riversides, and, rarely, around hot springs (Akhani, 2007)

The genus *Heliotropium* L., according to a recent survey, comprises nearly 300 species assigned to 19 sections (Al-Turki, 2001). Most species are distributed in tropical and temperate regions of both hemispheres in a variety of habitats, including drifting sands, hardened sandy plains, edges of cultivated or saline waste ground and steep rocky outcrops as high as 1.500 m (Collenette, 1999). Many researchers, in an effort to confirm the identity of *Heliotropium* taxa. *Heliotropium gruteri* that described by our study, is the new species described in Kayseri region. *Heliotropium gruteri*, seen as extensively in the region, have fragrant flowers. These plants are suitable for use in the refuges and ground cover plants. They are very attractive and actively grows up until the first frost by forming flowers. Research is needed for further information on the properties of this species.

MATERIALS AND METHODS

Kayseri province is natural habitat of *Heliotropium gruteri*. In this study, 50 genotypes were characterized for leaf and flower characteristics, two most important features of ornamental plants. Comparative morphological analyses were conducted at the sites where the plants naturally grow. The observation sites are in the elevation at 1000-1800 meters within Kayseri province of Turkey in Central Anatolia. Means of leaf sizes and flower diameters were analyzed with SPSS statistical software.

RESULTS AND DISCUSSIONS

Heliotropium greuteri belonging to *Boraginaceae*, is an annual and branched plant. The observations have been made in Kayseri region with hot summers and cold winters. The soil is mostly sandy-loam with desert pavement and coralline features. Flowering and fruiting occurs profusely from June to November. This study revealed valuable information on *H. gruteri's* leaf and flower structure of 50 genotypes. *Heliotropium gruteri* genotypes had high level of variation in this region. Leaf sizes ranged between 11 and 42 cm. These properties are given in Table 1 below. While leaf size of genotype 2, 9 and 17 were the largest, genotypes 37, 42 and 44 had the smallest leaves. Looking at the flower top diameter, genotypes had the values between 8.09 and 12.37 mm. While genotypes 25, 39 and 9 having the highest flower diameter, genotypes 31, 10, 48 had the smallest flowers. These properties are shown in Table 2 below. No significant differences were detected between genotypes in terms of flowers top diameter. There was no correlation between leaf length and diameter of flower top.

Literally, *H. gruteri* are 20-60 cm high herbs, with spreading and retrorse hairs, herbs annual and much branched. Petiolate, up to 11.5-42.0 cm long , gradually reduced in upper portion of the stem; narrowly ovate, undulate, spreading and retrorsely bristly hairy; hairs of 2 to 3 types, the basal portion of the hairs on the midrib and veins extremely swollen.

Infloresence terminal, of usually more than 2, spicate cymes, elongating in fruit. Flowers white, 5-10 mm long. Calyx 3 mm long, the lobes 2.5 mm long, densely bristly hairy, setose except at the bases. Corolla 5 mm long; tube slightly constricted near the mouth and below the stigma level; lobes obtuse, and spreading. Anthers 1 mm long, pointed, inserted at about the middle of the corolla tube. Stigma short, conical, apex obtusish, glabrous. Nutlets shortly winged, 3 mm long and wide including the wings, glabrous, with or without one or two spongy callosities on the outer surface. Shrubs of *H. gruteri* are given in Figure 1 below.

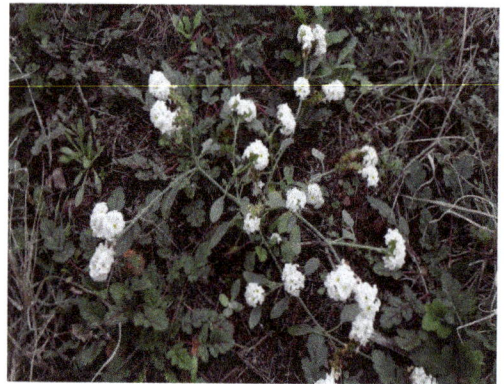

Figure 1. Pictures of shrubs of *Heliotropium greuteri* plants in August

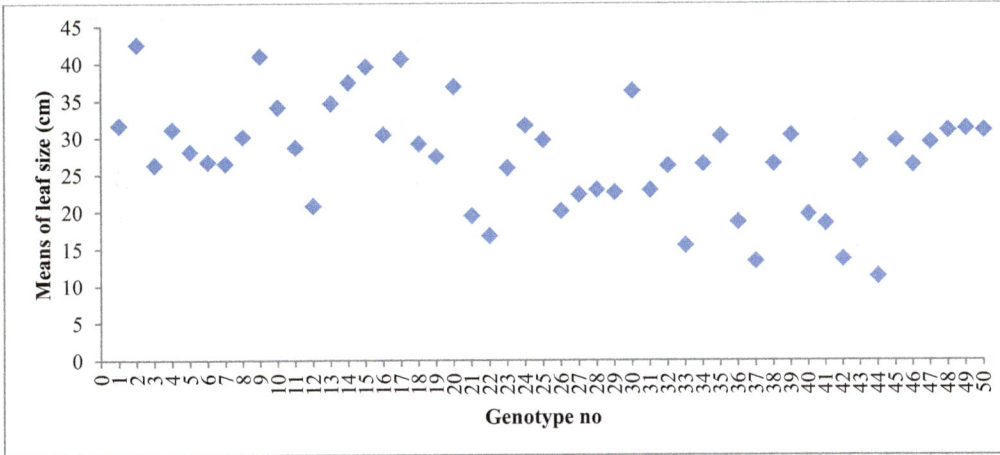

Figure 2. Means of leaf size

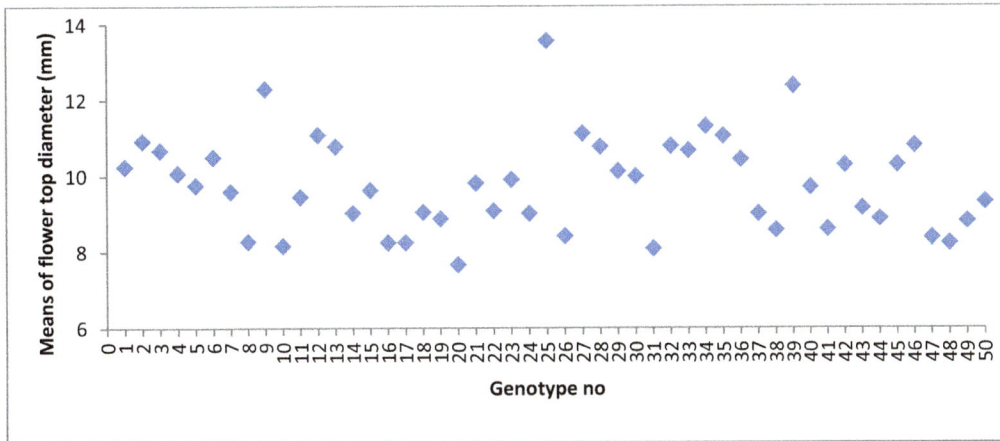

Figure 3. Means of flower top diameter

CONCLUSIONS

Turkey, because of its geographical location, high variable climates due to elevation differences, mountains ranging in short distance, geological, geomorphological structure, soil diversity and historical reasons, is one of the most important diversity centers in the world in terms of plant species.

As endemics of Turkey, *H. greuteri* may play a role in landscaping. We emphasis aesthetic value of natural species with their ornamental plant capacity. The usage of natural species play significant mission with ecological and providing identity for cities.

In this study, we indicated variation and potential of two important ornamentally important characteristics of 50 *H. greuteri* plants occurring in Kayseri, Turkey with elevation of 1200-1800 meters.

They are highly drought tolerant, attractive for fragrance, but having little size of flowers. Improving flower sizes may be better by breeding through selection, mutation breeding, etc.

Study results showed that *H. gruteri* have high potential for the use in landscape design.

REFERENCES

Al-Turki T. A., Omer S., Ghafoor A., 2001. Two new species of *Heliotropium* L. (*Boraginaceae*) from Saudi Arabia. Botanical Journal of the Linnean Society, 215-220.

Collenette S., 1999. Wild flowers of Saudi Arabia. Riyadh: National Commission for Wildlife Conservation and Development, 89-92.

Akhani H., 2007. Diversity, biogeography, and photosynthetic pathways of Argusia and *Heliotropium* (*Boraginaceae*) in South-West Asia with an analysis of phytogeographical units. Botanical Journal of the Linnean Society, 401–425.

Saad-Limama S. B., Nabli M. A., Rowley J. R., 2005. Pollen wall ultrastructure and ontogeny in *Heliotropium europaeum* L. (*Boraginaceae*). Review of Palaeobotany and Palynology, 135– 149.

30

EFFECT OF MULCHING ON WEED INFESTATION AND YIELDS OF LEEK (*ALLIUM PORRUM* L.)

Nina GERASIMOVA[1], MilenaYORDANOVA[2]

[1]Institute of Plant Physiology and Genetics, Bulgarian Academy of Sciences, Acad. G. Bonchev Street, Bldg. 21,1113, Sofia, Bulgaria, GSM: +359.889.968.339, Email:gerasimova_n@abv.bg
[2]University of Forestry, Faculty of Agronomy, 10 KlimentOhridski Blvd, 1756, Sofia, Bulgaria, GSM: +359.887.698.775, Email: yordanova_m@yahoo.com
Corresponding author email: yordanova_m@yahoo.com

Abstract

The aim of this field experiment was to study the effect of two mulching materials on weed infestation and yield of leek, cv. 'Bulgarian Giant'. The field experiment was carried out in the period 2010-2012 in the experimental field on University of Forestry – Sofia. The experimental design was the randomized block with four replicates. Two different mulching materials – barley straw mulch (BSM) and mulch from spent mushroom compost (SMCM) were compared with two control variants – non-mulching, but weeding control (WC) and non-mulching and non-weeding control (NWC). The mulching materials were spread manually in a 5 cm thick layer, one week after transplanting the seedlings of leek. On the 30th, 60th and 90th day after mulching were recorded the number of weeds on each plot. It was found out that mulching with BSM and SMCM have a significant depressing effect on weeds, especially on Echinochloa crus-galli L., Setaria glauca (L.) Beauv., Galinsoga parviflora Cav., Polygonum lapathifolium L. and Portulaca oleracea L. The yields were increased from 3.7 to 4 times when the leek was grown with mulches, compared with NWC. Data were subjected to statisticall analysis using dispersion method. Means were separated by application of Duncan's Multiple Range Test at p ≤ 0.05.

Key words: *barley straw mulch, spent mushroom compost mulch, weed infestation, leek.*

INTRODUCTION

Leeks (*Allium porrum* L.) are members of onion family, closely related to onion, garlic, shallots and chives. (Cholakov, 2009).
Weeds are competitors of most vegetable crops and can reduce their yields significantly. The main annual weeds that occur on arable land under cultivation of species of family Alliaceae are different types of amaranth, fat-hen, thorn-apple, pale persicaria, bristle-grass, cockspur, red finger-grass etc. Also, infestation of arable land with perennial weeds such as Johnson grass, creeping thistle, field bindweed, etc. has been observed (Tonev, 2000). Decrease of weed infestation depend on fact that leek is growing under irrigation and natural fertilizer.
One of alternative method for weed control is use of different kinds of mulch. In the integrated and ecological agriculture systemsmore attention is being paid to the longest possible periodof soil coverage with plant mulches and mulches from straw left aftercereal grain harvest (Szymona, 1993).

Organic mulch can block light to the soil surface,reducing the germination and growth of weeds (Anyszka, Dobrzański, 2008).A number of studies have documented that straw mulch isa good means of decreasing weed emergence and growth (Duppong et al., 2004; Grassbaught et al., 2004; Teasdale and Mohler, 2000). Covering or mulching the soil surface canreduceweed problems bypreventing weedseed germination or by suppressing the growth of emerging seedlings (Bond et al., 2003). Mulching decrease the numbers of hand-hoeing and mechanical cultivations for remove of weeds. The key factors that make straw mulchattractive are low cost and easy in availability and application(Ramakrishnaetal., 2006).
According to the data of experiments, straw mulchis best for weed control. In plots with straw mulch weeddensity wasestablished at2.8–6.4 times lower compared with weed density in plots without mulch (Sinkevičienė et al., 2009). According to Radwanand Hussein (2001) broad-leaved weeds were more susceptible than

grassy weed to mulching treatments. Mulching improves plant growth, increases yields their quality (Sharma, Sharma, 2003; Singh et al., 2007).

In studying of effect of different organic mulches on weed infestation was establish that mulching with spent mushroom compost, crushed corn cobs and long wheat straw reduced weed germination and weed growth. They suppressed bettermonocotyledonous than dicotyledonous weeds, except straw mulch (Yordanova, Shaban, 2007).

The aim of thepresentstudy was to evaluate the influence of different organic mulches on weed infestation and yield of leek.

MATERIALS AND METHODS

The studies were conducted in the period 2010-2012 in the experimental field on University of Forestry – Sofia, on the Fluvisol soil type.

The leek cultivar "Bulgarian giant" was grown through seedlings which were planted in the second half of June by scheme 60+25+25+25+25/15. The preceding crop was broccoli. The leek was cultivated by drip irrigation. Each trial was laid out in a randomized block-design with four replications (4x40), with protection zones.

The experiment was carried-out with four treatments: 1 - non mulching, but weeding control (WC); 2 - non mulching and non-weeding control (NWC); 3 – mulch from spent mushroom compost (SMCM); 4 – barley straw mulch (BSM). The mulching materials were spread manually in a 5 cm thick layera month after planted of leek.

The occurrence, extent and types of weeds werestudied at 30 and 60 daysafter mulching (DAM) at fixed sites of $1m^2$for each treatment and replicate. All weeds in eachquadrat were identified, counted and recorded forsubsequent data analysis.

The efficacy of the tested mulching materials was recorded by Abbot's formula:

WG% = (CA-TA/CA) x 100, where:

WG% - the percentage efficacy of the herbicides;

CA - living individuals in the control after treatment;

TA - individuals living in the variant after treatment.

The length and diameter of the false stem were measured on 10 plants and presented the average results. The total yield is established in tones per decare (t/da) in replications and variants.

Data were subjected to statistical analysis using dispersion method. Means were separated by application of Duncan's Multiple Range Test at p ≤ 0.05.

RESULTS AND DISCUSSION

The level of weed infestation in agrocenoses of leeks recorded on 30th and 60^{th}DAM is given in Table 1 - 3. In this agrocenoses the following weed species were established: cockspur (*Echinochloa crus-galli* L.), red finger-grass (*Digitaria sanguinalis* (L.) Scop.), green foxtail (*Setaria viridis* (L.) Beauv.), yellow foxtail (*Setaria glauca* (L.) Beauv.), galinzoga (*Galinsoga parviflora* Cav.), amaranth (*Amaranthus retroflexus* L.), common lambsquarters (*Chenopodium album* L.), purslane (*Portulaca oleracea* L.), and pale persicaria (*Polygonum lapathifolium* L.). In the variant with barley straw mulch (BSM) was recorded and barley (*Hordeum vulgare* L.).

The unmulched plots showed a greaterdiversity of weed species than the mulched plots in period 2010-2012. At 30DAM mulching from spent mushroom compost (SMCM) showed lower weed infestation than barley straw mulch (BSM) (table 1). In this variant weed species *Amaranthus retroflexus* had average number per square meter 10.5. The other weeds in agrocenoses of leek were with single numbers which didn't affect on leek. At 60DAM was establishing low increase of weed infestation in V3. *Amaranthus retroflexus* again was with the most numbers per square meter - 11.75. In treatment with BSM was established higher weed infestation with annual monocotyledonous weeds than dicotyledonous. Weed scores showedsignificant differences (p ≤ 0.05) in three experimental years.

Analogous results were obtained in year 2011 (table 2). The most effective weed control was recorded in the plots with mulch from spent mushroom compost (SMCM) except for amaranth (*Amaranthus retroflexus* L.) which average number was 18.75 per 1 m^2. In spite of this a significantdifferencewas observed between mulching by spent mushroom compost and non-mulching control. Mulching with barley straw (BSM) showed a slightly difference with SMCM, which is in little higher growth of annual monocotyledonous weed species greenfoxtail (*Setaria viridis* (L.) Beauv.). There were established single numbers in particular replications of weed species *Galinsoga parviflora* and *Portulaca oleracea* in which there was delay in its grow up.At 60DAM the applied organic mulches affected in high extent theweed species. The number of *Amaranthus retroflexus* in SMCM was unaffected but in BSM it was increase. In this variant has reported the single plants of barley (*Hordeum vulgare* L.), which is because of the presence of barley seeds in straw mulch. A significant differences in the average number of weeds in 1m^{2}was observed between NWC, SMCM and BSM.

Table 1. Average number of weeds in 1 m^2 after mulching (2010)

Weed species	NWC		SMCM		BSM	
	30 DAM	60 DAM	30 DAM	60 DAM	30 DAM	60 DAM
Echinochloa crus-galli	17.75 [a]	51.5 [a]	1.5 [b]	1.75 [b]	2.50 [b]	2.75 [b]
Digitaria sanguinalis	9.75 [a]	11.5 [a]	0 [b]	0.75 [b]	1.50 [b]	2.00 [b]
Setaria viridis	11.25 [a]	17.5 [a]	0 [b]	0 [b]	9.50 [a]	9.50 [a]
Setaria glauca	3.50 [a]	6 [a]	0 [b]	0 [b]	0 [b]	0 [b]
Hordeum vulgare	0	0 [b]	0	0 [b]	0	7.00 [a]
Galinsoga parviflora	11.00 [a]	19.5 [a]	0 [b]	0.75 [b]	1.25 [b]	2.25 [b]
Amaranthus retroflexus	43.50 [a]	52.5 [a]	10.5 [b]	11.75 [b]	2.75 [c]	7.50 [c]
Chenopodium album	2.25 [a]	3.25 [a]	0 [b]	0 [b]	0 [b]	0 [b]
Portulaca oleracea	18.75 [a]	21.5 [a]	0.5 [b]	0.50 [b]	0 [b]	0 [b]
Polygonum lapathifolium	1.75 [a]	3.5 [a]	0 [b]	0 [b]	0 [b]	0 [b]

Values with the same letter within years are not significantly different (Duncan's Multiple Range Test at p ≤ 0.05)

Table 2.Average number of weeds in 1 m^2 after mulching (2011)

Weed species	NWC		SMCM		BSM	
	30 DAM	60 DAM	30 DAM	60 DAM	30 DAM	60 DAM
Echinochloa crus-galli	27.75 [a]	45.75 [a]	1.25 [b]	1.25 [b]	1.25 [b]	1.50 [b]
Digitaria sanguinalis	10.25 [a]	21.50 [a]	0 [b]	0.25 [b]	1.50 [b]	1.50 [b]
Setaria viridis	9.75 [a]	9.75 [a]	0 [b]	0 [b]	9.50 [a]	9.50 [a]
Setaria glauca	0.75 [a]	2.00 [a]	0 [b]	0 [b]	0 [b]	0 [b]
Hordeum vulgare	0	0 [b]	0	0 [b]	0	9.00 [a]
Galinsoga parviflora	8.00 [a]	15.50 [a]	0.50 [b]	1.00 [b]	1.25 [b]	2.25 [b]
Amaranthu sretroflexus	24.50 [a]	32.50 [a]	18.75 [b]	18.75 [b]	1.75 [c]	6.00 [c]
Chenopodium album	0.25 [a]	0.25 [a]	0 [b]	0 [b]	0 [b]	0 [b]
Portulaca oleracea	4.25 [a]	4.25 [a]	1.00 [b]	1.25 [b]	0 [b]	0 [b]
Polygonum lapathifolium	0.25 [a]	0.25 [a]	0 [b]	0 [b]	0 [b]	0 [b]

Values with the same letter within years are not significantly different (Duncan's Multiple Range Test at p ≤ 0.05)

Table 3.Average number of weeds in 1 m^2 after mulching (2012)

Weed species	NWC		SMCM		BSM	
	30 DAM	60 DAM	30 DAM	60 DAM	30 DAM	60 DAM
Echinochloa crus-galli	10.50 [a]	32.75 [a]	0.75 [b]	1.75 [b]	1.25 [b]	2.50 [b]
Digitaria sanguinalis	3.75 [a]	6.50 [a]	0 [b]	0.50 [b]	0.75 [b]	1.75 [b]
Setaria viridis	7.50 [a]	13.50 [a]	0 [b]	0 [b]	4.75 [a]	5.00 [a]
Hordeum vulgare	0	0 [b]	0	0 [b]	0	6.50 [a]
Galinsoga parviflora	7.50 [a]	12.25 [a]	0 [b]	0.50 [b]	0.25 [b]	0.75 [b]
Amaranthu sretroflexus	22.75 [a]	24.25 [a]	0.50 [b]	0.75 [b]	0 [b]	0.50 [b]
Chenopodium album	1.50 [a]	3.50 [a]	0 [b]	0 [b]	0 [b]	0 [b]
Portulaca oleracea	9.75 [a]	10.50 [a]	0 [b]	0.50 [b]	0 [b]	0 [b]
Polygonum lapathifolium	1.50 [a]	2.75 [a]	0 [b]	0 [b]	0 [b]	0 [b]

Values with the same letter within years are not significantly different (Duncan's Multiple Range Test at p ≤ 0.05)

In year 2012 was established lower weed infestation than previous experimental years (Table 3). In SMCM treatmentwere monitored single plants in particular replications at 30 and 60 DAM of *Echinochloa crus-galli*, *Digitaria sanguinalis*, *Galinsoga parviflora*, *Amaranthus retroflexus* and *Portulaca oleracea*. In BSM treatment again was established higher growth of monocotyledonous weed species than dicotyledonous. Straw mulch's favourable effect on the limiting ofweeds infestation was also confirmed in the study by Ramakrishna et al. (2006).

The lowest weed infestation was recorded in mulching variants. This show the effectiveness of this method in the suppressing weed germination. The spend mushroom compost has a strong depressing effect on the development of annual monocotyledonous weeds, which has been found by other authors (Yordanova, Shaban, 2007). Lowerinfestation on the covered plots was due to the fast rateof crop plant growth and higher possibilities to competewith weeds compared to plants with non-mulching and non-weeding control.The results showed that two types of mulch caused a decreasein weed infestation, compared to the control plot. This was confirmedin the study by Kosterna (2014).

The efficacy of applied soil mulches on weeds is shown in Figure1-2. In year 2010 at 30 DAM mulching from spent mushroom compost (SMCM) showed higher efficacy than barley straw mulch (BSM). It range from 75,9%

against *Amaranthusretroflexus* to 100% against *Digitaria sanguinalis*, *Setaria viridis*, *Setaria glauca*, *Galinsoga parviflora*, *Chenopodium album* and *Polygonum lapathifolium*. The lowest efficacy in BSM was recorded against *Setaria viridis* – 15.6% (fig. 1). In year 2011 the lowest efficacy to *Amaranthus retroflexus* - 23.5% at 30 DAM and 42.3% at 60 DAM and to *Portulaca oleracea*– 76.5% at 30 DAM and 70.6% at 60 DAM was established in SMCM.The toxicity of mulch from spent mushroom compost on the other weeds of agrocenoses that interfere with leek production was above 93%. In BSM treatment the efficacy to monocotyledonous weeds were from 2.6% to *Setaria viridis* at 30 DAM to 96.7% to *Echinochloa crus-galli* at 60 DAM. Barley straw mulch shows lower efficacy for *Amaranthus retroflexus* than mulch from spent mushroom compost (92.9% at 30 DAM and 81.5% at 60 DAM). In year 2012 at 30 DAM was established higher efficacy of mulching from spent mushroom compost than barley straw mulch. There were only single numbers of weed species *Echinochloa crus-galli*and *Amaranthus retroflexus* in SMCM.

At 60 DAM the efficiency of mulching materials retained high (Figure 2). Mulching variants were characterized by low growth rate of existing weed species as they did not competed with the growth of leek plants. The used mulching materials showed good efficacy at 60 DAM against weed species in leek agrocenosis.

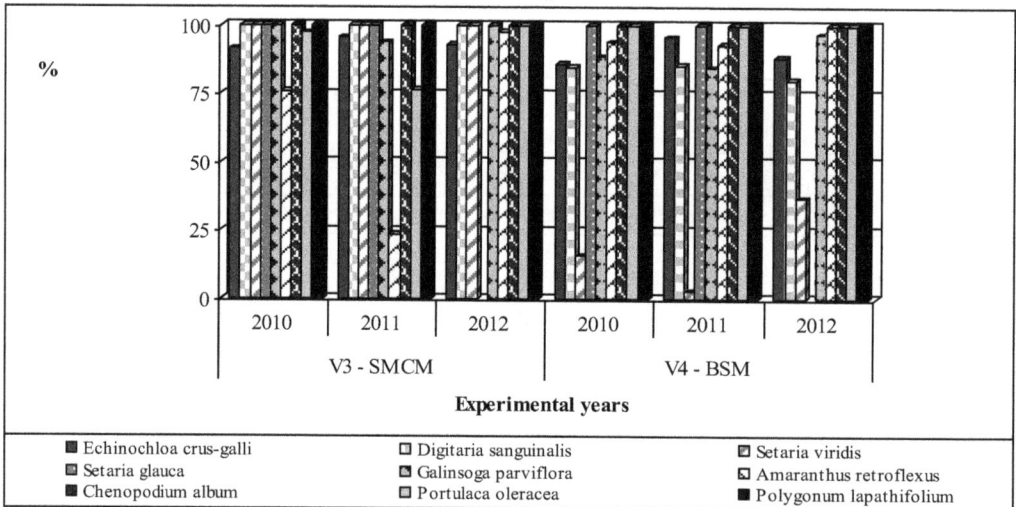

Figure 1. Efficiency of soil mulches compared to the control at 30 DAM (2010-2012)

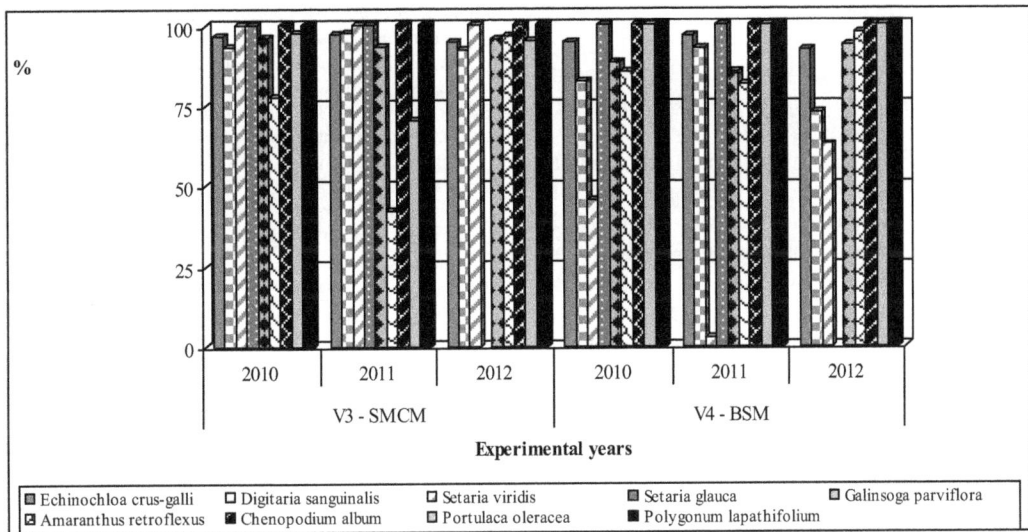

Figure 2. Efficiency of soil mulches compared to the control at 60 DAM (2010-2012)

The results obtained after gathering crop show that the yield of leek is lowest at the variant 2 – non-mulching and non-weeding control (NWC) (Figure 3). The yield obtained by the other variants is highest in year 2012 when the weed infestation was poorly developed compared to the other experimental years. During the three years of the field experiment the highest average yield was obtained in plots, mulched with barley straw mulch – 7.4 t/dain 2010, 7.7 t/da in 2011 and 7.8 t/da in 2012. In the variants mulching with spent mushroom compost the average yields were 6.4 t/da in 2010, 7.2 t/da in 2011 and 7.5 t/da in 2012. The lowest yield was obtained in plots from the second control, which is with non weeding plots (NWC).

The higher yield of mulching plots, compared with both controls – weeding control and non weeding control proves the efficiency of the mulches against weeds, but also in increasing the yields. These results were observed in studies made by other authors (Sharma & Sharma, 2003; Singh et al., 2007).

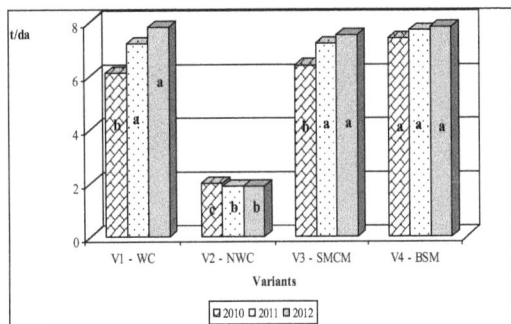

Values with the same letter within years are not significantly different (Duncan's Multiple Range Test at p ≤ 0.05)

Figure 3 Average yield (t/da) of leek

Differences between non weeding control (NWC) and other variants were very well statistically proven in the three years of field experiment.Weed infestation of non weeding plots decreased significantly the yield – from 3.7 to 4 times lower yield compared with mulching plots.

After the statistical analysis of data we can make the conclusion that yields obtained at mulching by spent mushroom and barley straw mulches differ statistically from the control.

CONCLUSIONS

It was found that growing leek by mulching with barley straw or spent mushroom compost reduces weed infestation.

It is proved that mulching leading to increased yields by 3.7 to 4 times in comparison with plots with weeds.The yields obtained in mulching plots with these studied mulches are similar or higher than those of the weeding plots. This indicates that the mulching is suitable for growing leek through reduced tillage.

The applied mulches can be used easily during the growing stage of leeks and they control efficiently the widespread monocotyledonous and dicotyledonous weed species.

REFERENCES

Anyszka Z., A. Dobrzański, 2008. Changesin weed infestation in transplanted leek grown in organicmulch. Prog. Plant Prot./Post. Ochr. Roślin 48 (4):1391–1395. (in Polish)

Bond, W., R. J. Turner, A. C. Grundy, 2003. A Review of Non-chemical Weed Management. HDRA, 20-24 (available at www.organicweeds.org.uk)

Cholakov, D. T., 2009. Vegetable growing. Plovdiv. (in Bulgarian)

Duppong L. M., Delate K., Liebman M., HortonR., Romero F., Kraus G.,Petrich J., Chowdbury P. K., 2004. The effect of natural mulches on crop performance, weed suppression and biochemical constituents ofCantip and St. John's Wort. Crop Sci. 44 (3): 861-869.

Grassbaugh E. M., Regnier E. E., Bennett M. A., 2004. Comparison of organic andinorganic mulches for heirloomtomato production. Acta Hort. 638: 171-177.

Kosterna, E., 2014. The effect of soil mulching with organic mulches, onweed infestation in broccoli and tomato cultivatedunder polypropylene fibre, and without a cover. Journal of Plant Protection Research, 54 (2): 188–198.

Radwan, S., H. Hussein, 2001.Response of Onion (Allium Cepa,L.) Plants and Associated Weeds to Biofertilization under Some Plant Mulches and Associated Weeds. Annals Agric. Sci., Ain Shams Univ, Cairo (Egipt), 46: 543-564.

Ramakrishna, A., H. Tam, S. Wani, T. Long, 2006. Effects of mulch on soil temperature, moisture, weed infestation and yield of groundnut in northern Vietnam. Field Crops Research, 95 (2-3): 115-125.

Sharma, R.R., V.P. Sharma, 2003. Mulch influences fruit growth, albinism and fruit quality in strawberry (Fragaria x ananassa Duch.). Fruits 58: 221–227.

Singh, R., S., R.R. Sharma, R.K. Goyal, 2007. Interacting effects of planting time and mulching on "Chandeler" strawberry (Fragaria x ananassa Duch.). Sci. Hortic. 111: 344–351.

Sinkevičienė A., D. Jodaugienė, R. Pupalienė, M. Urbonienė, 2009. The influence of organic mulches on soil properties and crop yield.Agronomy Research,7: 485-491.

Szymona, J.,1993. Soil cultivation. Ecological agriculture fromtheory to practice.Stiftung Leben und Umwelt,Warszawa: 131-137. (in Polish)

Teasdale J. R., Mohler CH. L., 2000. The quantitative relationship between weed emergence andthe physical properties of mulches. Weed Sci. 48: 385-392.

Tonev, T., 2000. Manual for integrated weed control and crop farming. Plovdiv, Agricultural University. (in Bulgarian).

Yordanova, M., N. Shaban, 2007. Effect of mulching on weeds of fall broccoli. Buletinul USAMV-CN, 64 (1-2): 99-102.

31

IDENTIFICATION AND PRESERVATION OF CULTURAL AND LANDSCAPE IDENTITY – THE PLANE TREE PARK OF BUCHAREST

Elisabeta DOBRESCU[1], Sanda PETREDEANU[2]

[1]University of Agronomic Sciences and Veterinary Medicine of Bucharest,
59 Marasti Blvd., District 1, Bucharest, Romania
[2]The National Bank of Romania, Bucharest, Romania
Corresponding author email: veradobrescu@gmail.com

Abstract

Urban green spaces are defined as city areas where complex interaction of environmental, human, socio-economic and cultural factors take place, that gives us a dynamic perspective, but with some stability, given by the customs and values of a society under transformation. The valuable urban landscape, both culturally and in terms of the quality of urban life, is a subject that can be approached among climate, social, and territorial changes attempt by the city in its evolution. Urban green spaces under increased real estate and social pressure are gradually diminished in terms of quality and value. This leads to focusing on elements that serve as memorial landmarks, heritage landscape that can define and characterize the evolution of a society. The secular trees, or those white aesthetic, historical, memorial and social value, are such cultural landmarks that require great attention, taking into account their value, doubled by the lack of a perspective to protect and preserve historical and landscape significance. In the current context, public or private green areas of Bucharest has a total of 110 specimens of protected trees, considered natural monuments, according to the List of protected trees, managed by the Romanian Academy. An urban green area less known but very valuable, containing a single area with almost a third of the total number of trees mentioned on this list, is represented by the area owned by the National Bank of Romania, located near the pier of Dambovita, better known by citizens as the place of sports competitions NBR Arenas. The 31 trees of the Platanus x acerifolia species present in this site require thorough research, are considered, in an advanced state of decay. Our study aims to raise awareness and advertise, this concentrated protected historic landscape site, in order to initiate a complex process of integrated and differentiated management of the tree vegetation of cultural identity and social values.

Key words: regeneration, rebirth of value, urban historical reintegration, trees protection.

INTRODUCTION

Urban green spaces have a large contribution in defining the quality parameters of social and cultural life, being a dynamic component of environment, under constant urban evolution at the same time with the society evolution.

Generally, we can talk about two basic senses of the quality notion:

-quality based on the understanding of the Aristotelian sense of the "quality-species" term, which defines the sense of being of a thing, as nature, the essence of the thing;

-quality defined by considering a thing like "good" or "bad" defining the "quality-value" term.

At the same time, we can talk about the primary and secondary qualities of landscaping, defined in terms of philosophy, the primary ones being intrinsic qualities, of real things and the secondary ones being those that have the power to bring us certain feelings (Stănescu, 2008). All these relations determine another category of quality, more special for a city, namely its inner energy, which defines and characterizes a city (Sandu, 1992).

In a city, urban energy and quality of life is supported and determined in a large scale by the dynamic component of the vegetation too, especially by tree vegetation, playing its part in regulating the urban ecosystem. The world becomes richer due to the memorable places where people design their life and constructions (Simonds, 1967). The ecological and cultural-historical principles of the building and analyzing urban landscapes lead to conservation and preservation measures, in case of valuable urban landscaping.

The management, based on the analysis and control of urban landscaping development, allows the controlled evolution of vegetal assemblies, based on the principles of globality and continuity of the analysis, assessment and intervention over the vegetation.

Provided that the urban space has special features, given by the historical, memorial, ecological or social value, the urban vegetal assembly becomes heritage which can help in the interpretation of the society evolution, by understanding social, aesthetic and cultural concerns.

In the heart of Bucharest, very close to the "heart of the city", on the right side of Dâmbovița River, we have a very valuable landscape, dominated by some special historical plane trees, of the *Platanus × acerifolia* species, which actually gave the name of the land - the Plane Tree Park. The piece of land suffered progressive changes, currently belonging to the National Bank of Romania.

The current study aims to investigate the evolution of the site, analysis of the history and continuity of the site, identification of trees with historical, ecologic, social and memorial value, control and update of information included in the List of protected trees, related to the *Platanus × acerifolia* trees existing in the Plane Tree Park. After the heritage trees have been identified, the study aims for the future to continue the researches, to achieve a complex analysis of the dendrometric characteristics of each tree, determining the functional, aesthetic and memorial value in order to preserve and protect the heritage of this site.

All researches and analysis on the field and in the archives will lead to specific measures for the preservation and regeneration of the historical landscaping in the Plane Tree Park.

MATERIALS AND METHODS

Urban landscaping can be considered a symbol, a sensible unit, which interpretation is determined by the cultural experience, respectively of practices and events specific to communities (Majuru, 2012). At the same time, it is a specific layer accumulation of several practices, policies, customs, events which builds the memory of the place (Tudora, 2009). Amid cultural accumulation and territorial changes, the site investigation method was focused on two important issues:

- studies and historic analysis of the landscape structures of the site - analysis of the urban area;
- studies dendrometric and visual analysis of the secular plane trees on the studied site.

The historical analysis was based on the archival studies of the place history, completed with the study and analysis on the field of the existing structure. The historical study analyses the evolution of the current territory of the Plane Tree Park, in order to identify the specific moments during its existence, as well as in order to raise awareness on the landscape heritage value of Bucharest.

The historical plans of Bucharest have been analyzed (1846, 1852, 1871, 1895, 1899 and 1911), together with the literature in terms of evolution of the public urban green space and of gardens of Bucharest.

Historical research started with the analysis of the most important structures - roads, benchmarks, nodes, limits, specific typology areas - which defined the mental map (Lynch, 1960), but also the physical and evolutional map of the place.

The historical analysis also indicated how the secular vegetation of the Plane Tree Park started and developed in terms of urban and social conditions on this location.

Nevertheless, the complex landscaping analysis methods conducted on the field referring to the historical vegetation were focused on dendrometric measurements, analysis to identify the aesthetic, historical and landscape values, on considering the state of health of the analysed trees.

A precise survey was conducted with the position of all plane trees, each tree analyzed was assigned a unique registration code, according to the vegetation inventory method (basic tool in Management of Urban Vegetation).

Tomography analyses were conducted for the internal structure of each historical plane tree log existing in the analyzed site. This complex analysis helped in identification of the internal gaps, of wood firmness, and finding hidden cavities which affect trees' stability and health.

All assessments and analysis performed were scientifically construed and they were considered as the first database organized on scientific grounds of the Plane Tree Park. This particularly important document is the main tool used to prepare the Management and Integrated Plan of the Plane Tree Park.

RESULTS AND DISCUSSIONS

Analysis and historical studies

The analysis and interpretation of the historical plans indicated the evolution of the analysed space, from the peri-urban area to the modern area, emblematic of today's society.

In terms of historical period evolution, we can see the stratification of at least six determined period, with insignificant overlaying of years, when the current territory undergone significant changes:

- the first period -until 1870, indicates us a peri-urban area, affected by the constant flooding of Dâmbovița. The area is left somehow not systematized because the city not included Dâmbovița River. Starting with 1800, Bucharest inhabitants begin to capitalize the river's adjacent spaces, for leisure purposes. The land under study was overlapping the semi-agricultural piece of land existing almost two centuries ago in this location (Figure 1).

Figure 1. Bucharest Plan in 1846 - taken over from the Borroczyn Plan - according to the redo of drawings in 1911 coordinated by Cincinat Sfințescu
(Source: www.ideiurbane.ro)

-the second period - from 1962 until 1895, when the first systematization of Dâmbovița river occurs and the land is designed as target range. The project coordinated by Major Papassoglu shows a space organized with an alley, bordered by a bilateral tree alignment, leading to the shooting gallery - "Tiru Gherman" (Figure 2). Dâmbovița is regularized, drained, registered in the land of

the "Societatea de Dare la Semn", established by the ruler Cuza in 1862, the access was done by Notagiilor Street, part of the current Dr. Staicovici Street.

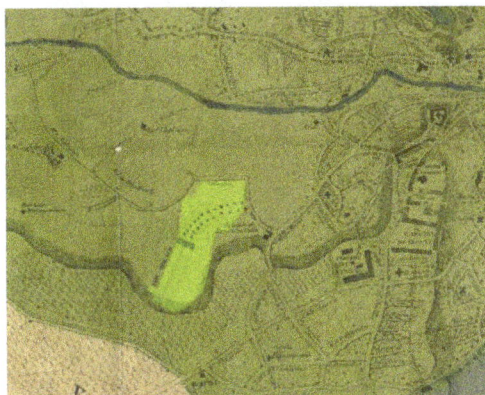

Figure 2. Bucharest Plan in 1871 - taken over from the Papassoglu Plan - detail of Green Color
(Source: www.ideiurbane.ro)

-the third period - since 1890 until 1911, when final rectification of Dâmbovița takes place and when the main boulevards are drawn in the urban structure. We can see that the land of the "Societatea de Dare la Semn" is organized with alleys, pavilions, plantations and a pond. The streets and lots of the new district, Cotroceni are drawn (Figure 3 and Figure 4).

-the fourth period – 1912, the area is partially rethought in terms of functionality, the park is shared by Tenis Clubul Român (Romanian Tennis Club) and Societatea de Tir (Shooting Gallery Company), the previous landscape is changed.

-the fifth period - the interwar period - it does not bring any radical changes, the neighborhood is defined as street scanning field, Dr. Lister Streets appear, when the main entrance of Romanian Tennis Club is moved. In 1939, the tennis courts were taken over by ANEF, whipped out by the new political regime, nine years later.

Instead of the pond, we can see now a modern pool, on the east side, which will be opened until 1962.

-the sixth period from 1945 until 1990 – define the current configuration of the land, the Cotroceni Stadium is built (1950), the new

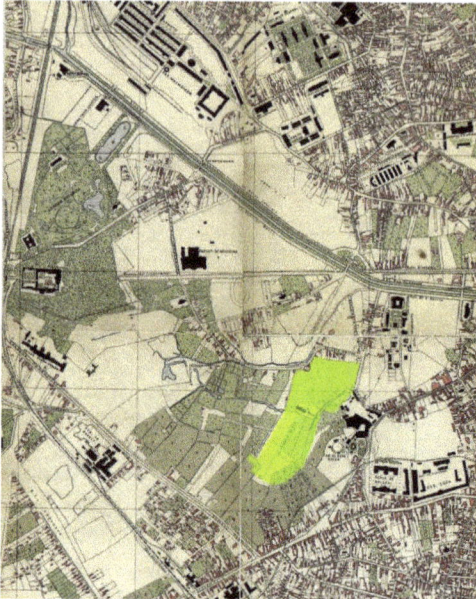

Figure 3. Bucharest Plan in 1895 – 1899
(Source: www.ideiurbane.ro)

Figure 4. Bucharest Plan in 1911
(Source: www.ideiurbane.ro)

center of Victoria Socialismului is drawn and built now (1980).

Currently the piece of land can be accessed on the two main entrances (from Dr. Lister Street and Dr. Staicovici Street), a complex scanning field of roadway and pedestrian walkways, which serves the tennis courts, football fields, office spaces, sports halls, restaurant, guest house, outbuildings and annexes, seating areas, pavilion, gazebo, leisure areas.

Vegetation analysis

Valuable tree vegetation, of historic trees - *Platanus × acerifolia,* the species that named

the place - the Plane Tree Park - it has 31 monumental trees, of different health stages.

The List of Protected Trees, under the administration of Romanian Academy, holds 39 historic trees of *Platanus × acerifolia* out of the total of 110 trees listed and inventoried in Bucharest. Out of the total of 39 protected plane trees, 31 are to be found in the Plane Tree Park, counting for almost 80% (79.8%). At the same time, out of the total of protected trees in Sector 5, more than 85% are located in the Plane Tree Park (Figure 5 and Figure 6).

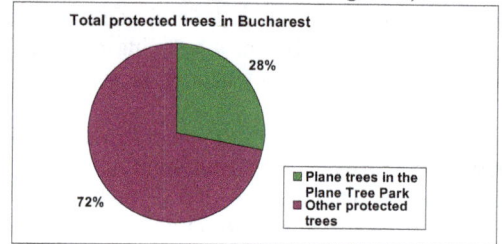

Figure 5. Percent of plane trees out of the total of protected trees in Bucharest

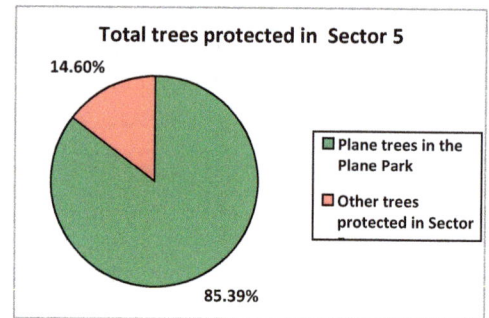

Figure 6. Percent of plane trees out of the total of protected trees in Sector 5 Bucharest

The analysis started in 2014 generated a Specialized Technical Expertise Report which indicated the health condition of the protected plane trees. Three major categories were identified for the physical health condition of the trees identified in the topographic plan: good, average and poor (Figure 7).

Each tree was analyzed individually following a scheme drawn according to the principles of Sustainable Management, issuing an Individual Analysis Sheet.

The following elements were analyzed individually: tree morphology (basic anatomical elements: stem, bark, crown), internal structure of the stem (considered following the tomography analysis conducted), the health condition and aesthetic appearance.

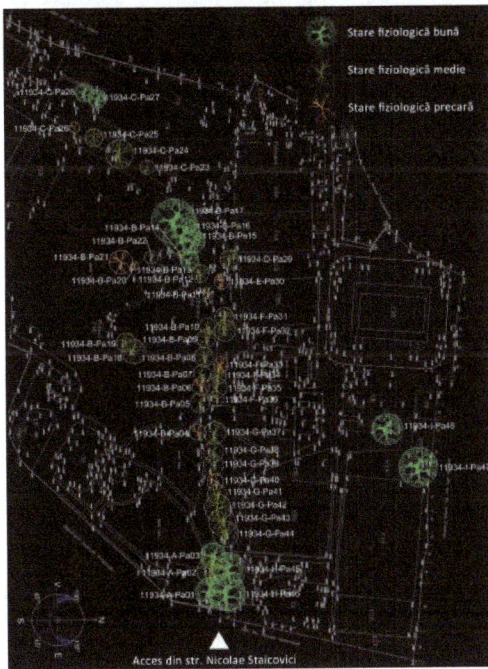

Figure 7. Identification plan of the protected trees, assessment of their physical health condition

Figure 8. Individual Specialized Technical Expertise Sheet with dendrometric recording and aesthetic analysis

Each *Individual Analysis Sheet* includes a series of recommendation and interpretations of the results analyzed and the measurements conducted (Figure 8 and Figure 9).

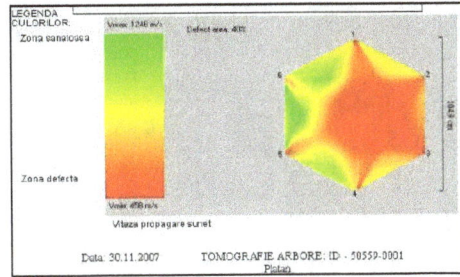

Figure 9. Individual Tomography Expertise Sheet

The historical trees were analyzed in full, all existing diseases and affections, growing and development anomalies were identified, as well as the condition and balance of the morphological elements. All information was recorded in hard support format and photographic documents, in order to set a customized diagnosis, for a tree to be given the proper treatment to improve its health condition (Figure 10).

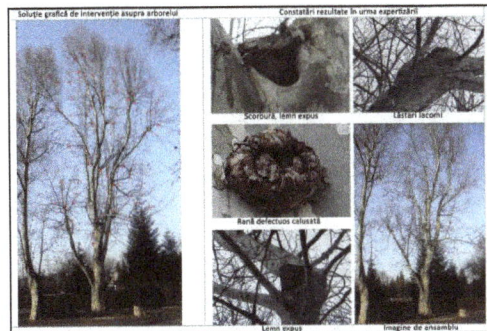

Figure 10. Analysis, diagnosis and individual treatment sheet

A part of the trees analyzed were seen in an advanced state of decay, as the main parts supporting it were not present. It meant they had to be removed, being considered as imminent hazard of safety for the buildings and pedestrians (Figure 11).

Protected trees, with significant health problems, in an advanced physiological state of decay (but with great historical value) were mentioned in special sheets (but not recorded as vegetal elements) under provisional preservation, with mandatory constant monitoring and they can be cut at first sign of possible hazard. The individual analysis sheet indicates actions to be taken to increase their safety and stability (Figure 12).

Figure 11. Hazardous Trees Identification Sheet

Figure 12. Individual sheet of protected trees with special monitoring condition

Out of the total of 48 plane trees analyzed, only 31 trees are noted in the List of Protected Trees. The individual analysis of the trees present in the studied site indicated that there are trees with the same landscape and historical value, but which are not mentioned in this list (Figure 13).

Figure 13. Individual analysis of the trees with similar value as the protected trees

Careful analysis of each tree provided primary information, centralized in complex inventory sheet. They include data related to the surface code, analyzed tree code (unique and individual), diameter of the trunk (measured at

1.3 m from base), crown inserting height, diameter of the crown, total height of the tree, number of branches (main, secondary), aesthetic value, intervention required, general health condition (pruning, treatment), type of intervention recommended, root system, approximate age (Figure 14 and Figure 15).

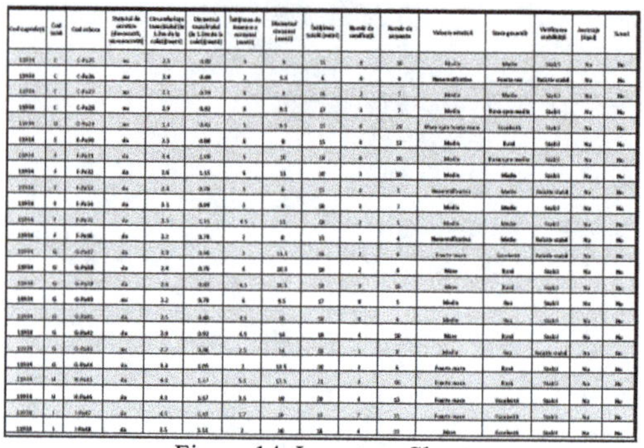

Figure 14. Inventory Sheet

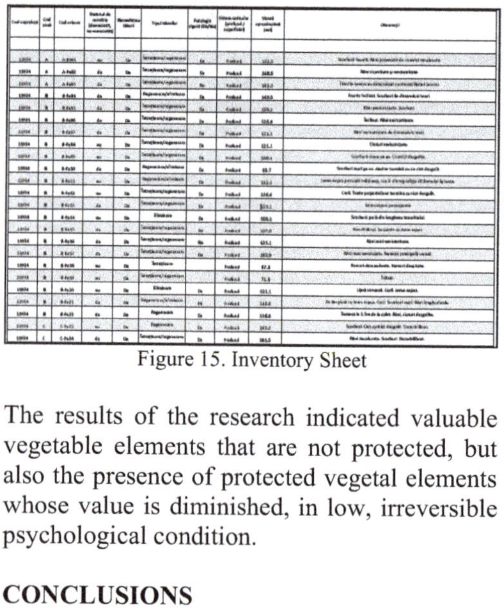

Figure 15. Inventory Sheet

The results of the research indicated valuable vegetable elements that are not protected, but also the presence of protected vegetal elements whose value is diminished, in low, irreversible psychological condition.

CONCLUSIONS

The impressive number of historical trees grouped in the land assessed can give us an idea about the importance and the heritage value of the analyzed site.

The valuable tree vegetation in the Plane Tree Park counts for almost 1/3 of the total of trees listed in the Protected Tree List (31 trees, out of the total of 110).

Nevertheless, another 17 trees of *Platanus × acerifolia* species found and analyzed in the site, have aesthetic values and dendrometric consideration as the protected ones (Figure 16).

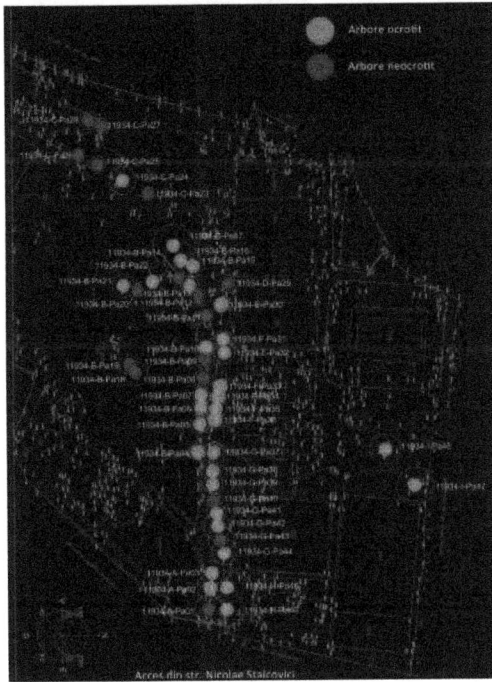

Figure 16. Plan for the identification of the protected trees

In total, the number of trees that can be considered protected, could be of 48, not 31, as they are currently listed in the Protected Trees List.

The classification of the trees into the Protected Trees List could follow a similar procedure listed in the methodological norms for the classification and inventory of historical monuments. This procedure is based on classification criteria, which assess the cultural significance and importance of the immovable assets, establishing the legal category and valuable group of the national cultural heritage which these assets belong to, respectively group A or B of historical monuments (landscape). Therefore, the classification of an immovable asset is conducted based on the following criteria:

    a. Seniority criterion.
    b. Criterion related to the artistic, urban (landscape) value.
    c. Criterion related to the memorial and symbolist value.

These criteria, applied in line with the legislation of architectural historical moments could be slightly adjusted in order to be applied for landscaping of historical monuments.

The general health condition of the trees analyzed, protected or not protected, is poor, most of them indicating early signs of aging. A part of the trees existing, mainly historical plane trees, was presented even in the first stage of the site analysis (starting with the 19th century). The plane tree is a long-lasting species and the estimated age of the oldest trees (max. 180-200 years), does not explain the advanced grade of their physiological decay.

Related to the phenotype description of the species, significant differences were registered between the typo species and the species analyzed. Most of the trees present of the land indicate mainly vertical growing, while the typo species has large crowns, with horizontal growing.

The trees stability analysis indicated as main issue the internal cavities or even open cavities, some of them noticed on the field, others indicated following the tomography analysis.

Previous interventions (infilling, cutting) were severe, the trees reacted in uncontrolled growing of greedy springs, concurrent branches, terminal ends, with the internal wood exposed. All these interventions lead to the worsening of the plane trees physiological health condition.

All the trees examined were diagnosed and recommendations were issued for regeneration pruning and balancing of crowns, of disinfection and infilling of hollows and open cavities.

The analysis and measures were taken and organized in the first Tree Inventory Register, a basic document in planning the interventions and Landscaping Sustainable management. The integrated landscape management survey indicated the importance of constant monitoring of valuable plane trees, but also the program of required interventions, as a matter of emergency or immediate operations. The scientific system for the registration and inventory of each tree as settled. The results of the survey indicated 8 trees of *Platanus × acerifolia* species in a very poor condition, some of them should be cut immediately as there are facing a constant hazard of falling. In the following current management stage, we will update the analysis of vegetation - once in 5 years - and the tomography analysis will be redone for the constant monitoring of the health condition of the historical plane tree in the Plane Tree Park.

ACKNOWLEDGEMENTS

This survey was conducted with the support of the National Bank of Romania, through the project "Integrated Management and Administration of the Plane Tree Park -owner NBR."

REFERENCES

Lynch K., 1960. The image of the city, The MIT Press, page 20-60.
Sandu Al., 1992. Structura urbană, Curs, cap. III, Ed. UAUIM, București, 18.
Simonds J.O., 1967. Arhitectura peisajului, Ed. Tehnică, 53.
Stănescu A., 2008. Aspecte ecologice ale calității spațiilor verzi - implicații în peisagistica urbană, Ed. AmandA Edit, București, 16.
Tudora I., 2009. Teoria peisajului. Definirea interdisciplinară a peisajului. USAMV, 112.
***www.ideiurbane.ro
***The methodological norm for classification and inventory of historical monuments dated April 18, 2008.

Permissions

All chapters in this book were first published in SPSBH, by University of Agronomic Sciences and Veterinary Medicine of Bucharest; hereby published with permission under the Creative Commons Attribution License or equivalent. Every chapter published in this book has been scrutinized by our experts. Their significance has been extensively debated. The topics covered herein carry significant findings which will fuel the growth of the discipline. They may even be implemented as practical applications or may be referred to as a beginning point for another development.

The contributors of this book come from diverse backgrounds, making this book a truly international effort. This book will bring forth new frontiers with its revolutionizing research information and detailed analysis of the nascent developments around the world.

We would like to thank all the contributing authors for lending their expertise to make the book truly unique. They have played a crucial role in the development of this book. Without their invaluable contributions this book wouldn't have been possible. They have made vital efforts to compile up to date information on the varied aspects of this subject to make this book a valuable addition to the collection of many professionals and students.

This book was conceptualized with the vision of imparting up-to-date information and advanced data in this field. To ensure the same, a matchless editorial board was set up. Every individual on the board went through rigorous rounds of assessment to prove their worth. After which they invested a large part of their time researching and compiling the most relevant data for our readers.

The editorial board has been involved in producing this book since its inception. They have spent rigorous hours researching and exploring the diverse topics which have resulted in the successful publishing of this book. They have passed on their knowledge of decades through this book. To expedite this challenging task, the publisher supported the team at every step. A small team of assistant editors was also appointed to further simplify the editing procedure and attain best results for the readers.

Apart from the editorial board, the designing team has also invested a significant amount of their time in understanding the subject and creating the most relevant covers. They scrutinized every image to scout for the most suitable representation of the subject and create an appropriate cover for the book.

The publishing team has been an ardent support to the editorial, designing and production team. Their endless efforts to recruit the best for this project, has resulted in the accomplishment of this book. They are a veteran in the field of academics and their pool of knowledge is as vast as their experience in printing. Their expertise and guidance has proved useful at every step. Their uncompromising quality standards have made this book an exceptional effort. Their encouragement from time to time has been an inspiration for everyone.

The publisher and the editorial board hope that this book will prove to be a valuable piece of knowledge for researchers, students, practitioners and scholars across the globe.

List of Contributors

Ana Cornelia Butcaru and Florin Stănică
University of Agronomic Sciences and Veterinary Medicine of Bucharest, 59 Mărăşti Blvd., 011464, Bucharest, Romania

Gabi-Mirela Matei and Sorin Matei
National Research-Development Institute for Soil Science, Agrochemistry and Environment - ICPA, 61 Mărăşti Blvd., 011464, Bucharest, Romania

Arda Akçal
Çanakkale Onsekiz Mart University, Faculty of Agriculture, Terzioglu Campus, 17020, Çanakkale, Turkey

ÖzgüR Kahraman
Çanakkale Onsekiz Mart University, Faculty of Architecture and Design, Terzioglu Campus, 17020, Çanakkale, Turkey

Marian Burducea, Andrei Lobiuc, Naela Costică and Maria-Magdalena Zamfirache
"Alexandru Ioan Cuza" University of Iasi, Carol I Bld., 20 A, 700505, Iasi, Romania

Ana Maria Bădulescu and Florina Uleanu
University of Pitesti, Department of Science Romania, 1 Târgu din Vale str., 110040, Pitesti, Romania

Anca Stănescu
University of Agronomic Sciences and Veterinary Medicine of Bucharest, 59 Marasti Blvd, District 1, Bucharest, Romania

Alexandru Mexi
"Vasile Goldiş" Western University of Arad, 94-96 Revoluţiei Avenue, Arad, Romania

Smaranda Comănescu
Landscape architect Freelance

Violeta Răducan
University of Agronomic Sciences and Veterinary Medicine of Bucharest, 59 Marasti Blvd, District 1, Bucharest, Romania

Cristina Petrişor, Alexandru Paica and Florica Constantinescu
Research and Development Institute for Plant Protection, Ion Ionescu de la Brad Blvd., No.8, District 1, Bucharest, Romania

Oana-Alexandra Drăghiceanu and Liliana Cristina Soare
University of Piteşti, Târgu din Vale Street, No 1, 110040, Piteşti, Argeş County, Romania

Costel Vînătoru, Bianca Zamfir and Camelia Bratu
Vegetable Research and Development Station Buzău, No. 23, Mesteacănului Street, zip code 120024, Buzău, Romania

Adrian Peticila
University of Agronomic Sciences and Veterinary Medicine of Bucharest, 59 Mărăşti Blvd, District 1, 011464, Bucharest, Romania

Andreea Coşoveanu, Samuel Rodriguez Sabina and Raimundo Cabrera
UDI Fitopatología, Facultad de Ciencias, Sección Biología, Universidad de La Laguna (ULL), ES-38206 La Laguna, Tenerife, Spain

Cristina Emilia Niţă
University of Agronomic Sciences and Veterinary Medicine of Bucharest, Faculty of Horticulture, 59 Marasti Blvd, District 1, Bucharest, Romania

Beatrice Michaela Iacomi
University of Agronomic Sciences and Veterinary Medicine of Bucharest, Faculty of Agriculture, 59 Marasti Blvd, District 1, Bucharest, Romania

Yavuz Bağci and HüSeyin Biçer
Department of Biology, Faculty of Science, Selçuk University, Konya, Turkey Ardıçlı Mh., Alaaddin Keykubat Kampüsü, Diş Hekimliği Fakültesi Kampüs, Merkez/Konya, Turkey

Vladimir Ionuţ Boc and Robert Mihai Ionescu
University of Agronomic Sciences and Veterinary Medicine of Bucharest, Department of Landscape Architecture, Biodiversity and Ornamental Horticulture. 59, Marasti Bd., 011464, Bucharest, Romania

Vladimir Ionuţ Boc
University of Agronomical Sciences and Veterinary Medicine of Bucharest, Department of Landscape Architecture, Biodiversity and Ornamental Horticulture. 59, Marasti Bd., 011464, Bucharest, Romania

Alexandru Ciobotă and Smaranda Bica
Politehnica University of Timişoara, 2 Vasile Pârvan Blvd, 300223, Timişoara, Romania

Ion Roşca, Elisabeta Onica, Alexei Palancean
Botanical Garden (Institute) of the Academy of Sciences of Moldova, 18 Padurii Street, Chisinau, Republic of Moldova

Elisabeta Dobrescu and Mihaela Ioana Georgescu
University of Agronomic Sciences and Veterinary Medicine of Bucharest, 59 Mărăşti Avenue, District 1, 011464, Bucharest, Romania

Daniel Constantin Potor, Mihaela Ioana Georgescu and Dorel Hoza
University of Agronomic Sciences and Veterinary Medicine of Bucharest, 59 Marasti Blvd., District 1, Bucharest, Romania

Marinela Vicuţa Stroe and Cristinel Ioana
University of Agronomical Sciences and Veterinary Medicine of Bucharest, 59 Mărăşti, 011464, Bucharest, Romania

Elisabeta Dobrescu and Claudia Fabian
University of Agronomic Sciences and Veterinary Medicine Bucharest, 59 Marasti Blvd., District 1, Bucharest, Romania

Aleksandra Stanojković-Sebić, Radmila Pivić, Zoran Dinić and Dragana Jošić
Institute of Soil Science, Teodora Drajzera 7, 11000 Belgrade, Serbia

Snežana Pavlović
Institute for Medicinal Plant Research "Dr Josif Pančić", 11000 Belgrade, Serbia

Mira Starović
Institute for Plant Protection and Environment, 11000 Belgrade, Serbia

Zorica Lepšanović
Military Medical Academy, 11000 Belgrade, Serbia
Aurora Dobrin, Roxana Ciceoi and Vlad Ioan Popa
University of Agronomic Sciences and Veterinary Medicine of Bucharest, Laboratory of Diagnosis and Plant Protection of Research Center for Studies of Food Quality and Agricultural Products, 59 Marasti Blvd, District 1, Bucharest, Romania

Ionela Dobrin
University of Agronomic Sciences and Veterinary Medicine of Bucharest, 59 Marasti Blvd, District 1, Bucharest, Romania

Arif Turan and Yusuf Ucar
Süleyman Demirel University, Agricultural Faculty, Farm Structure and Irrigation Department, 32260, Isparta-Turkey

Soner Kazaz
Ankara University, Agricultural Faculty, Horticulture Department, 06100, Dışkapı-Ankara-Turkey

Monica Luminiţa Badea, Aurelia Dobrescu, Elena Delian, Ioana Marcela Pădure and Liliana Bădulescu
University of Agronomic Sciences and Veterinary Medicine of Bucharest, 59 Marasti Blvd., District 1, Bucharest, Romania

Corina Bubueanu, Alice Grigore and Lucia Pîrvu
National Institute for Chemical-Pharmaceutical R&D (ICCF-Bucharest), Vitan Road 112 Sector 3, Bucharest, ROMANIA

Constantina Chireceanu, Andrei Teodoru and Andrei Chiriloaie
Research and Development Institute for Plant Protection Bucharest, 8 Ion Ionescu de la Brad, District 1, 013813 Bucharest, Romania

Ileana Maria Panțu
University of Agronomic Sciences and Veterinary Medicine of Bucharest, 59 Marasti Blvd., District 1, Bucharest, Romania

Akife Dalda Şekerci, Tuğçe Tecirli and Osman Gülşen
Erciyes University, Department of Horticulture, Kayseri, Turkey

Nina Gerasimova
Institute of Plant Physiology and Genetics, Bulgarian Academy of Sciences, Acad. G. Bonchev Street, Bldg. 21,1113, Sofia, Bulgaria

Milenayordanova
University of Forestry, Faculty of Agronomy, 10 KlimentOhridski Blvd, 1756, Sofia, Bulgaria

Elisabeta Dobrescu
University of Agronomic Sciences and Veterinary Medicine of Bucharest, 59 Marasti Blvd., District 1, Bucharest, Romania

Sanda Petredeanu
The National Bank of Romania, Bucharest, Romania

Index

www.ingramcontent.com/pod-product-compliance
Lightning Source LLC
Chambersburg PA
CBHW062002190326
41458CB00009B/2947